Tasmania's Offshore Islands:
seabirds and other natural features

Nigel Brothers
David Pemberton, Helen Pryor, Vanessa Halley

Tasmania

Tasmanian Museum and Art Gallery,
Hobart, Tasmania.

First published in Australia in 2001 by
Tasmanian Museum and Art Gallery,
19 Davey Street, Hobart
Tasmania 7000
Australia

Copyright © 2001 Nature Conservation Branch of the Department of Primary Industries, Water and Environment.

All rights reserved. No part of this publication may be reproduced, stored in a retrieval system or transmitted, in any form or by any means, electronic, mechanical, photocopying, recording or otherwise, without the prior written permission of the publisher and copyright holder.

ISBN 0 7246 4816 X

Typesetting and pre-press: Show-Ads Hobart
Original design: Hannah Gamble
Editor: Clodagh Jones
Index: Clodagh Jones
Printer: Printing Authority of Tasmania

Contents

Maps of Islands, islets and rocks	iv
Acknowledgements	vi
Foreword	vii
Introduction	viii
Stages of book's development and methods used	ix
Descriptions and Distribution of Seabirds	1
Table 1 – Seabird Populations on Tasmanian Offshore Islands	40
Island Industry	41
Island-friendly Visitation	42
Regional Map	44
North West Islands – Region 1	45
North Coast Islands – Region 2	131
North Bass Strait Islands – Region 3	139
Furneaux Islands – Region 4	195
North East Islands – Region 5	373
East Coast Islands – Region 6	409
South and West Coast Islands – Region 7	501
Appendix 1 - Status of Fauna	605
Appendix 2 - Islands with no Breeding Seabirds	606
Appendix 3 - Land Tenure Status	608
Appendix 4 - Common and Scientific Names of Fauna Species	611
Appendix 5 - Common and Scientific Names of Plant Species	618
References and Bibliography	622
General Subject Index	625
Index to Islands, Isles & Rocks & Regions	632
Index to seabirds and terrestrial birds	638

Maps of Islands, islets and rocks

(Islands and rocks represented by photographs: Black Pyramid Rock; South Black Rock; Craggy Island; Wright Rock; Bass Pyramid; The Nuggets; Pedra Branca; Eddystone Rock; Sidmouth Rock)

North West Island - Region 1 45	**North Bass Strait Islands - Region 3** 139	**Long Island Group** 226
New Year Island Group 46	**Kent Group** 140	Long Island 227
Christmas Island 47	Judgement Rocks 141	Big Black Reef 229
New Year Island 50	South West Isle 144	Key Island 231
Councillor Island 53	Erith Island 147	Key Reef 233
Hunter Island Group 56	Dover Island 149	Boxen Island 234
Steep Island 57	Deal Island 151	**Badger Island Group** 236
Nares Rocks 60	North East Isle 154	Badger Island 237
Bird Island 62	**Hogan Group** 157	Little Badger Island 239
Stack Island 65	Hogan Island 158	Mount Chappell Island 241
Dugay Islet 68	Twin Islets (North) 161	North West Mount Chappell Islet 243
Edwards Islet 70	Twin Islets (South) 163	Inner Little Goose Island 245
Penguin Islet 72	Long Island 165	Little Goose Island 247
Bears Island 76	Round Island 167	Goose Island 249
Three Hummock Island 77	East Island 169	Beagle Island 252
Albatross Island 80	**Curtis Island Group** 171	**Tin Kettle Island Group** 254
Black Pyramid Rock 84	Curtis Island 172	Anderson Island 255
South Black Rock 87	Cone Island 174	Little Anderson Island 257
Trefoil Island Group 89	Sugarloaf Rock 176	Mid Woody Islet 259
Doughboy Island East 90	Devils Tower 178	Tin Kettle Island 261
Doughboy Island West 93	**Rodondo Island Group** 180	Oyster Rocks 263
Trefoil Island 96	Rodondo Island 181	Oyster Rocks West 265
Little Trefoil Island 99	West Moncoeur Island 184	Neds Reef 266
Seacrow Islet 101	East Moncoeur Island 186	Doughboy Island 268
Henderson Islets 103	**Bass Pyramid Group** 188	**Great Dog Island Group** 270
Harbour Islets 105	Craggy Island 189	Little Dog Island 271
Murkay Islet (East) 108	Wright Rock 191	Little Dog Island (rock to north) 273
Murkay Islets (Middle) 110	Bass Pyramid 192	Great Dog Island 275
Murkay Islets (West) 112	**Furneaux Islands - Region 4** 195	South East Great Dog Islet 278
Shell islets 114	**Passage Island Group** 196	Briggs Islet 280
Petrel Island Group 116	Gull Island 197	Billy Goat Reefs 282
South West Petrel Island 117	Passage Island 199	Little Green Island 284
Little Stony Petrel Island 119	Forsyth Island 202	Spences Reefs 286
Big Stony Petrel Island 121	Low Islets (east island) 205	Samphire Island 288
Big Sandy Petrel Island 124	Low Islets (west island) 207	Fisher Island 290
Half Tide Rock 126	Moriarty Rocks 209	Fisher Island Reef 292
Kangaroo Island 127	Spike Island 211	**Vansittart Island Group** 293
Howie Island 129	Battery Island 213	Ram Island 294
North Coast Islands - Region 2 131	**Preservation Island Group** 215	Apple Orchard Point Island 296
North Coast Group 132	Night island 216	Puncheon Island 298
Sisters Island 133	Preservation Island 218	Puncheon Islets 300
Egg Island & Horseshoe Reef 135	Preservation Islets 222	Pelican Island 302
Penguin Island 136	Rum Island 224	Pelican Island Reef 304
The Carbuncles 137		

Tucks Reef	305
Vansittart Island	307
Cooties Reef	310
Babel Island Group	**311**
Babel Island	312
Cat Island	316
Storehouse Island	320
Sisters Island Group	**322**
Inner Sister Island	323
Outer Sister Island	325
Shag Reef	327
Sentinel Island Group	**329**
Sentinel Island	330
Little Island	332
Gossys Reef	334
Pasco Island Group	**335**
Roydon Island	336
Middle Pasco Islands	338
North Pasco Island	340
South Pasco Island	342
Marriot Reef	344
Prime Seal Island Group	**346**
Prime Seal Island	347
Low Islets	350
Wybalenna Island	353
Bird Island	355
Big Green Island Group	**357**
East Kangaroo Island	358
Big Green Island	361
Little Chalky Island	365
Chalky Island	367
Mile Island	369
Isabella Island	371
North East Coast Islands - Region 5	**373**
Waterhouse Island Group	**374**
Tenth Island	375
Ninth Island	377
Waterhouse Island	380
Little Waterhouse Island	383
Maclean Island	385
Baynes Island	387
Cygnet Island	389
Foster Island	391
Little Swan Island	394
Swan Island	397
Bird Rock	400
George Rocks	402
St Helens Island	405
Paddys Island	407
East Coast Islands - Region 6	**409**
Schouten Island Group	**410**
The Nuggets	411
Schouten Island	414
Diamond Island	417
Governor Island	419
Picnic Island	421
Taillefer Rocks	423
Ile des Phoques	425
Refuge Island	428
Little Christmas Island	429
Maria Island Group	**431**
Ile du Nord	432
Maria Island	434
Lachlan Island	437
Tasman Island Group	**439**
Hippolyte Rocks	440
The Lanterns	443
The Thumbs	445
Tasman Island	447
Wedge Island	450
Sloping Island Group	**452**
Spectacle Island	453
Little Spectacle Island	455
Barren Island	457
Hog Island	459
Sloping Island & Sloping Island Reef	460
Smooth Island	462
Fulham Island	464
Visscher Island	466
Betsey Island Group	**468**
Betsey Island & Little Betsey Island	469
Iron Pot	472
Green Island	474
Partridge Island Group	**476**
Huon Island	477
Arch Rock	479
Charity Island	481
Partridge Island	482
Curlew Island	484
Actaeon Island Group	**486**
Courts Island	487
The Friars	489
Southport Island & Southport Island Reef	491
Blanche Rock	493
Actaeon Island	495
Sterile Island	498
The Images	500
South and West Islands - Region 7	**501**
Pedra Branca Group	**502**
Mewstone	503
Pedra Branca	506
Eddystone and Sidmouth Rocks	510
Maatsuyker Island Group	**512**
Chicken Island	513
Hen Island	516
Ile de Golfe	518
Louise Island & adjacent islets	521
De Witt Island	524
Flat Witch Island	527
Western Rocks	530
Maatsuyker Island	531
Walker Island	534
Flat Top Island	537
Round Top Island	540
Needle Rocks	542
Mutton Bird Island Group	**545**
Inner Rocks, New Harbour	546
Wild Wind Islets	548
South East Mutton Bird Islet	550
South West Mutton Bird Islet	552
Mutton Bird Island	553
Sugarmouse Island	556
East Pyramids	557
Sugarloaf Rock	559
Wendar Island	561
Swainson Island Group	**563**
Big Caroline Rock	564
Swainson Island	566
Hay Island	568
Shanks Islands	570
Lourah Island	572
Breaksea Island Group	**574**
Breaksea Islands	575
Kathleen Island	578
Fitzroy Islands	580
Mavourneen Rocks	582
Trumpeter Islet Group	**584**
The Coffee Pot	585
West Pyramid	587
Trumpeter Islets	589
Hobbs Island	591
Hibbs Pyramid Group	**593**
Montgomery Rocks	594
Leelinger Island	596
Hays Reef	598
Hibbs Pyramid	599
Entrance Island	601
Bonnet Island	603

V

Acknowledgements

The former Parks and Wildlife Service of the Department of Primary Industries, Water and Environment funded a 20 year survey program primarily conducted by Nigel Brothers to gain the information which forms the basis of this book.

Appreciation is extended to Neil Smith and *Wild Wind* not just for the logistics of transport but also for an immensely handy extra pair of eyes ashore, when conditions permitted – the greatest rower of a Purdon and Featherstone 10 footer ever. Top job Louie!!

Also thanks to Nick Mooney, trusty captain of the survey dinghy, who only managed the odd near-disaster amongst countless crazy dashes to deposit the shore party.

Of particular note amongst others who assisted are Peter Atkins, Jayne Balmer, the late Viv Careless, Jack Chesterman, George Davis, Rosemary Gales, Dave Gatenby, Sheryl Hamilton, Steve Harris, Mark Holdsworth, Jim Luddington, Pete Mooney, Nat Murphy, Bob Plummer, Lloyd Robson, Chris Short, Irynej Skira, Al Wiltshire, Chas Wessing, Mick Brown for plant collection identificaton, Max Banks for geological enthusiasm and Bob Green for his interest in and curation of reptiles.

Some photographs of seabirds were kindly supplied by Dave Watts, Dave James and Bill Wakefield.

The authors also express their gratitude for the efforts of the staff at the Tasmanian Museum and Art Gallery who have readily embraced this project recognising its importance as a key reference work for the State. Their drive and enthusiasm together with their capacity to explore a range of funding methods has enabled the work to be published. In particular publication of this work has been made possible with the direct assistance of the Office of Vocational Education and Training (OVET) Tasmania through the Environmental Tourism Program and through the support of the Department of State Development (DSD). Special thanks are due to Jeff Kelly, Chief Executive (DSD) and Therese Taylor (OVET) for recognising the significance of this major work and ensuring that it could be made widely available.

Foreword
Tasmania's Offshore Islands: seabirds and other natural features

Premier of Tasmania, Jim Bacon, MHA

Tasmanians are privileged. We are fortunate enough to share our State with unique animals and ancient plants, to breathe unpolluted air and have World Heritage listed wilderness on our doorstep. The 334 islands that make up the archipelago of Tasmania are among the last truly unspoiled natural attractions in the world.

Tasmania's Offshore Islands: seabirds and other natural features highlights the uniqueness and importance of 280 of our islands as significant seabird breeding refuges and our wealth of natural resources.

This extensively researched book provides an invaluable addition to the body of knowledge on Australia's rich and diverse wildlife heritage. The seabirds and their habitats documented in this book not only showcase the richness and diversity of Tasmania's seabird population but highlight the fragility of their breeding habitat.

It draws attention to the special duty of care Tasmanians have to ensure our diverse seabird population continues to enjoy safe and unspoilt breeding refuges and that Tasmania continues to play host to such a spectacular and diverse wildlife population.

I commend Nigel Brothers for his dedication as a wildlife biologist with the Department of Primary Industries, Water and Environment. His research conducted between 1978 and 2000 forms the basis of this beautifully illustrated and informative book.

I also congratulate everyone involved in its publication. This book is another shining example of what is being achieved by Tasmanians working together for Tasmania.

Jim Bacon, MHA
Premier

Introduction

South of Australia, the largest island continent on Earth, is this continent's largest offshore island and only island state - Tasmania, whose natural beauty, highland landscapes, cool temperate rainforests, unique and abundant wildlife, spectacular coastal scenery of clear waters, white sands and towering cliffs are of world renown. Often overlooked due to their size, as Tasmania can be alongside the rest of Australia, are the numerous islands around its shoreline. They are equally spectacular and have their own individuality.

The notoriety of Tasmania and its offshore islands, in recent history, originated from 1642 when Dutchman Abel Jansz Tasman achieved the objective of his voyage: discovery of the unknown Southland. In fact, the Maatsuyker group of islands off Tasmania's south coast was the first locality of this 'new' continent, along with Tasmania itself, to be encountered and named.

Thousands of years before Tasman's 'discovery', however, Tasmanian Aborigines recognised the importance of the islands for the abundant food resources, such as seals and seabirds, they provided and with their maritime skills, were able to access even some of the more remote islands.

From the early 1800s white settlers used the islands that were 'fenced' by water as the bases for growing their sheep, as well as their seal and seabird harvesting industries. However despite the vital role they have played in shaping our history, many of these islands have remained a mystery, until the surveys that comprise the content of this book.

Times have changed with the islands once farmed no longer viable, the depredations of the past now largely prohibited and many of the islands having been assigned a conservation status in recognition of their significance as refuges for a diversity and abundance of wildlife. The seabird populations that frequent the majority of these islands are one of their most significant attributes, and as such were the primary focus of the surveys undertaken.

Lying within us all is an enchantment, an inexplicable attraction to these remote specks on Earth, perhaps fostered by the tales of old such as Daniel Defoe's 'Robinson Crusoe'. Few of us, however, take or are given the chance to actually live out our dream of visiting such places, or perhaps wisely choose not to do so because it can be fraught with difficulty, discomfort and unwelcome or disruptive to their fragility. But through the information and pictures in this book, everyone is given an opportunity to appreciate such inaccessible, spectacular and romantic localities.

This book contains information about 280 of the 334 islands that belong to Tasmania, the island state. Most omitted are those not in the marine environment, so of little interest to seabirds. The omissions do not presume or imply any level of importance. The surveys have revealed just how important the region is, being the stronghold both locally and globally for many species. Tasmanians have, therefore, a considerable responsibility of care for the islands and their inhabitants. The information in this book will hopefully contribute to ongoing efforts to preserve these unique, precious and vulnerable places.

Nigel Brothers
April, 2000

The Stages of the Book's Development and Methods Used

From 1978 to 1997 Parks and Wildlife Service biologist Nigel Brothers spent many months during the spring and summer seasons visiting Tasmania's offshore islands, principally to document the breeding distribution and abundance of seabirds and to record all observations of other birds, mammals, reptiles, topography, vegetation, geological features and any items of potential historical significance. The purpose of the surveys was to provide basic information to assist better understanding of Tasmania's flora and fauna and their management. Long gone are the days when to simply leave an island and its inhabitants alone is adequate protection. Information on species' distribution and abundance is fundamental to conservation biology.

The trading ketch, *Wild Wind* and skipper Neil Smith were chartered for much of the survey work. Otherwise a 4 metre aluminium dinghy with two motors was used to access islands from a centrally-sited base camp in a particular region.

The trading ketch, *Wild Wind*

In most cases the islands were surveyed by walking everywhere and recording all the surface-nesting seabirds including Sooty Oystercatchers, Pied Oystercatchers, Silver Gulls, Pacific Gulls, Kelp Gulls and terns. The burrowing birds were surveyed using a transect method. Transects were a metre wide and randomly placed through the colonies. All active burrow entrances were counted within the area and it was assumed that an active burrow represented one potential breeding pair. This assumption does not apply in all instances as several pairs or more can occupy nests with a common burrow entrance and not all active burrows are actually used for breeding.

A density of birds was calculated using the length and width of the transect and number and spacing of birds counted. The density was extrapolated to the total area of the colony. The total area was measured in situ when the size of the colony allowed, or via the mapping facilities of the Geographic Information Systems (GIS) mapping system. The number of transects varied according to time available and size of colony. Usually more than three were undertaken. The position of all surface-nesting birds and colonies was noted in the hand-drawn maps which utilised aerial photographs, topographic maps and nautical charts. The objective was, wherever possible, to survey an entire island group in the breeding season. The purpose of this was to accurately determine total population estimates for most species, a task complicated by the fact that some species shift breeding sites between years. So to complete an island region over several seasons could result in sectors of populations being

overlooked or perhaps counted twice but at different localities. What can be missed with this survey technique is knowledge of all breeding sites that, for species such as Black-faced Cormorants, can be strictly of traditonal use, but infrequently occupied. Generally species do not undertake long-distance breeding site shifts between years, which minimised the risk of over-counting, as different island regions were assessed in different years.

While the main focus of the surveys was to assess the distribution and abundance of nesting seabirds, timing was important. To increase the likelihood of encountering most species whilst they were breeding, surveys were conducted at the optimum time in November and December. This timing was, however, not ideal for all species as some 'uncooperative' species will choose to breed at any time of the year. For other species such as Little Penguins there is no ideal time; they can be difficult to survey accurately, having a particularly extended breeding season, with burrows well-disguised amongst thousands of burrows of other species.

Botanical collections were made on most islands, with these being deposited in the State Herbarium for preservation and identification. Similarly rock samples considered of potential interest were collected for Dr. Max Banks of the Geology Department of University of Tasmanian. Skinks were caught, wherever possible, for identification and inclusion in the Queen Victoria Museum reference collection.

Meticulously hand-drawn maps to identify breeding sites and other significant features accompanied Nigel's field notes.

During the past twenty years, Nigel's field notes have been used as the foundation for a variety of reports and studies, including the National Estate Grants Program Report on Seabird Islands, 1995, *A Reconnaissance Inventory of Sites of Geoconservation Significance on Tasmanian Islands*, 1996 and the *National Oil Spill Response Atlas*, 1999. Also accounts of a number of islands were used in the excellent Australia-wide series produced by the late Bill Lane for the journal of the Australian Bird Study Association, *Corella*.

Zoologist David Pemberton's enthusiasm for ensuring that these important data were recorded

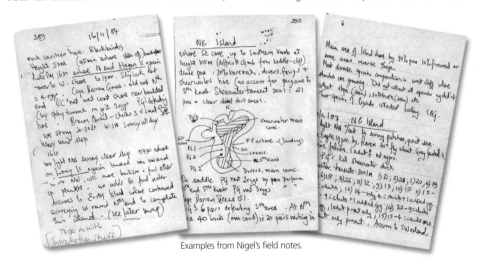

Examples from Nigel's field notes.

for perpetuity was one of the catalysts for the production of this book, which he foresaw as the first inventory of Tasmania's offshore islands.

The timely adoption of GIS technology meant that it was now possible not only to record the seabird population and distribution data, but to also effectively map them. Vanessa Halley was employed to undertake this significant task and Helen Pryor was employed to translate the field notes and transect data.

The information from Nigel's field notes and hand-drawn maps was captured by GIS and computer maps were generated. Small variations may occur between the locations shown on the map and given in the text. The location in the text is taken from the Tasmanian Nomenclature Gazette. Map grid references have been centrally derived using the GIS system.

Integral to the surveys were Nigel's extensive photographic records, which form the basis for most of the visual material in the book.

At the beginning of the book the salient characteristics of each of the breeding seabirds, particularly in relation to their breeding and migration, are summarised. Distribution patterns are also discussed. Table 1 give the total populations of breeding seabirds on Tasmania's offshore islands. Appendix 1 give the definitions of status as defined by the Tasmanian *Threatened Species Protection Act 1995*.

To compartmentalise the book, the islands have been divided into geographic regions and groups within those regions, defined by the largest island, unless nomenclature-listed group names such as the Kent Group exist. For example, the group of Wendar Island, East Pyramids, Sugarloaf Rock and Mutton Bird Island becomes the Mutton Bird Island Group. Whilst the original objective was to undertake surveys of all Tasmania's offshore islands, some of the larger islands – King, Flinders, Bruny, Cape Barren, Hunter, Clarke, Robbins and Walker – remain to be done and a few others were insufficiently complete to include. In addition, the islands listed in Appendix 2 have not been described in detail because they do not appear to be used by seabirds for breeding. Generally, for seabirds in the region, larger islands have fewer on them, however there are exceptions such as King Island, which is known to have large numbers of Short-tailed Shearwaters and Little Penguins. For the sake of completeness, population estimates, where available for islands not covered in the book, are included in the table of total

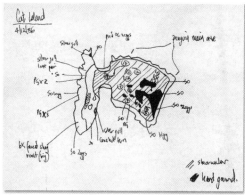

Hand drawn map from Nigel's field notes.

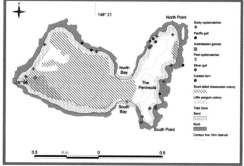

GIS – drawn map.

population size estimates. Any land mass surrounded by water, irrespective of how small, whether named or not, was included. There are some places (Sidmouth Rock for example) that are incapable of supporting terrestrial life, as they are always, or frequently, awash.

For each map there is a legend with standard symbols. Common Diving-Petrels, for instance, are indicated by a blue star. However in may cases their colonies are large therefore the text should be consulted for detailed distribution information.

Maps of several islands are not included because they have not yet been digitised. Aerial photographs have been substituted.

At the beginning of each island description is a summary table showing the island's location, survey date, area and land tenure category, referred to as status. Where more than one survey date is given, information is drawn from the most recent survey, unless otherwise indicated. Generally, the location of breeding sites tends to remain relatively consistent for burrowing species, but for surface-nesting species, it may vary considerably from season to season or year to year. Crested Terns and Black-faced Cormorants are known to move between islands from year to year.

An explanation of the terms used to define status is given in Appendix 1. Most islands lying adjacent to the South West World Heritage Area have been automatically encompassed by that status, which may not necessarily represent their values. An arrow next to the island's status (↑↓) indicates the authors' recommendations for land tenure changes, which are discussed in the Comments section for each island.

Each island's topography, landscape features, geology and human-made structures are briefly described in the initial island summary.

Breeding seabird species are then listed, with population size, densities (where known or significant) and distribution. Lists of non-breeding seabird species, mammals, other birds and reptiles follow. The native and introduced species of mammals and birds have been differentiated. Common names are used, with a table indicating corresponding scientific names in Appendix 4.

The Royal Australasian Ornithologists Union format for bird names (common and scientific) has been adopted throughout the text.

The dominant vegetation of each island is summarised, using scientific names to avoid possible confusion over common names. A table of corresponding scientific and common names is in Appendix 5.

Under 'Comments', the authors discuss each island's significance in terms of the representativeness or uniqueness of its natural values in relation to seabirds. Factors affecting the status of each island, such as existing or potential disturbance, are also canvassed, with suggested changes to the status if considered desirable for the protection of the island's seabirds.

For most of the islands these are the first published surveys and a number of the islands had never been landed on before.

To detect and monitor small mammals such as mice and *Antechinus* and to confirm the presence and more accurate population estimates for some seabird species (especially those occurring in small numbers) would have required considerably more time and effort than was possible during these surveys.

Description and Distribution of Seabirds

A summary of the characteristics of the major breeding seabird species and their distribution on Tasmania's offshore islands.

Little Penguin
Eudyptula minor

Other names:	Fairy penguin, blue penguin
Average length:	40 – 45 cm
Average weight:	About 1 kg
Breeding season:	August to February. Lay September – November. They can also breed outside these seasons.
Breeding habits:	Colonially
Habitat:	Occur in temperate seas; they land wherever they are able to climb – sandy beaches, rocky shores and rock ledges. Breed in coastal areas and offshore islands, mostly near the sea in grasslands and herbfields with good depth of soil for burrowing. They can climb to quite high altitudes, in some cases 100 metres or more. They are also found in cliff crevices and caves.
Movements:	Adults are present at colonies throughout the year, though numbers are lowest between completion of moulting in April and start of breeding in August, with a 3 week cycle of attendance. Young birds disperse widely after fledging, but most return to their natal colonies after 3 years to moult. Adults are generally close to the colony.
Other features:	Flightless. Both sexes dig burrows and build nests. Lay 2 eggs. Incubation shared equally, shifts vary. Both parents share in feeding, brooding and guarding of young.
Threats:	Dogs, cats, oil spills, habitat destruction, gill nets, diet overlap with commercial fisheries species.

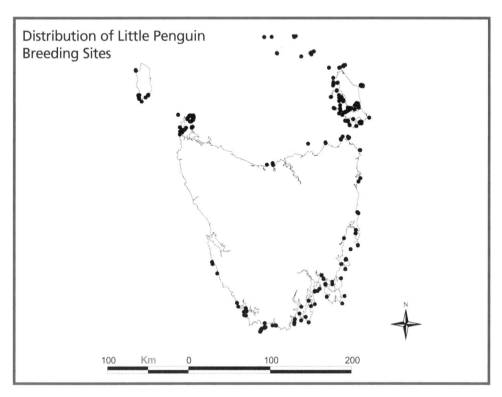

Distribution of Little Penguin Breeding Sites

This species' Australian stronghold is in the Tasmanian region, with birds nesting on virtually all islands where it is possible for them to gain access from the sea. They occupy a wide variety of habitats including rock crevices, deep burrows and beneath a dense vegetation canopy. In fact, they will nest anywhere that provides concealment, their habitat often extending from near sea level to over 100 metres above sea level. Some competition for nest sites is likely to occur between Little Penguins and Short-tailed Shearwaters as they occupy similar habitats. Little Penguins tend to breed in relatively low numbers everywhere, but with some exceptions. At such places as Councillor Island, Ninth Island, Betsey Island and some places on King Island, they occur in very large numbers. Although this species is secure, being widespread and abundant throughout the Tasmanian region, this could easily change as they are particularly vulnerable to predation by cats and dogs, to fire and to drowning in gill nets. As the *Iron Baron* disaster demonstrated, this species is also especially vulnerable to environmental pollutants such as oil.

They generally occupy their burrows year round but with a small proportion vacating islands entirely for several months during the winter. Their nesting habits, (widespread season and mixing with Short-tailed Shearwaters) makes this species particularly difficult to count.

Short-tailed Shearwater
Puffinus tenuirostris

Other names:	Muttonbird, moonbird, yolla
Average length:	40 – 45 cm, wingspan 95 –100 cm
Average weight:	480 – 800 g
Breeding season:	A classic example of synchrony. Return to same burrow around September 20, lay eggs around December 26 and chicks are hatched from 16 – 23 January.
Breeding habits:	Colonially
Habitat:	Occur mainly over continental shelf waters both inshore and offshore but also found in pelagic waters. Breed on islands and on headlands and promontories, avoiding areas of very dense vegetation and very steep slopes. Burrows may be conspicuous and unprotected to well-hidden amongst vegetation, dug anywhere in stabilised dunes of soft stable soil of at least 23 – 31 cm depth. Occasionally tunnels are made in dense vegetation without burrowing.
Movements:	Fully migratory, moving through the Pacific Ocean north to north west, extending to 71°N. Adults leave breeding colonies mid April. Non-breeding immature birds leave as early as February. Chicks leave late April/early May and return to natal colony. When breeding, they generally return to their burrows every night, but some feed 150 – 200 km from colonies.
Other features:	Gregarious, feeding in small groups or flocks of thousands. Burrow dug by pair. Lay only one egg per year. Both sexes incubate alternately. Chicks brooded or guarded usually by the female for 2 – 3 days, then deserted by day and visited each night for up to 10 days. Chicks fed by both parents alternately. Mate for life. Return to the same burrow every year. Highly synchronised breeding timetable.
Threats:	Harvesting, habitat destruction particularly fires, gill nets, ingestion of plastic debris.

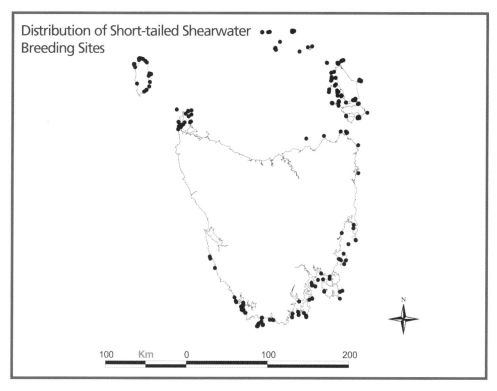
Distribution of Short-tailed Shearwater Breeding Sites

This is undoubtedly the most abundant and widespread seabird species in the region, with Tasmania being the most important stronghold world-wide. There are few islands upon which this species does not nest. Adequate depth of soil, it seems, is its only prerequisite for breeding, as all types of islands – flat and steep, forested and sparsely vegetated – are occupied.

Typically a Short-tailed Shearwater colony is dominated by thick *Poa poiformis*, with deep but loosely-compacted, friable peat soils that facilitate easy burrow excavation. Although subjected to the commercial and recreational harvesting of nestlings, livestock grazing, particularly in the recent past, and fire are considered the species' greatest onshore threats. But offshore, vast numbers are likely to be killed in various fisheries, particularly in the northern hemisphere, during their annual migration to the Bering Sea and north Pacific region.

They co-exist with Little Penguins, Fairy Prions and Common Diving-Petrels.

Short-tailed Shearwaters provide a classic example of synchronised breeding, often regarded as a feature of many species of seabirds. They return each year to the same burrow around 20 September, lay eggs around the 26 December, hatch chicks between 16 – 23 January and all fledge in April/May.

Sooty Shearwater
Puffinus griseus

Other names:	King muttonbird, New Zealand muttonbird
Average length:	40 – 46 cm, wingspan 94 – 105 cm
Average weight:	650 – 950 g
Breeding season:	October to May. Eggs laid mid November to early December, fledge late April.
Breeding habits:	Generally colonially, but in Tasmania more likely solitary nests and sparsely distributed.
Habitat:	Marine, pelagic; breed mainly on subtropical and subantarctic islands and mainland New Zealand. They nest in burrows or rock crevices on coastal slopes, ridges and cliff tops mainly in tussock-dominated areas.
Movements:	Migrate to the north Pacific and Atlantic Oceans in the non-breeding season. Most adults depart breeding grounds by the second week of April. Chicks remain for up to a month later, before leaving. During the breeding season they may travel as far as the north Ross Sea in Antarctica.
Other features:	Gregarious, migrating in huge continuous flocks.
	Both sexes incubate and feed chicks, but after the first week, chicks spend most of their time alone in their burrows.
	Lay one egg per year.
Threats:	Introduced predators, particularly feral cats, habitat destruction and commercial harvesting in New Zealand.

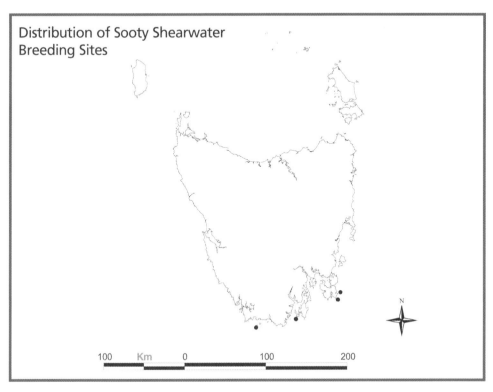

Distribution of Sooty Shearwater Breeding Sites

A species with a stronghold elsewhere, particularly in New Zealand, Sooty Shearwaters only occur at 4 localities in the Tasmanian region with few elsewhere on the Australian continent.

At each locality, they occur intermixed with Short-tailed Shearwaters, whose abundance hampers the detection of the less populous species. Despite having a call that is clearly discernible and characteristic, this as a means of detection is reliant on night time searches and not all localities have been visited overnight. So the distribution and abundance of Sooty Shearwaters is likely to be considerably greater than what is currently known, especially on the south coast islands which harbour large numbers of Short-tailed Shearwaters. Their abundance at sea indicates the possibility of larger numbers than are currently known to occupy burrows in the region.

Their breeding timetable is similar to that of the Short-tailed Shearwater.

Fairy Prion
Pachyptila turtur

Other names:	Whale bird, oil bird
Average length:	25 cm, wingspan 56 cm
Average weight:	90 – 175 g
Breeding season:	September to early March.
Breeding habits:	Colonially – few pairs to tens of thousands, depending on extent of habitat.
Habitat:	In subtropical and subantarctic seas. Breed on islands and rock stacks. Burrow in soil or use crevices, caves or rock falls. Sometimes nest in scrub or tussock, but avoid dense vegetation.
Movements:	Generally, migratory or dispersive, probably travelling to subtropical waters during winter, returning late August, but in Tasmania adults maintain burrows all year.
Other features:	Gregarious when feeding at sea and when breeding. Both parents incubate and feed young. Brooded by one parent for 1 – 5 days then fed at night by both parents. Monogamous. Lay one egg.
Threats:	Habitat destruction, cats.

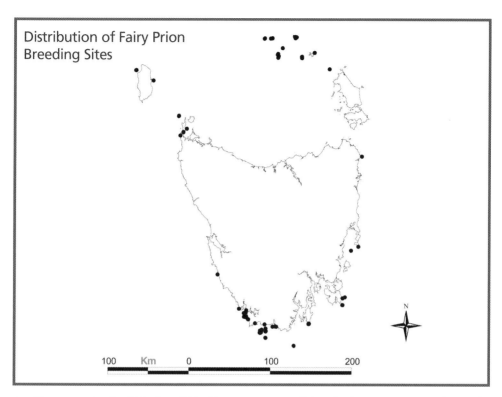

As with many species, the distribution of Fairy Prions may largely be determined simply by the existence of an island. Oceanic conditions, including food availability, must also be suitable within the species' foraging capacity, but most importantly, it requires a safe refuge. As can be seen from the distribution map, Fairy Prions occur throughout the Tasmanian region. Unlike many species of prions and also perhaps this species in other breeding localities, Fairy Prions in this region are not migratory. They can be found ashore occupying burrows and courting at any time of the year.

They will use a wide variety of nesting habitats but the majority nest in rock crevices, burrows amongst rocky habitat or on steep slopes with sparse soil cover. Soil depth can limit their burrowing distribution: competition for nest space when soils are sufficient for larger more aggressive species such as Short-tailed Shearwaters, precludes prion occupation. Although there are comparatively large islands such as Tasman Island and Ile du Golfe with vast numbers of prions nesting, typical Prion islands are small, rocky and steep. In many localities burrows can be interconnected in soft, unstable, shallow soils, making accurate census of their population difficult.

Apart from their natural range of predators such as Forest Ravens, Pacific Gulls and Peregrine Falcons, Prions are subjected to predation by a non-indigenous predator, the feral cat, at only one locality - Tasman Island.

This species' breeding habits and survivorship have been little studied in this region.

Common Diving-Petrel
Pelecanoides urinatrix

Other names:	None
Average length:	20 – 25 cm, wingspan 33 – 38 cm
Average weight:	110 – 150 g
Breeding season:	Broadly August, when eggs are laid, to April.
Breeding habits:	Colonially
Habitat:	Circumpolar usually between 35° and 55° S. Breed in burrows on islands with peat or stable soil, under vegetation on slopes with direct access to the sea. Their burrows are concentrated around the coast, but some may occur inland, often amongst tussock grass.
Movements:	In Tasmania, burrows are occupied in all months. Maximum foraging range about 360 km during the breeding season.
Other features:	Gregarious when breeding. Birds are very active and noisy upon arrival at breeding colony. Compete with Fairy Prions for nest sites, often unsuccessfully.
Threats:	Habitat destruction, oil, cats.

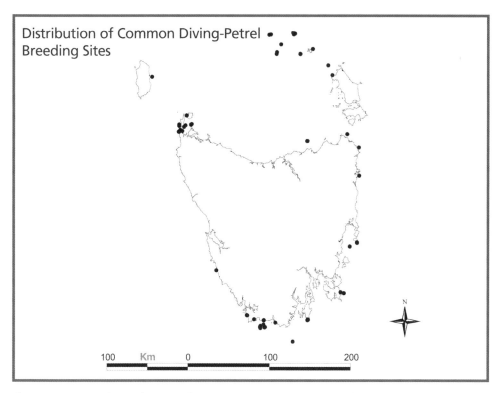

Distribution of Common Diving-Petrel Breeding Sites

There are many similarities between the nesting habits of this species and Fairy Prions. In fact, seldom on Tasmanian islands does one nest without the other. An important exception is Steep Island, where Common Diving-Petrels occur in great abundance but where Fairy Prions are absent. Where both species occur, there is evidence that they compete for nesting sites, with prions being the more aggressive. Such competition is likely to produce a contour effect, as on the Doughboys, where Common Diving-Petrels occupy the least suitable areas, Fairy Prions the moderately suitable areas and Short-tailed Shearwaters the prime habitat. However, unlike the other species, Common Diving-Petrels actually favour the most steep, vegetated, ledges for burrow construction. This habitat is preferred because they are particularly ungainly once ashore and find departure from flat terrain difficult. Such habitat is fragile, with honeycombed burrows and sparse vegetation cover, thus providing easy predation opportunities for gulls and Forest Ravens. Whilst this species is unusual in that it is asynchronous and also commences egg laying in mid winter, eggs can also be found in December, apart from on the south coast islands. It is possible that a second species of Common Diving-Petrel does, in fact, frequent this region, but has as yet not been discovered. Unusual for such species, diving-petrels come ashore to rest at localities where they do not actually breed, such as on Pedra Branca and Albatross Island. There are also apparently suitable breeding localities, where they do not breed, perhaps excluded by the sheer abundance of Fairy Prions that occupy the available nesting habitat. Whilst the Common Diving-Petrel breeds in all areas of the region and in large numbers, south coast islands are undoubtedly its stronghold. Few breed outside the Tasmanian region in continental Australia.

White-faced Storm-Petrel
Pelagodroma marina

Other names:	Mother Carey's chicken
Average length:	18 – 21 cm, wingspan 42 – 43 cm
Average weight:	40 – 70 g
Breeding season:	Laying starts before December
Breeding habits:	Colonially
Habitat:	Temperate and subtropical regions of Atlantic, Indian and south Pacific Oceans. Breed on low granite islands covered by sandy loam, sandy limestone islands, steep rocky outcrops or low sand hummocks. In eastern Bass Strait they nest among deep *Poa* or under mats of succulents, in other places in *Atriplex, Tetragonia* or other succulent herbs.
Movements:	Migratory from temperate breeding sites to tropical and subtropical waters, fledging mid February to mid March, returning to colonies late September to early October. Foraging range when breeding about 250 – 370 km.
Other features:	Gregarious in small parties of 2 to 50. Larger flocks, several hundreds reported near breeding colonies. Strictly nocturnal when visiting land. Lay one egg.
Threats:	Habitat destruction, disturbance by humans, rats.

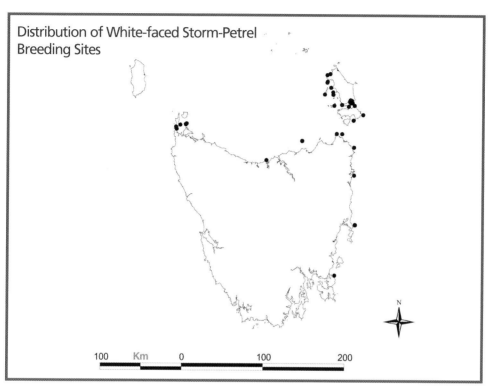

Distribution of White-faced Storm-Petrel Breeding Sites

This species breeds chiefly on islands in Bass Strait, particularly in the Furneaux group of eastern Bass Strait. Several colonies, only, can be found south of Bass Strait, all on the east coast and no further south than Visscher Island.

This is a particularly delicate species, which perhaps accounts for its tendency to prefer soft, sandy soil for burrow excavation: anything other than such soils would prohibit it from digging. Also, wherever White-faced Storm-Petrels nest, soils are insufficiently deep or too friable to permit the successful occupation by other larger, more aggressive species.

So, at most breeding localities, disturbance of the colonies of this species must be avoided at all cost, as burrows will readily collapse under foot. Burrows are often long, but dimensions are unable to be determined due to their very small entrances and narrowness. This species should be considered vulnerable, nesting at only 32 localities in relatively small numbers, with specific, but fragile habitat requirements.

Pacific Gull
Larus pacificus

Other names:	Mollymawk
Average length:	50 – 67 cm, wingspan 131 – 169 cm
Average weight:	1.6 kg male, 1.1 kg female
Breeding season:	September to January Lay in October to December.
Breeding habits:	Singly or in loose colonies.
Habitat:	Endemic to Australia, mostly south and west coasts and infrequently east coast. Breed mainly in Bass Strait. Prefer elevated sites on rocky outcrops, headlands, small hillocks, ridges, cliffs and islands. Also beaches among stipa and *Poa*.
Movements:	Partly resident and partly dispersive, particularly in winter. Generally stay close to breeding sites during breeding season.
Other features:	Largest gull. Nests are either scrapes or depressions in the ground unlined or lined with gravel, or a neat shallow bowl constructed of grass, sticks, seaweed and feathers. Adults are usually seen singly or in pairs, rarely in large groups. Lay 2 – 3 eggs.
Threats:	Alteration to natural distribution and feeding patterns through artificial food supplies at tip sites or inappropriate handling of bulk food scraps.

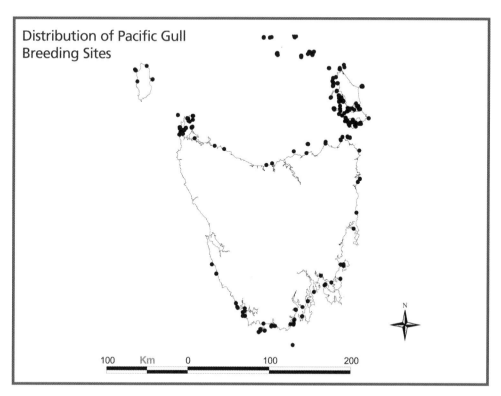

Distribution of Pacific Gull Breeding Sites

Although widespread throughout the whole region, this species does not occur in large numbers. Generally each island has a population of Pacific Gulls, but on many islands, only one or two pairs occur. Largest numbers traditionally occupy very small islands which are adjacent to a larger island, island group or mainland Tasmania. This provides the gulls with a suitable nesting refuge and a nearby larger foraging area for the whole population.

They are a predator of the small burrow-nesting species, especially the Common Diving-Petrel. Nests are generally positioned adjacent to a stipa clump or similar vegetation.

Not especially abundant, but widespread in low numbers, its population is considered secure, except perhaps from displacement by Kelp Gulls, a relatively recent arrival, which is increasing in distribution and abundance.

Silver Gull
Larus novaehollandiae

Other names:	Seagull
Average length:	36 – 44 cm, wingspan 91 – 96 cm
Average weight:	265 – 315 g
Breeding season:	Mainly spring/summer, but can be at all times. Usually lay September to December.
Breeding habits:	Usually in colonies, but may be single pairs.
Habitat:	Widespread in Australia in coastal areas and inland waters. Breed on islands, rocks or small peninsulas.
Movements:	Partly migratory, partly dispersive and partly resident. In south-east Australia birds that breed in spring and summer generally leave colonies when breeding is complete and return to breeding areas in winter and spring. Generally stay close to breeding sites during breeding season.
Other features:	Nest is a scrape in the ground lined with rootlets and seaweed amongst low vegetation. They will often roost nocturnally and communally. Incubation is by both sexes with rotations of 1 to 9 hours. Lay 2 – 3 eggs.
Threats:	Alteration to natural distribution and feeding patterns through artificial food supplies at tip sites or inappropriate handling of bulk food scraps.

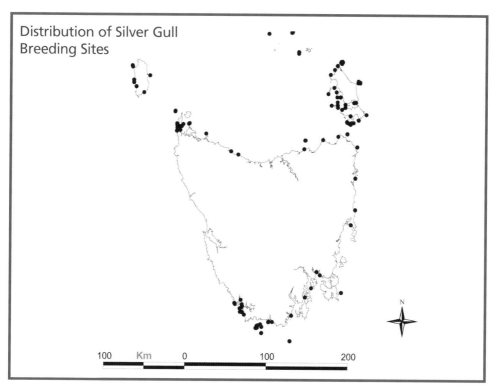

Distribution of Silver Gull Breeding Sites

Although the Silver Gull has exploited urban opportunities for its food supply perhaps to a greater extent than any other seabird, it has likely to have always been, relative to other gull species in the region, the most abundant. In the more remote island groups, it can still be found at natural levels in quite small, scattered colonies consisting of 20 to 50 pairs.

Often Silver Gulls nest in association with other seabirds, especially the tern species and are quick to exploit opportunities to rob their nests of eggs and chicks, when human presence induces disturbance. It has a wide variety of habitats from steep cliff ledges to flat bare rock or vegetated areas, and, regardless of habitat, will exploit whatever material is locally available to build its nests – vegetation, feathers, seaweed, marine debris. In the vicinity of larger colonies, vegetation can be largely destroyed by bird activity.

It is generally traditional to a specific nesting locality on an island and resident year round.

Proclaimed a pest in some situations such as when it occurs adjacent to airports, roadways, refuse disposal sites, town water supplies and research sites (such as Fisher Island), the Silver Gull has been subjected to poisoning at some localities, which has had significant impact in the short term on local populations but unknown longer-term impact.

Kelp Gull
Larus dominicanus

Other names:	Dominican gull
Average length:	49 – 62 cm, wingspan 106 – 142 cm
Average weight:	1.05 kg male, 830 g female.
Breeding season:	Lay late August to mid November. Hatch mid November to late January.
Breeding habits:	Usually in colonies, but sometimes as single pairs.
Habitat:	Antarctic to subtropical zones. In Australia, almost exclusively coastal, mostly in sheltered harbours, bays, inlets and estuaries or on sandy or rocky beaches and rock platforms. Roost on islands, rock stacks, beaches and in estuaries.
Movements:	Generally dispersive, but in many parts of Australia they are resident all year. Stay close to breeding site during breeding season.
Other features:	Similar to the Pacific Gull but a more slender bird with a slimmer bill and, when adult, without the black tail band. Occur mainly in pairs but in flocks during the winter. Self-introduced in the 1960s, so expected to increase in abundance. Lay 2 – 3 eggs.
Threats:	Potential alteration to natural distribution and feeding patterns through artificial food supplies at tip sites or inappropriate handling of bulk food scraps.

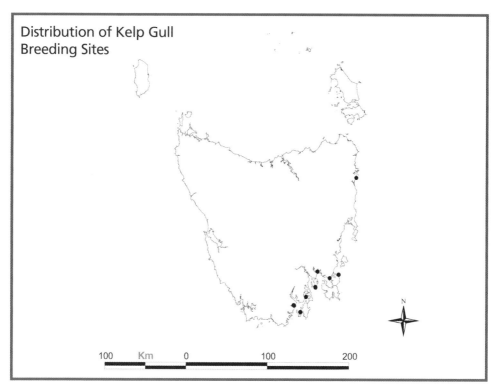

Distribution of Kelp Gull Breeding Sites

A relatively recent arrival in Australia and the Tasmanian region (possibly as recent as the 1960s), it is now well-established and believed to be still increasing in distribution and abundance, a tribute to its ability to exploit urban food sources, especially refuse disposal sites. Unlike the Pacific Gull, it tends to be more a colonial nester with several quite large breeding sites such as Green Island in the D'Entrecasteaux Channel and Visscher Island off the north-east of Tasman Peninsula. It has a similar nesting habitat to the Pacific Gull and is resident year round. It will fly long distances from its more traditional and larger nesting colonies such as Green Island to exploit urban food sources such as Margate and Hobart refuse disposal sites, where minimal effort is made to exclude such scavenging birds – a potentially serious problem.

Sooty Oystercatcher
Haemotopus fuliginosus

Average length:	40 – 52 cm, wingspan about 100 cm
Average weight:	About 750 g
Breeding season:	September to January.
Breeding habits:	Solitary territorial pairs, nests a minimum of 50 metres apart. On islands, generally one breeding pair unless size permits more eg. Bird Island.
Habitat:	Endemic to Australia. Coastal areas, mainly rocky islands, reefs and headlands, usually within 50 metres of the shore above the high tide mark. Breed on offshore islands and rock stacks. Bass Strait islands are a stronghold.
Movements:	Some resident, but mostly move between March and September to wintering sites.
Other features:	Occur as single birds or in pairs, frequently with Pied Oystercatchers. Flocks in winter. Nest is a scrape in the ground amongst rocks, shingle, pigface or seaweed. Lay 2 eggs.
Threats:	Human disturbance, cats, dogs, off road vehicles.

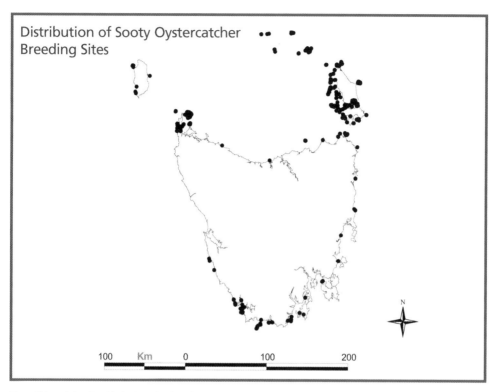

Distribution of Sooty Oystercatcher Breeding Sites

The offshore islands with their rocky shorelines, are the breeding stronghold for this species in the Tasmanian region. Usually no island, regardless of size, is without at least one breeding pair. Seldom do more than only a few pairs occur on any one island with the exceptions being such localities as Bird Island in the Hunter Island group, where a combination of suitable rock shoreline with adjacent large expanses of low tide foraging habitat make high nesting density possible. The extent of the inter-tidal feeding zone correlates with the nesting density of this species.

Sooty Oystercatchers frequently nest within a metre or two of a Pacific Gull nest, seemingly to exploit the gull's aggressive defence strategy when nesting, thus assisting their own protection.

There is perhaps a symbiotic relationship with the Pacific Gull, which in its predator role, chooses to exploit the early warning alarm of the oystercatchers for its own protection.

Although they are considered to be resident year round, Sooty Oystercatchers tend to vacate many of the smaller, more remote offshore islands during the winter, presumably for more favourable foraging shores, which are less influenced by frequent high seas.

Pied Oystercatcher
Haemotopus longirostris

Average length:	42 – 50 cm, wingspan 85 – 95 cm
Average weight:	650 – 750 g
Breeding season:	Early September to late January.
Breeding habits:	Solitary nesting in well-defined and strongly-defended territory.
Habitat:	Coastal, preferring intertidal mudflats and sandbanks in bays, also sandy ocean beaches. Breed on beaches, shores of lagoons, estuaries on sandy spits and islands, usually just above high water mark.
Movements:	Generally sedentary but winter flocking occurs.
Other features:	Occur singly or in pairs, frequently with Sooty Oystercatchers.
	Nest is a scrape in the ground, unlined or lined with shells, seaweed, dry grass, twigs, bark and leaves.
	Both sexes incubate, brood and guard their chicks.
	Lay 2 eggs.
Threats:	Disturbance by humans, off road vehicles, dogs and cats.

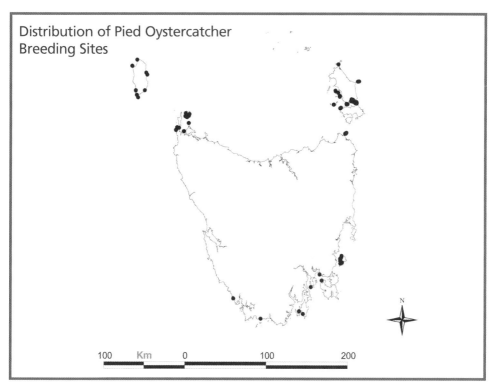

Distribution of Pied Oystercatcher Breeding Sites

With sandy beaches seeming to be a prerequisite for this species to breed, they tend not to depend on the offshore islands. On the islands they occur in small numbers, with only a few pairs at any one locality. Bass Strait islands have a tendency for sandy beaches and are therefore more frequently used than the southern islands with their rocky shores. Although not officially listed, the Pied Oystercatcher is possibly vulnerable given its predilection for sandy beaches prone to disturbance by summer visitors and its small numbers.

Caspian Tern
Sterna caspia

Average length:	47 – 54 cm, wingspan 130 – 145 cm
Average weight:	About 680 g
Breeding season:	July to March. Eggs laid November to December.
Breeding habits:	In isolated pairs or small colonies
Habitat:	Mostly sheltered bays, lagoons, inlets and estuaries. Breed mainly on low, offshore islands on spits, banks, ridges and beaches.
Movements:	Partially resident and partially dispersive. Breed irregularly in some areas, sometimes deserting breeding sites after disturbance.
Other features:	Nests are slight hollows scraped in the ground, bare or well-lined, with grass, a few twigs, seaweed, feathers, small stones and shells.
	Incubation and guarding of chicks by both parents.
	Act defensively at nest site and very noisy towards intruders.
	Lay 2 eggs.
Threats:	Human disturbance, dogs, cats, off road vehicles.

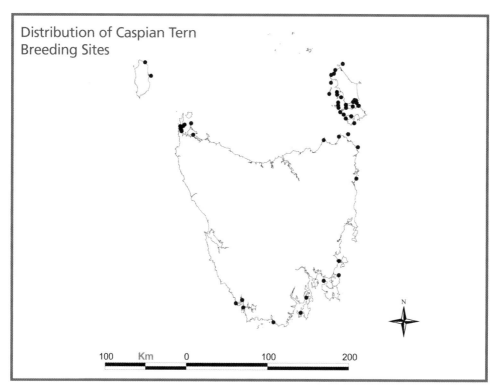

Distribution of Caspian Tern Breeding Sites

This species is sparsely-scattered throughout the region, generally occurring as solitary pairs, but occasionally colonies of up to fifteen to twenty pairs can be encountered. Caspian Terns have a nesting preference for small isolated islands, where they breed adjacent to shorelines or on poorly-vegetated areas or bare rock.

They remain faithful to nesting sites and defend them noisily and aggressively.

They are particularly susceptible to disturbance and as they are present in such low numbers they should be classified as 'vulnerable' under the Tasmanian *Threatened Species Protection Act, 1995*.

Crested Tern
Sterna bergii

Average length:	40 – 50 cm, wingspan 90 – 115 cm
Average weight:	Approx. 310 g
Breeding season:	September to March.
Breeding habits:	Nest in dense colonies, rarely in isolated pairs, colonies ranging from small groups to thousands of pairs.
Habitat:	Mostly sheltered bays, lagoons, inlets and estuaries. Breed mainly on islands, banks of sand, shells or rock, any area of relatively level ground above high water. Rarely on reefs or in dunes. They prefer areas without tall vegetation.
Movements:	Considered sedentary, resident, dispersive and partly migratory. Seem to be faithful to local area rather than specific colony.
Other features:	Nests are slight hollows scraped in the ground, usually unlined or sometimes with a little grass, guano or shells. Incubation and guarding of chicks by both parents. Lay 2 eggs.
Threats:	Human disturbance, dogs, cats, off road vehicles.

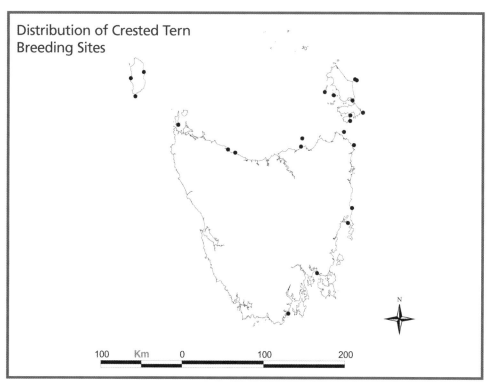

Distribution of Crested Tern Breeding Sites

Whilst this is by far the most abundant tern species of the region, and the most frequently observed fishing off beaches, it is by no means common. It is likely that no more than about 15 breeding colonies exist on the islands in any season. Like many surface-nesting species, the crested Tern remains faithful to specific nesting sites although in each island group, there will be several alternative sites that may be occupied in any season. They tend to select flat, relatively open spaces close to the shoreline for nesting. Interestingly, most of the few colonies contain relatively large numbers of pairs, reaching over one thousand on Cat Island, East Kangaroo Island and Ninth Island. Their nests are often adjacent to a Silver Gull nesting site, which is a problem, especially if human disturbance provokes the terns to leave nests exposed for gulls to raid.

The status of this species should be upgraded to 'vulnerable', because of the existence of so few breeding sites, with the majority prone to disturbance. Also the consequence of colony site selection on flat ground near the sea can precipitate storm destruction.

White-fronted Tern
Sterna striata

Average length:	35 – 43 cm, wingspan 79 – 82 cm
Average weight:	Approx. 130 g
Breeding season:	Lay October, hatch November to early December, fledge January or February.
Breeding habits:	Colonially, ranging from as few as 10 – 20 pairs to hundreds.
Habitat:	Coastal seas and exposed rocky coasts, often sandy beaches of sheltered coasts, especially bays with spits. In Australia, they breed on exposed rocky islets, stacks or exposed reefs, vegetated with grass and mats of succulents.
Movements:	Dispersive, possibly partly migratory. Some stay in nest areas during winter. May return to nest at same site in next season, as whole colony or just a few pairs, or may change sites after a successful season or after a few years. In some areas, breeding is intermittent. Where they breed annually, numbers of breeding birds vary greatly between years. They appear to return to general area rather than specific breeding site.
Other features:	Nests are shallow scrapes or depressions in sand, soil in cracks between granite slabs or in pigface. Often no nest is made, birds laying on bare substrate. Occasionally nests are lined with fragments of shells, pebbles or soft vegetation. Incubation, feeding and guarding of chicks by both parents. Lay 2 eggs. A relatively recent arrival from New Zealand.
Threats:	Human disturbance, habitat destruction, cats, dogs, off road vehicles.

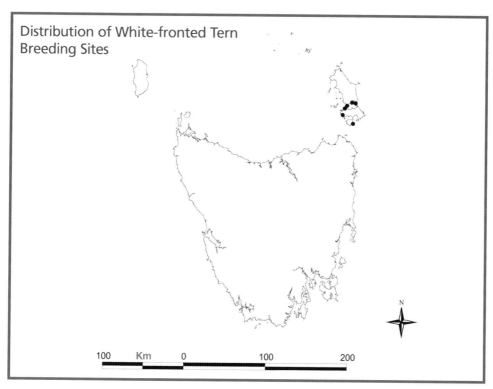

Distribution of White-fronted Tern Breeding Sites

Believed to be a very recent coloniser of Australian shores, having come from New Zealand in the 1960s, their stronghold seems to be in the southern sector of the Furneaux group, where 35 pairs are known to breed at up to 6 different island locations. Although faithful to specific nesting sites, alternative sites are used. Colonies are small with up to a maximum of 20 pairs known for any one location.

Although the majority of breeding sites are at relatively remote, seldom-visited locations, they are very vulnerable to potential visitation, as gulls will exploit any opportunity to plunder eggs or chicks left unattended. Nest sites, which are situated near sea level on flat ground, are also often vulnerable to ill-timed adverse sea conditions.

Their status under the Tasmanian *Threatened Species Protection Act, 1995* is rare given their low numbers in the Tasmanian region and unknown population trends. There is a presumption that with a recent colonising species, its abundance will increase over time, but if this is the case for this species, progress has been very slow so far.

Fairy Tern
Sterna nereis

Average length:	22 – 27 cm, wingspan 44 – 53 cm
Average weight:	70 g
Breeding season:	Lay early October to late December and in late January. Young in late January and early February.
Breeding habits:	Usually colonially, but may be solitary. Colonies can vary from 2 to 400 pairs.
Habitat:	Mostly sheltered coasts and some on inshore and offshore islands. Nests are usually just above high water mark on sheltered beaches, spits, bars, banks and ridges, usually of sand but also of shell grit. Prefer low sparse vegetation.
Movements	Partly migratory. Generally leave Tasmanian breeding sites between February and August with a few records in midwinter. Populations tend to breed within same local area, but not at fixed sites.
Other features:	Nests are scrapes in substrate, sometimes beside driftwood or other debris, usually on lee side. Incubation, feeding and guarding of chicks by both parents. Lay 2 eggs.
Threats:	Human disturbance, off road vehicles, dogs, cats.

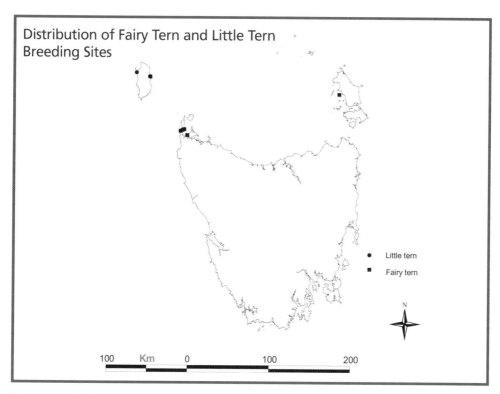

Distribution of Fairy Tern and Little Tern Breeding Sites

These two species have very specific nesting habitat requirements, which leave them extremely vulnerable to the summertime surge of beach-going holiday makers, off road vehicles and dogs. They are not species particularly associated with offshore islands, preferring nest sites on sandy beaches and spits often adjacent to a river or creek outflow. There were only 5 recorded breeding sites for Fairy Terns on offshore islands, with 112 pairs. By far the majority (89 pairs) were recorded on the sandy beaches of Chalky Island in the Furneaux group. Up to 50 individuals were recorded fishing offshore from several of the islands, usually in pairs.

The Fairy Tern is listed as rare under Tasmania's *Threatened Species Protection Act 1995* and the Little Tern as endangered under both the Tasmanian and Commonwealth Acts. Both species are extremely susceptible to disturbance and will become extinct unless their preferred nest sites are made secure through exclusion of visitors throughout the breeding season.

Black-faced Cormorant
Phalacrocorax fuscescens

Other names:	Black-faced shag
Average length:	61 – 69 cm, wingspan 93 – 102 cm
Average weight:	No data
Breeding season:	At any time of year, but eggs are laid mainly from September to December.
Breeding habits:	Colonially from 2 to 3 nests to thousands. Rarely, solitary.
Habitat:	Endemic to Australia, confined to the coasts. Marine and estuarine, nesting on rocky islands, stacks and reefs on coastal slopes and shores with rock platforms, slabs or boulders or on cliff tops and ledges up to 100 metres. Roost on islands, offshore rocks, sandbanks, navigation beacons and jetties. Vegetation is sparse or absent at most sites but they occasionally nest amongst shrubs.
Movements:	Largely sedentary with some dispersal of juveniles. During breeding, they forage close inshore.
Other features:	Solitary or gregarious, congregating where food is copious. May form large cooperative foraging flocks. Nests are large, round and open, made of seaweed, driftwood, pigface and debris, sometimes lined with grass, 40 – 50 cm in diameter. Usually lay 2 eggs. Colonies may move between islands.
Threats:	Human disturbance, high seas destroying breeding sites, gill nets.

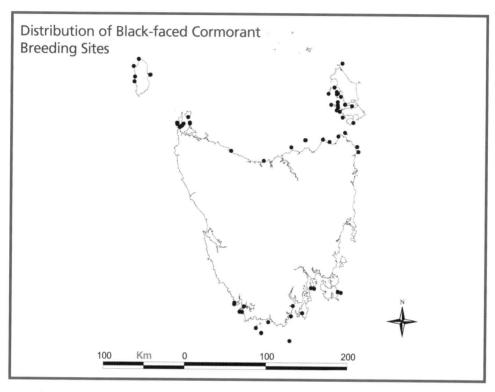

Distribution of Black-faced Cormorant Breeding Sites

Australia's only true 'oceanic' species of cormorant, it is probable that higher numbers breed in the Tasmanian region than elsewhere in Australia. Colonies are widespread in all areas of the region, but in each area there is generally only up to three nesting sites. While each colony site is traditionally used, the birds at each site invariably have one or two alternative sites within a kilometre or two, with no specific pattern of use. A site may be used in consecutive years over a period of many years, or can be vacated in alternate years or for many years, before eventually being used again. Where colonies are sited in vegetated areas, the vegetation is rapidly destroyed leaving sites conspicuously bare and white with guano. Although colonies are usually very close to sea level, at some localities on cliff ledges, they can be 20 – 40 metres above.

This species is not a particularly close-sitter and will leave eggs and chicks if disturbed. It is also at considerable risk from entanglement in gill nets, especially around high-use gill netting areas that are close to colonies, such as in Frederick Henry Bay and Storm Bay. Because of their precarious breeding habitats, some colonies are also at risk from destruction by high seas.

Australian Pelican
Pelecanus conspicillatus

Other names:	Pelican
Average length:	1.6 – 1.8 m, wingspan 2.3 – 2.5 m
Average weight:	4 – 6.8 kg
Breeding season:	Can breed all through the year.
Breeding habits:	Colonially
Habitat:	Australia, PNG, Indonesia and Fiji. Accidental to New Zealand. In terrestrial wetlands, estuarine and marine habitats, extending into arid zone. Breed on low, secluded sandy islands, islets or shores, where they nest on bare ground or amongst low or patchy vegetation. Need undisturbed site with abundant and assured food supply.
Movements:	Dispersive in wetter areas, but no regular seasonal variation in southern areas. Forage close to nest site during breeding season.
Other features:	Both sexes incubate alternately. Nests are hollows or shallow depressions, sparsely lined with material consisting of plant stems, sticks, seaweed, leaves, feathers and bones, if available. Lay 1– 3 eggs, mostly 2.
Threats:	They are particularly nervous breeders and, when disturbed, leave nests vulnerable to predation or weather extremes.

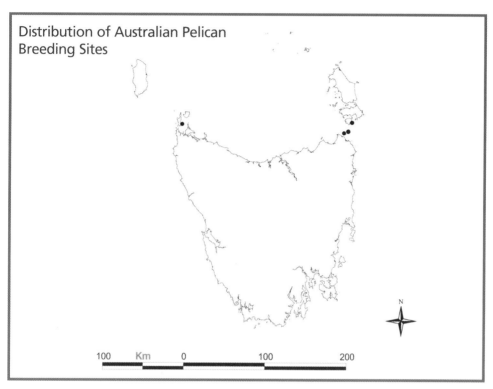

Distribution of Australian Pelican Breeding Sites

Whilst this species is occasionally encountered in most Tasmanian coastal waters, it only breeds at four localities in Bass Strait. In western Bass Strait there is one breeding site on Penguin Islet, whereas in eastern Bass Strait there are three islands where they are known to breed. However, in any one year they are likely to be found at only two of these sites. Perhaps the most frequently used sites are Low Islets to the south of Cape Barren Island and Little Swan Island. They breed at all times on Penguin Islet.

Of all the seabirds that breed in Tasmania the Australian Pelican is the most vulnerable to disturbance.

A feature common to their breeding sites in this region is easy access for this relatively cumbersome species.

Australasian Gannet
Morus serrator

Other names:	Gannet
Average length:	84 – 91 cm, wingspan 170 – 200 cm
Average weight:	2.3 kg
Breeding season:	July to February. Eggs laid August to December, fledge March/April.
Breeding habits:	Colonially
Habitat:	Marine, mostly within limit of continental shelf. Breeding sites are usually on small rocky islands using all types of terrain.
Movements:	Generally migratory/dispersive, but in Tasmania they are in colonies all year round. In breeding season foraging range is about 268 km.
Other features:	Loosely gregarious when feeding. Both sexes incubate, feed and guard chicks. Lay one egg per year.
Threats:	Habitat destruction, human disturbance, oil spills, potential fishing interaction, diet overlap with commercial fisheries species.

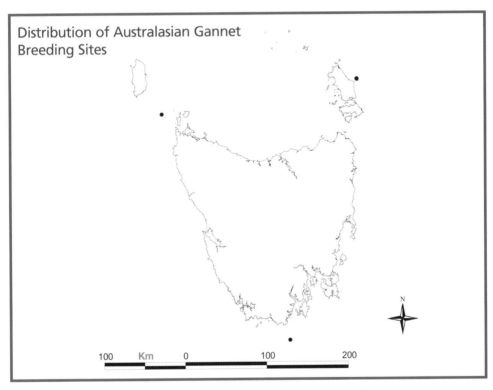

Distribution of Australasian Gannet Breeding Sites

Now only three colonies of this species remain in the Tasmanian region, with the original largest breeding site at Cat Island used as a breeding site in the early 1990s, with birds still visiting. Fortunately the other three sites at Black Pyramid Rock, Pedra Branca and Eddystone Rock were not as readily accessible and so avoided human visitation and destruction. Monitoring of these three islands has shown that the Australasian Gannet populations are steadily increasing.

Gannets are particularly vulnerable to human disturbance. They have nervous dispositions and this, in combination with their steep, rocky nesting habitat, makes eggs and chicks vulnerable to being dislodged as disturbed adults take fright and leave hastily.

On Pedra Branca, with little soil and vegetation, birds rely on seaweed and other flotsam such as plastic net and rope fragments, for nest construction. In the process of collecting such material, they are susceptible to being entangled. Currently at Black Pyramid Rock, there is sufficient soil and vegetation for nest construction, but this may change as the population continues to expand, using what is available.

Being essentially a relatively flat island, Cat Island was atypical of a gannet colony, whereas Pedra Branca and Black Pyramid Rock have the more common terrain, being steep and rugged. There are 7 Australasian Gannet breeding localities in Australia and 23 in New Zealand.

Shy Albatross
Thalassarche cauta

Other names:	White-capped albatross, mollymawk
Average length:	90 – 100 cm, wingspan 212 – 256 cm
Average weight:	4 – 4.4 kg male, 3.4 – 3.8 kg female
Breeding season:	August. Lay mid September, hatch late November, fledge April/May.
Breeding habits:	Colonially
Habitat:	Marine, breeding on only 3 Tasmanian islands, Albatross, Mewstone and Pedra Branca. They nest on level or gently sloping ledges or summits of rocky islets and stacks, usually in broken terrain with little soil and vegetation.
Movements:	Adults sedentary at all times, some adults visit in winter. Some immature birds move as far as South Africa.
Other features:	Mate for life. Lay one egg annually. Breed at the same site each year.
Threats:	Fishery interaction, diet overlap with commercial fisheries species.

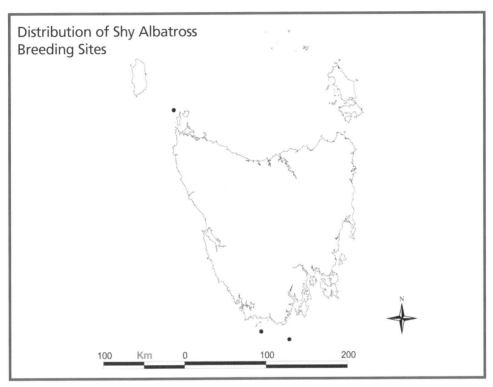

There are only three nesting localities in the world, all in Tasmania, at Albatross Island in the north and Pedra Branca and Mewstone in the south. Unusual for an albatross, this species occupies its nesting islands all year round. Egg laying commences in the last days of August and finishes in the first days of October (most are laid in mid September). Chicks hatch in December and fledge throughout April and May. Pairs generally build a high pedestal nest of soil and vegetation.

Despite the world-wide destruction of albatrosses by longline fisheries, the population on Albatross Island is recovering in numbers after near extermination at the hands of feather harvesters in the late 1800s.

Table 1 – Seabird Populations on Tasmanian Offshore Islands

Numbers represent breeding pairs

Seabird species	NW	North	Bass Strait	Furneaux	NE	East	South	TOTALS
Little Penguin	30 400	200	6000	272 500	5000	31 000	20 000	365 100
Short-tailed Shearwater	2 457 300	20	1 977 000	7 050 000	22 000	465 000	2 800 000	14 771 320
Sooty Shearwater	0	0	0	0	0	1000	5	1005
Fairy Prion	43 910	0	38 120	0	12 500	530 000	510 000	1 134 530
Common Diving-Petrel	37 180	0	27 400	1	0	275 000	25 200	364 780
White-faced Storm-Petrel	1680	10	0	60 800	19 500	10 200	0	92 190
Pacific Gull	340	2	120	570	80	170	40	1322
Silver Gull	1020	0	100	1000	800	1500	550	4970
Kelp Gull	0	0	0	0	0	400	1	401
Sooty Oystercatcher	200	1	50	350	40	20	60	721
Pied Oystercatcher	30	0	0	50	5	20	5	110
Caspian Tern	10	0	0	30	5	20	5	70
Crested Tern	25	0	0	5000	2000	400	0	7425
White-fronted Tern	0	0	0	40	0	0	0	40
Fairy Tern	1	0	0	90	0	0	0	91
Black-faced Cormorant	920	550	0	2050	700	450	300	4970
Australian Pelican	10	0	0	30	5	0	0	45
Australasian Gannet	12 400	0	0	0	0	0	7000	19 400
Shy Albatross	5000	0	0	0	0	0	7750	12 750

Island Industry

The trends towards ecotourism in the 21st century

Throughout the last two centuries, the prevailing attitude to the islands around Tasmania was to exploit them if at all possible. This occurred with little regard, perception or ability to carry it out in a sustainable manner. Certainly any natural attributes of these places have been of inconsequential importance until relatively recent times. The first evidence of a change in attitude came with the recognition that natural attributes deserved to be preserved and so islands were set aside for conservation.

For most islands little more is required to assure their protection than for them to be left to their own devices, not visited, not managed. Island habitats and island inhabitants are often extremely vulnerable to human disturbance. But in present times there are ever-increasing demands on localities that are remote or unique and the more fragile places such as islands have tended to become the greatest attraction to visitors. These visitors – ecotourists – have the potential to be as destructive to islands and their inhabitants as those involved in industries of the past – sealers, farmers, fishermen, feather pluckers and guano diggers. But unlike industries of the past, we are now equipped with sufficient knowledge to ensure that this new development proceeds responsibly.

Despite the rough sea conditions often encountered around Tasmania, by comparison to other localities around the world, Tasmanian island habitats and the species they harbour are relatively accessible. For instance, where else within an hour or two from a city the size of Hobart, Tasmania's capital, can visitors encounter a variety of species of albatrosses, view immense colonies of these breeding, watch fur seals basking on rock platforms and observe penguins and shearwaters returning to their colonies, all amidst such diverse and spectacular coastal scenery?

With appropriate consideration during the development of ecotourism, there are opportunities to further enhance, not degrade, conservation values. Public education is vital in the conservation of our marine species and their unique habitats.

Island-friendly Visitation

Guidelines for preventing and minimising impacts

Every island is different with respect to potential visitor impact because of the variety of island size, shape, topography and other natural features. To cater for this variety, visitor guidelines and codes of conduct have, thus far, been relatively generic. As visitor patterns and impacts become better understood, island-specific guidelines and management plans can be created. This process is already underway with the Small Bass Strait Islands Management Plan, which sets out procedures and actions to protect twelve of the most vulnerable seabird and seal breeding colonies in Bass Strait. Planning is also underway for the most vulnerable east and south coast islands.

Similarly, guidelines have been drawn up for sustainable ecotourism practices for visitors to seal and seabird islands. These are listed below.

Another approach is to encourage the implementation of codes of conduct for the recreational and professional groups that utilise the islands. A sea kayaking code of practice is currently being developed. Similar codes could be implemented for yachtspeople and fishers, who are likely to visit Tasmania's islands.

To maximise island visits, the following is provided as a set of guiding principles:

- Think about why you want to land on an island. Do you really need to? Try to limit your impact by restricting your time and movement ashore.
- Do not land on islands with seals, pelicans, White-fronted Terns, fairy terns or Australasian Gannets.
- Check the island's status and vulnerability to visitation. Nature reserves will often require visitor permits, which may require weeks to process.
- Visit at dusk to watch the arrival of birds from sea. Position yourself so you are not on their flight path or landing areas (they are soon apparent).
- Visit when chicks are big or fledged, as disturbance is less likely.
- Notify somebody you have gone to an island as very few have telephones!
- Avoid islands with snakes as, when present, they are in large numbers.

The greatest threats to islands are:

- the introduction of feral plants and animals;
- fire; and
- disturbance to nesting birds which results in desertion of nests and death.

Guidelines for visiting a seabird colony

When visiting a seabird colony the following actions will minimise your impact on the resident birds and their habitat:

- Always report your visit to the Parks and Wildlife Service (PWS) and obtain the relevant permits.
- If possible, plan for a day trip with no overnight camping.
- If an overnight stay is unavoidable, preferably stay on your boat or set up camp as far away from the bird colonies as possible.

- Ensure that there are no feral pests such as rats and mice aboard your boat. They could devastate a seabird colony either by eating or destroying the eggs and carrying ticks, fleas or other harmful parasites.

- Prior to arrival, thoroughly wash your shoes, tent pegs and other soil-contacted equipment in salty water to avoid transporting *Phytophthora cinnamomi* and other potentially damaging weed seeds or fungal spores.

- Where possible, walk on the rocky shoreline to ensure that you do not trample on burrows and nests.

- Look out for birds being disturbed. When agitated they fly in circles or squawk. Ensure that you keep well away, if they are displaying any agitated behaviour.

- Look out for small nesting birds, particularly between October and March. Some, like terns, are often difficult to see, so stay alert.

- Do not light fires. Smoke can distress some birds and there is always a risk of escape, no matter how careful you may be. Always carry a fuel stove.

- Avoid setting gill nets, within 500 metres of a bird colony and in particular along shorelines that penguins visit to breed. Many diving birds are killed in nets, especially during their breeding season when they forage nearer their colony.

- Record in as much detail as possible information about dead wildlife and unusual occurrences you may witness during your visit. Report them to PWS rangers on your return.

Seal watching guidelines

- Approach a seal colony quietly. Seals are sensitive - fast boats, noisy engines, clattering sails and rattling anchor chains frighten them, making them flee into the water.

- During the breeding season, disturbance may cause stampedes, where pups are crushed or forced off the colony. Lower your sail or reduce speed to under 10 knots within 200 metres and 5 knots within 100 metres of the colony.

- Approach downwind of the colony slowly and quietly - seals have an acute sense of smell and will be wary of your intrusion into their territory.

- In November and December when pups are born, boats should not moor or pass closer than 100 metres to a seal colony.

- At all other times of year it is best to go no closer than 50 metres to a seal colony.

- Do not swim with seals, as sharks often share the same habitat.

- Never land at a seal colony. Disturbed seals will stampede, endangering their pups. In cases of severe disturbance, they may abandon them.

- Never discard plastic material overboard. Seals can be killed by swallowing it or, more likely, from becoming entangled in it.

- Avoid setting or constantly attend gill nets within 500 metres of the bird colony. Many diving birds are killed in nets, especially during their breeding season when they forage nearer their colony.

- Record as much detail as possible information about dead wildlife and unusual occurrences you may witness during your visit. Report them to PWS officers on your return.

Tasmania's Offshore Islands by Region

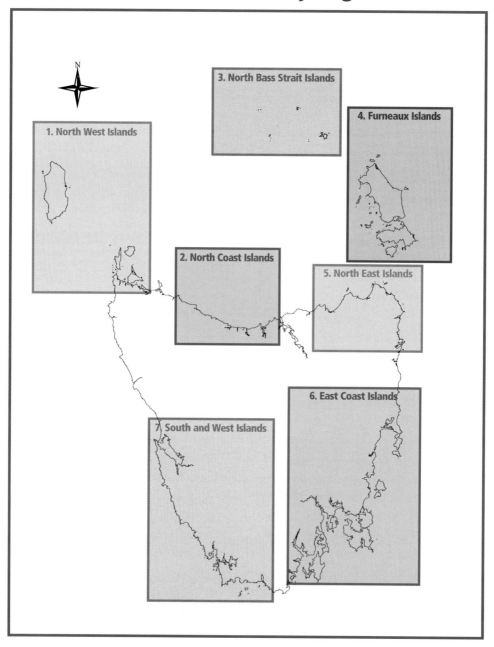

North West Islands – Region 1

(Regions, page 44)

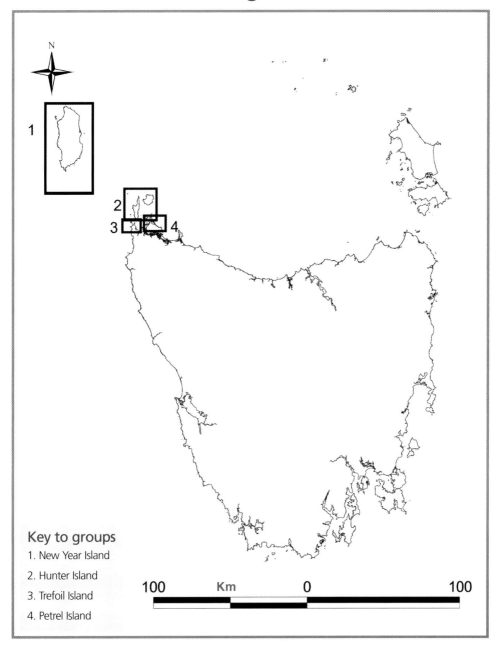

Key to groups
1. New Year Island
2. Hunter Island
3. Trefoil Island
4. Petrel Island

New Year Island Group

(Region 1, page 45)

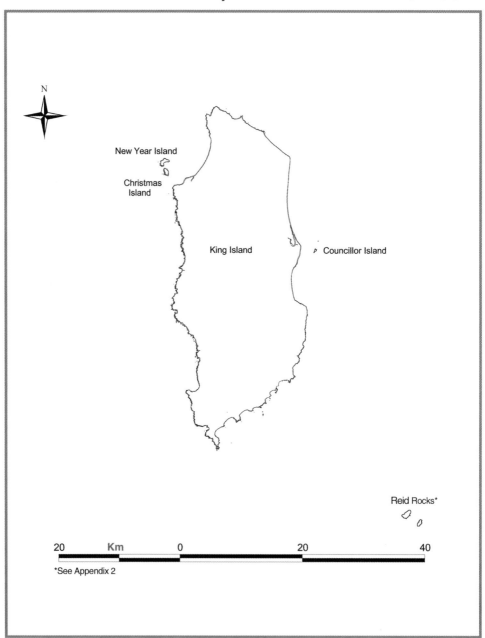

Christmas Island

(New Year Island Group, page 46)

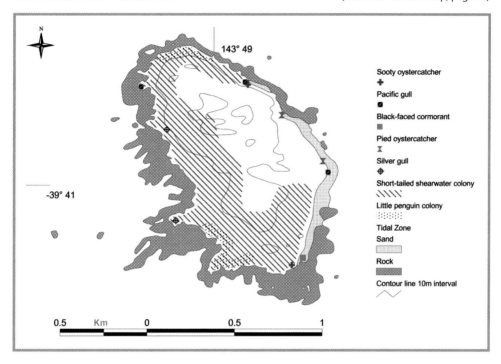

Location: 39°41'S, 143°50'E
Survey date: 26/11/84
Area: 63.49 hectares
Status: Nature Reserve

Christmas Island is similar to its neighbour, New Year Island, but with extensive north-east to south-east orientated beaches with steep sand banks leading up to scrub at the mid to north eastern side. All other shores are rugged granite rock with boulder and occasional grit beaches, which slope gently up from the rocks to the highest point in the north. There are remains of early habitation on the island.

BREEDING SEABIRD SPECIES

Little Penguin

An estimated 11 883 pairs, with eggs and small chicks, breed on the island. The majority are in the northern area, where they come ashore on rocks and beaches with well-worn tracks up steep dunes in the north-east. They nest either in shallow burrows and ditches under very thick high saltbush or under densely matted *Rhagodia*.

Short-tailed Shearwater

An estimated 412 732 pairs breed in an extensive area covering two-thirds of the island in *Rhagodia*, *Atriplex cinerea* and grasses in very sandy soil.

New Year Island to Christmas Island.

Pacific Gull
3 pairs, one with 2 eggs, one with one chick and one egg and one with 3 chicks, were located around the northern coastline of the island.

Silver Gull
650 pairs, with very small to fully-fledged chicks, were counted in the 1984 survey.

Many nests had been destroyed and dead chicks found, probably from very strong winds and rain of 3/11/84. There are two distinct Silver Gull colonies at the bare rock edge on the west coast, near saltbush and *Carpobrotus rossii*.

Sooty Oystercatcher
3 pairs, 2 with eggs were located around the east coast.

Pied Oystercatcher
2 pairs, one with 2 eggs, the other with no eggs, were located on the sandy beach on the mid-eastern coast.

Black-faced Cormorant
21 pairs, most with fleglings, were located on the rocky reef at the south end of the island, which is sometimes wave-washed.

No other seabirds were recorded.

MAMMALS
Mouse – possibly New Holland Mouse

BIRDS
Native:
Swamp Harrier – nest with 3 eggs, freshly laid

Ruddy Turnstone

Sanderling and other waders

Red-capped Plover

Hooded Plover

Masked Lapwing – 2 pairs

Silvereye

Introduced:
Skylark – nest with 4 eggs on sandspit

REPTILES
Tiger Snake – many

Lizards – unidentified

VEGETATION
Dominant vegetation is *A. cinerea*, *Rhagodia candolleana*, *C. rossii* and grasses such as *Poa poiformis* with scattered patches of scrub with boobyalla *Myoporum insulare*.

COMMENTS
This island has a high diversity of seabirds and intact natural vegetation.

Burrow density transect in thick vegetation of Christmas Island west coast.

New Year Island

(New Year Island Group, page 46)

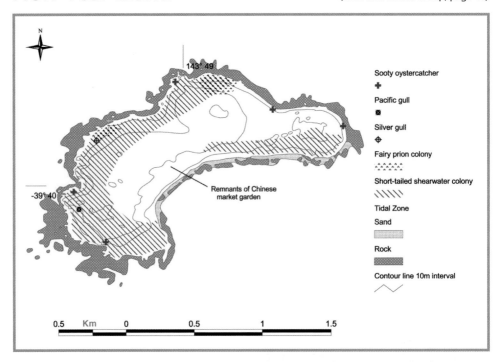

Location: 39°40'S, 143°50'E
Survey date: 1/12/86
Area: 98.22 hectares
Status: Game Reserve ⇑

It is a very sandy island with the central area badly wind-blown. The east side is characterised by sandy beaches broken by rocky outcrops with a steep sandy slope to the beach at the south end. The rest of the shoreline is granite rock and boulders edged by thick *Atriplex cinerea* patches and *Rhagodia*.

BREEDING SEABIRD SPECIES
Short-tailed Shearwater

71 464 breed in the south-east colony and 508 812 in the main western colony. The south eastern colony is in very sandy soil,

Mouse (species uncertain) burrow in sand blow entrance protected by bone arrangement.

dominated by a cover of *Rhagodia*, *Atriplex cinerea*, introduced cabbage and *Tetragonia*. The major colony along the western coast is dominated by 80% *Rhagodia* in sandy soil.

Fairy Prion
100 pairs were located in two small areas in rock outcrops in a limited number of crevices.

Pacific Gull
3 pairs, two with eggs and one with a downy chick, were nesting on the north west shoreline.

Silver Gull
35 pairs were defending but not nesting, in two areas on the western coast.

Sooty Oystercatcher
6 pairs were located scattered around coast.

VEGETATION
A. cinerea is the dominant vegetation with *R. candolleana*, *Tetragonia implexicoma* with introduced cabbage, remnants of the Chinese market garden which existed in 1861. (Hooper, 1973)

New Year Island, north side toward Christmas Island.

No other seabird species were recorded.

MAMMALS

Many mammal bones were found in the sandblow – Common Wombat, Tasmanian Pademelon, Red-necked (Bennett's) Wallaby, Spotted-tailed Quoll, Australian Fur Seal and pig (introduced in the 1800s either by sealers or the Chinese market gardeners)

Mouse – species not identified, common.

BIRDS

Native:

Swamp Harrier

White-faced Heron

Hooded Plover

Masked Lapwing

White-fronted Chat

Silvereye

Introduced:

Skylark

REPTILES

White's Skink

Metallic Skink

Blue-tongue Lizard

Tiger Snake – common

OTHERS

Cape Butterflies – many

COMMENTS

Its Holocene dune system, which supports a range of mammals' bones, is considered of significance for the local region. The terrestrial mammals at the site probably became extinct as the sea level rose, creating what is now New Year Island, an area insufficient in size and without fresh water to support the herbivores once present (Dixon, 1996).

Whether the wind blows present today are the consequence of interference to the habitat is uncertain. The historical significance of the island as a Chinese market garden in the 1860s and its diversity and abundance of seabirds warrants its status being upgraded to nature reserve.

Councillor Island

(New Year Island Group, page 46)

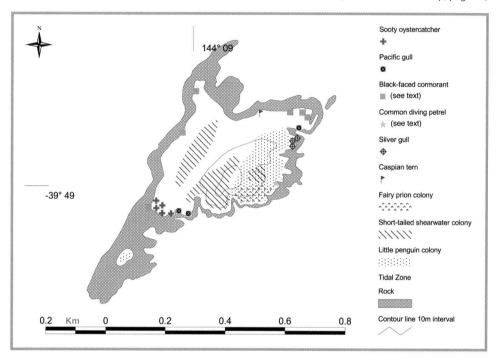

Location: 39°50'S, 144°09'E
Survey date: 19/11/87
Area: 10.53 hectares
Status: Non-allocated Crown Land ⇑

It is a low island with gently undulating topography and large expanses of flat ground on the west coast to the north. The north-east and south-west slopes are steep and extend to the bare rocks of the shoreline. Much of this shoreline is composed of small round black pebbles. The eastern shoreline is rocky and steep to the sea. Soil is insufficient over most of the island for burrowing birds: penguins mainly nest under vegetation. A small islet to the south-west is connected at low tide by a rocky reef. Its topography is flat with shores of round pebbles.

BREEDING SEABIRD SPECIES

Little Penguin

On the small islet, 26 pairs nest under the *Rhagodia* and on the main island an estimated 7990 pairs were found all over in average densities of $0.5/m^2$ but in higher densities down the eastern slope under *Rhagodia* and on the summit and its slopes. They come ashore all around the island and there are several well-worn pathways. The majority surveyed were on eggs and some were with small chicks.

Short-tailed Shearwater

Shearwaters were found in average densities of $1/m^2$ over the whole island. The main colony of 23 168 pairs runs north-east to south-west across

the base of the northern tip extension with short burrows mainly occurring in hollows or depressions, where there is sufficient soil. The other two colonies are on the mid-east side slopes, predominantly in dense *Poa*. The larger has an estimated population of 17 321 pairs and the smaller 4194 pairs.

Fairy Prion
1500 pairs, most with eggs, were interspersed with common diving-petrels, largely confined to eastern and south-eastern slopes in *Poa*, covered with *Carpobrotus rossii* patches and *Senecio* sp.

Common Diving-Petrel
An estimated 5000 pairs inhabit the island. With the exception of the few in the *Rhagodia* on the southern islet and a few scattered along the south-west edge of the main island, this species is mainly on the steep *Poa*-covered slopes in a 20 metre wide strip from the southern tip two-thirds of the way to the north-east tip. They are also interspersed with fairy prions. At the time of the survey some had eggs and some were with chicks, both down-covered, and fully-fledged.

Pacific Gull
27 pairs, all with eggs and one pair also with one chick, were counted. Most of the population is in two concentrated colonies, one at the beach edge alongside *Rhagodia*, with the main colony on the north face of the summit hill in low grassed clearings surrounded by *Poa*.

Silver Gull
27 pairs were located in a tight colony following a low bare rock ridge surrounded by scattered *Poa* and *Senecio* sp.

Sooty Oystercatcher
5 pairs were nesting on a rock ledge on the edge of the *Carpobrotus rossii*.

Black-faced Cormorant
50 pairs were located. Three obvious nesting sites were located at the edge of the vegetation on bare basalt rock surfaces. The southernmost colony is on steep rock slopes. The other two are on flat ground. 300 were also roosting at the far north-east tip with 6 juveniles.

Caspian Tern
1 pair, with 2 quarter-grown downy chicks, was located on the far north-west tip in an old cormorant colony.

VEGETATION
The vegetation of the main island is dominated by *Poa* from the summit to the coast, where *Rhagodia candolleana* is found in a 5 metre wide strip, particularly along the east coast. *C. rossii* and *Senecio lautus* dominate the rocky shore. The dominant vegetation of the south-west islet is *R. candolleana* (90%) with scattered *Poa poiformis* and *Senecio lautus*.

No mammals were recorded.

OTHER SEABIRD SPECIES
A White-faced Storm-Petrel landed on the deck of the boat off the western coast. Suitable habitat on shore is available, but no birds were found.

BIRDS
Native:
Swamp Harrier
Fairy Tern – 2
Welcome Swallow
Little Grassbird

REPTILES
Blue-tongue Lizard

COMMENTS

With such a high diversity of seabirds, the status of Councillor Island should be upgraded to nature reserve to afford it full protection. The possible presence of White-faced Storm-Petrels and mammals should be investigated further.

Hunter Island Group

(Region 1, page 45)

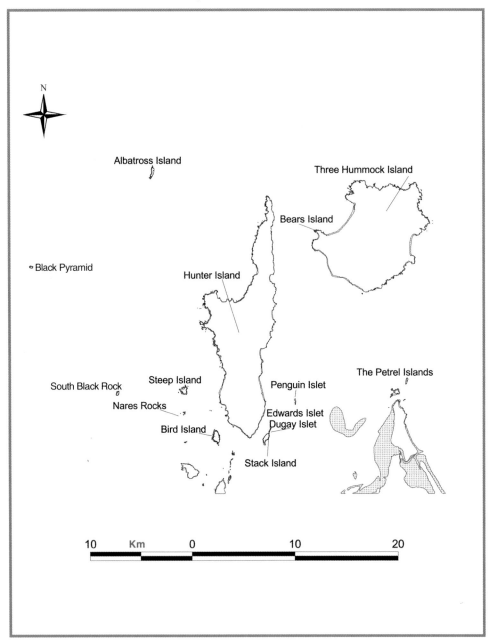

Steep Island
(Hunter Island Group, page 56)

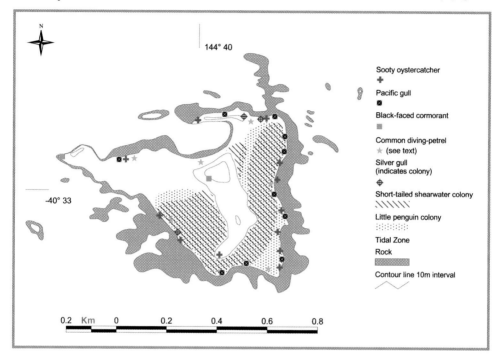

Location: 40°34'S, 144°41E
Survey date: 12/12/82
Area: 21.6 hectares
Status: Private property

Sometimes referred to as Steep Head, it is a steep, spectacular island, with cliffs dominating the coastline and an impressive crater-like structure on the north-west side. Pebble beaches, possibly basaltic beach cobble deposits, begin on the south-west side behind the rock platform and extend to the northern end. The north face is mostly an 80 metre wide rock platform with a pebble beach in the crater to the west. A large sea tunnel cave (20 m x 10 m) pierces the western headland. The offshore reef makes landing difficult. Buildings include huts associated with shearwater harvesting. There were vehicle tracks around the shearwater colony. This island was once used for sheep grazing with the last few still present in 1982.

BREEDING SEABIRD SPECIES
Little Penguin
2000 – 3000 pairs, with eggs and fully grown chicks, were estimated to be scattered all over the island amongst Short-tailed Shearwater colonies. They appear to be concentrated on the lower slopes, particularly the south-east to east, below the cliff face, where the habitat is most suitable.

Short-tailed Shearwater
An estimated 250 100 burrows, in average densities of 1.16/m^2, were recorded on the island. The highest densities occur in thick *Poa*, although some burrows are also in gravelly soil.

Common Diving-Petrel

An estimated 23 000 – 25 000 burrows were located in a 10 to 20 metre continuous strip around the cliff face all round the island. Burrows are often tunnels beneath the *Disphyma crassifolium*. On the main western ridge, burrows are on steeper parts in clay soil, often winding amongst sharp rocks away from the short-tailed shearwaters' territory. Carcasses were frequently found in the Pacific Gull breeding territory.

Pacific Gull

10 pairs, most on 1 - 2 eggs but a few with newly-hatched chicks and one with an almost fully-fledged chick, were found all around the island, apart from the west coast, on the base of cliffs and pebble beaches. A few were also on high ridges.

Silver Gull

33 empty nests, 19 with eggs, 12 with chicks and 2 with runners were recorded in three distinct colonies – one on the south-west coast, one on the north coast on the cliff face and one on the north-east on the cliff ledges up to 25 metres on *Sarcocornia* sp. and bare soil.

Sooty Oystercatcher

12 pairs, with eggs and small chicks, were mostly at the base of slopes at the edge of the pebble beaches with one nest at the top of the cliff, 30 metres up on the south-west side.

Black-faced Cormorant

80 – 100 pairs were nesting mostly in *Poa*-dominant areas on the western cliff ledges, which are virtually inaccessible. One group of nests was found in a cave 25 metres up the cliff.

VEGETATION

Poa poiformis is the dominant vegetation. The cliff edges right around the island are fringed by a strip of *D. crassifolium*, with myrtaceous scrub co-dominant in some places but generally not as extensive. *Sarcocornia* sp is also co-dominant in small patches.

Unique and spectacular topography of Steep Island with shoreline huts at the north-east corner.

Western gulch with diving petrel habitat in foreground.

No mammals or other seabird species were recorded.

BIRDS
Native:

Brown Quail – 1

Cape Barren Goose – flock of 6 , plus 2 pairs on the north of the island

Nankeen Kestrel – 1 pair with 3 fledged chicks on the cliff ledges of the crater

Forest Raven – 1 pair

Introduced:

Common starling – flock of 30+

REPTILES

Tiger Snakes, introduced irresponsibly, are now present on the island and potentially devastating to the seabirds.

COMMENTS

The ownership of Steep Island was recently transferred back to the Aboriginal community. Its unique geomorphology is of outstanding significance at a State level (Dixon, 1996).

Steep Island, west side.

Nares Rocks

(Hunter Island Group, page 56)

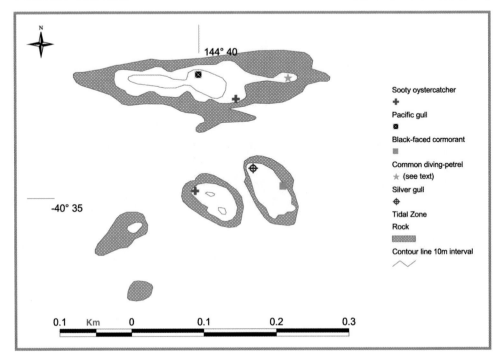

Location: 40°35'S, 144°41'E
Survey date: 10/12/82
Area: 0.76 hectares
Status: Non-allocated Crown Land

Nares Rocks, a group of three rocks, are steep and predominantly bare, with little suitable soil depth for burrowing birds.

BREEDING SEABIRD SPECIES
Common Diving-Petrel
Up to 20 pairs were located on the east side of the northern rock, about half way up, just above where the vegetation line starts.

Pacific Gull
1 pair, with a chick, was found on the northern rock, near the summit, hidden in a rock hollow in *Disphyma crassifolium*.

Silver Gull
6 pairs, some with eggs, some with chicks and some empty nests, were located on the eastern rock on ledges in the north west corner in *Poa poiformis*.

Sooty Oystercatcher
2 pairs, one with an empty nest and one with one chick, were located. One was on the northern rock with the nest tucked well into the rock hollow, low down at the vegetation edge of the south-east corner. The other was on the western rock on the east side in *D. crassifolium*.

Black-faced Cormorant
8 unused nests, constructed of *Tetragonia*, were located on the east side on a cliff ledge.

Nares Rocks, east side

VEGETATION
D. crassifolium and *Tetragonia implexicoma* are dominant.

No other fauna was recorded.

BIRDS
Welcome Swallow – one at the edge of the north rock

COMMENTS
Being relatively isolated and very small in area makes these rocks secure from disturbance.

Bird Island

(Hunter Island Group, page 56)

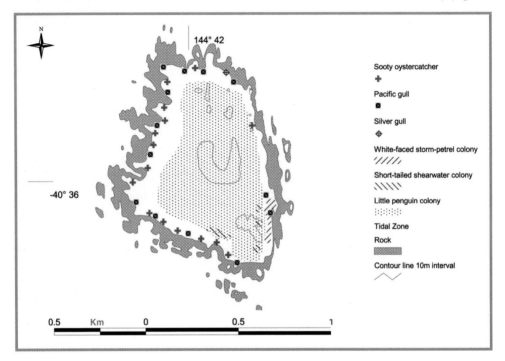

Location: 40°36'S, 144°43'E
Survey date: 12/12/82.
Area: 43.92 hectares
Status: Game Reserve ↑

It is a rectangular island orientated north to south, tapering at the southern end to form a small peninsula. Most of the island has either shallow soil or rocks. The only areas with adequate soil are occupied by burrowing petrels. The coastline from the southern tip to halfway up the west side comprises small shingle beaches alternating with rocky areas forming reefs out to sea, with extensive tidal flats. At the northern end there are a few small sandy beaches. The eastern side is mainly rocky with one small beach.

BREEDING SEABIRD SPECIES

Little Penguin

An estimated 3000 pairs, with eggs and small chicks, were located in burrows intermingled with the Short-tailed Shearwater colonies. They are also sparsely scattered over the island, extending into the thick scrub at the centre.

Short-tailed Shearwater

An estimated 4940 pairs were located in four distinct colonies at the southern end of the island, where the vegetation is dominated by *Poa poiformis* and *Senecio* sp. Where the *P. poiformis* forms dense mats, the burrow density is less than in the more open area. The soil is soft and sandy. The colonies end abruptly where the scrub commences.

White-faced Storm-Petrel
20 pairs, most with eggs, were located on the peninsula at the southern end of the island. The colony is covered by thick *P. poiformis*. The burrows are on average about 0.8 m long.

Pacific Gull
13 pairs, at nests with eggs and down-covered runners, were nesting on the shoreline all around the island, but concentrated on the western shore along the edge of the vegetation. They generally nest in association with stipa and bare rocks.

Silver Gull
23 pairs, mostly at nests with eggs and chicks, were nesting on low rocks 30 m from vegetation at the northern end of the island. The nests were constructed of *P. poiformis*.

Sooty Oystercatcher
30 pairs, some without nests but acting territorially, some on new empty nests and some with small and large chicks, were located all around the shoreline, but concentrated particularly on the western side. The nests were positioned on small pebbly beaches near rocks

Dense scrub in the central area of Bird Island.

Eastern shoreline with jagged granite formations

and *P. poiformis* or *Atriplex cinerea*. There was a strong association between the locality of Pacific Gull and Sooty Oystercatcher nests, with some only a metre apart and usually within 5 m of each other. The higher density of birds on the western side may be related to the extensive reefs and rock pools particular to this area, which are exposed at low tide, providing an abundant food source.

VEGETATION
The vegetation over most of the island is stunted, with thick, tangled scrub dominated by myrtaceous shrubs. The vegetation along the coastline is dominated by a strip of *A. cinerea*, stipa, *Disphyma crassifolium* and *Sarcocornia* sp. The vegetation at the eastern end is dominated by stipa.

OTHER SEABIRD SPECIES

Common Diving-Petrel – remains were found in several Pacific Gull territories, but it is suspected they do not breed here

Black-faced Cormorant – 3 were roosting at a regular site at the southern tip

Australian Pelican – 4 were seen ashore on beaches and a dead one was found

Crested Tern – 23 on rocks.

Fairy Tern –1 pair was seen feeding in rock pools on the western side

No mammals were recorded.

BIRDS

Native:

Brown Quail – 2 seen

Cape Barren Goose – 1 pair

Red-capped Plover – 4 seen

Ruddy Turnstone – 18 seen

Masked Lapwing – 1 nest with 1 large chick

New Holland Honeyeater – 1 seen

White-fronted Chat – 2 seen and 1 nest in *Sclerostegia arbuscula*

Welcome Swallow – 6 seen

Little Grassbird

REPTILES

Tiger Snake – abundant

Metallic Skink

COMMENTS

The thick scrub is vulnerable to fires, which result from activities associated with the harvesting of short-tailed shearwaters, one partly burning the island in the mid 1980s. Despite the island having a low seabird species diversity, its status needs upgrading to protect the high diversity of vegetation, which is rare for such a small island and to conserve the breeding site of the White-faced Storm-Petrel. Its large population of Sooty Oystercatchers is also unique.

Stack Island

(Hunter Island Group, page 56)

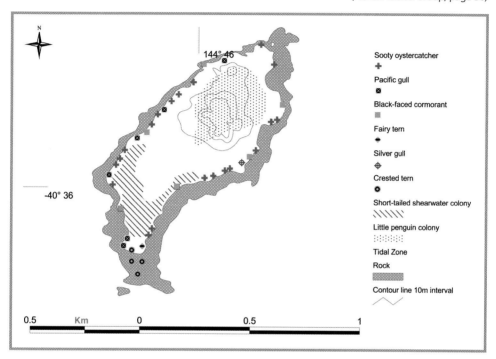

Location: 40°36'S, 144°46'E
Survey dates: 25/1/82-8/2/82, 15/12/82
Area: 23.7 hectares
Status: Game Reserve

It is an elongate island running north to south, with the summit of 54 metres in the north. It has an extensive sandy beach shoreline on the east and west sides of the southern sector and a granite rock shoreline elsewhere.

BREEDING SEABIRD SPECIES
Little Penguin

200 pairs, mostly with half-grown to fully-fledged chicks, were located, scattered amongst the Short-tailed Shearwater colonies. They were also nesting under boulders right to the summit and under rock clumps of the summit. Small numbers were also nesting beneath boulders, lightly scattered along the slopes.

Short-tailed Shearwater

63 342 pairs, generally with small chicks, were nesting in an average density of $0.51/m^2$ in sandy soil, mainly along the south-west and south-east coasts. They were also nesting beneath tea tree in a small strip just in from the beaches.

Pacific Gull

6 pairs, 3 accompanied by juveniles, were located on the southern and western shores.

Silver Gull
10 pairs, only one with eggs, were located on the central eastern shore at the vegetation/rock interface.

Sooty Oystercatcher
20 pairs, 3 with eggs were located all around the shore of the island.

Black-faced Cormorant
Several nests were recorded and up to 10 were seen fishing offshore.

Crested Tern
25 pairs were nesting on the southern sandbar.

Fairy Tern
1 pair was recorded as breeding at the southern end of the island.

VEGETATION
Leptospermum sp. scrub dominates the southern end of the island, with *Poa* and stipa around the rocky coastline.

OTHER SEABIRD SPECIES
Australasian Gannet – flew by

Australian Pelican – 2 were seen regularly fishing and flying around.

Little Pied Cormorant – a few

Great Cormorant – a few

Caspian Tern – 1 adult and 1 dead fledged juvenile

MAMMALS
Rabbit – in small numbers

Water Rat

BIRDS
Native:
Brown Quail – 2 at south end

Black Swan – 6 were seen off the west shore

Pacific Black Duck – regular flocks of up to 15

Chestnut Teal – usually in small parties of 3 - 5 but on several occasions flocks of up to 30 - 40 were also seen feeding on grit of west beach

White-faced Heron – 1

Cape Barren Goose – 1 on shoreline and tennis court area.

White-bellied Sea-Eagle – 1

Swamp Harrier – several seen

Brown Goshawk – hovering around the summit

Brown Falcon – 2

Ruddy Turnstone – 5 recorded on the south tip of the rocks

Hooded Plover – in pairs and parties of up to 5

Masked Lapwing – 1 pair

White-throated Needletail – 20+

Crescent Honeyeater

White-fronted Chat

Black Currawong – 2 nests were found in teatree thicket, the chicks had just left

Forest Raven – a nest in tea tree was found

Welcome Swallow

Silvereye – 5 eggs in nests in teatree – eggs were gone 3 days later

Introduced:
Common Starling – 1 flew by

No reptiles were recorded.

Stack Island vegetation.

COMMENTS

Stack Island has a particularly high diversity of bird species and should be given Nature Reserve status to reflect its ecological significance in the region. The reservation should include Dugay and Edwards Islets. Rabbit eradication is desirable to maintain the integrity of the vegetation. It is interesting to note the absence of reptiles on an island of this size and habitat, particularly as nearby Hunter Island has a large Tiger Snake population.

Dugay Islet

(Hunter Island Group, page 56)

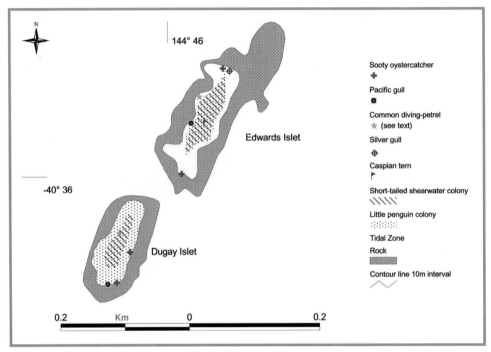

Location: 40°36'S, 144°47'E
Survey date: 15/12/82
Area: 0.44 hectares
Status: Non-allocated Crown Land ↑

It is a small, elongate, narrow, granite islet with a pebble beach at the southern end and rocky shores. The eastern side slopes more gently to the sea than the western.

BREEDING SEABIRD SPECIES
Little Penguin
An estimated 50 pairs breed all over the island in burrows amongst the stipa and *Carpobrotus rossii*, interspersed with short-tailed shearwaters.

Short-tailed Shearwater
An estimated 200 pairs were recorded all over the island in all available burrowing habitat, amongst rocks. Highest densities were found on the eastern side.

Common Diving-Petrel
2 burrows were found amongst the *C. rossii*, where the wind has formed sandy tunnels amongst the vegetation. However, there was no recent evidence of breeding.

Pacific Gull
One pair, with a small chick, was located at the southern end of the island.

Sooty Oystercatcher
2 pairs were found on the eastern side of the island.

Penguins evening arrival

VEGETATION
The vegetation is dominated by stipa., *C. rossii* and *Tetragonia* sp.

COMMENTS
Dugay and Edwards Islet should be included as part of a newly-created Stack Island Nature Reserve.

No mammals, reptiles or other seabirds were recorded.

BIRDS
Cape Barren Goose – 1 adult seen

Edwards Islet

(Hunter Island Group, page 56)

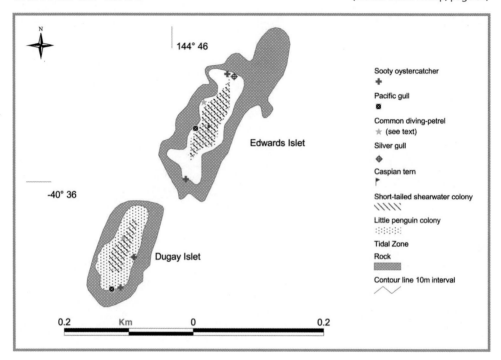

Location: 40°36'S, 144°47'E
Survey date: 15/12/82
Area: 0.58 hectares
Status: Non-allocated Crown Land ↑

It is a small, granite island, elongate in shape, increasing in width towards the northern end. The shoreline is rocky with a small shingle beach at the southern end. There is a ridge running down the centre of the island.

BREEDING SEABIRD SPECIES
Little Penguin
20 to 30 pairs are found all over the island with approximately half the birds nesting in burrows and half nesting under *Poa poiformis*.

Short-tailed Shearwater
An estimated 20 pairs were found all over the island nesting in burrows and under *Poa* tussocks.

Common Diving-Petrel
3 burrows were located on the western side of the island amongst *P. poiformis*. An adult carcass was also found. It is thought that up to 10 pairs may breed here.

Pacific Gull
1 pair, at an empty nest, was located on the western side of the island.

Silver Gull
An estimated 25 pairs were found breeding in a colony at the northern end amongst *Carpobrotus rossii* and *P. poiformis*.

Sooty Oystercatcher
2 pairs, with eggs, were located at opposite ends of the island.

Caspian Tern
A pair, with a large chick, was located on the central ridge midway along the island.

No mammals, birds, reptiles or other seabird species were recorded.

VEGETATION
The vegetation is dominated by *Atriplex cinerea* and *Rhagodia candolleana* with *P. poiformis*. stipa dominates the coastal strip and on the western side *Sclerostegia* sp. and *Sarcocornia* sp. are dominant.

COMMENTS
As for Dugay Islet, this islet should be incorporated into a Nature Reserve with Stack Island.

Penguin Islet

(Hunter Island Group, page 56)

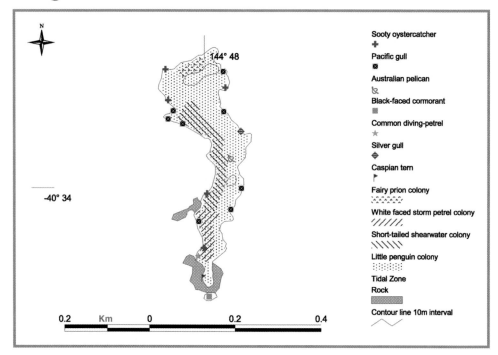

Location: 40°35'S, 144°49'E
Survey date: 11/12/82
Area: 3.46 hectares
Status: Nature Reserve

Penguin Islet is long and narrow, orientated north to south, tapering to a point at the southern end. It rises to a rocky knoll at the northern end and is surrounded by beaches and rocks. The largest beach is at the southern end. Small dunes occur towards the south-west end of the island. The soil is sandy.

BREEDING SEABIRD SPECIES
Little Penguin

50 pairs, some with empty nests, most with eggs and some with chicks of varying sizes, were located in burrows over most of the island amongst Short-tailed Shearwater burrows and in rock crevices along the shoreline.

Short-tailed Shearwater

An estimated 3060 burrows are concentrated towards the centre of the island in an area just behind the northernmost beach on the west side, stretching across to the east side. Burrows were approximately 0.8 m long and 0.5 m deep in sandy soil dominated by *Poa poiformis*. Burrowing was sometimes in association with White-faced Storm-Petrels and perhaps Common Diving-Petrels.

Fairy Prion

An estimated 100 pairs were located, almost exclusively in rock crevices, on the northern side of the rocky knoll, at the northern end. All birds

observed were on eggs or with very small chicks.

Common Diving-Petrel

There are possibly up to 200 pairs breeding on the island. They are intermixed with White-faced Storm-Petrel burrows and in crevices, so the numbers are difficult to estimate. 30 carcasses, most pre-fledged young, were found in Pacific Gull nests.

White-faced Storm-Petrel

Up to 1000 burrows are estimated to exist in the sandy soil in the southern half of the island, where vegetation is dominated by *Poa poiformis* and *Tetragonia* sp. They do not extend further north of the west coast beach. On average, burrows were 0.8 m long and 0.5 m deep. The birds were on eggs in unlined, fragile burrows, which could easily collapse due to the sandy soil.

Pacific Gull

11 pairs, at nests with eggs, small chicks and runners, were located nesting on the shoreline all around the island. Concentrations occur at the southern end where a colony of 4 pairs was nesting on the west side and 4 pairs were on the east side.

Silver Gull

12 pairs, at freshly-built nests, were defending sites on the eastern side of the island.

Sooty Oystercatcher

5 pairs, on new empty nests, were located on the western side of the island on the shoreline and 1 pair was on the eastern side.

Black-faced Cormorant

28 pairs at newly-constructed nests with only one nest with an egg, were located on a rocky reef at

Aerial photography is used to monitor Australian Pelican breeding without the disruption that ground visits can cause.

Toward the southern end of Penguin Island.

the south end of the island, less than 1 m above high water. 200 to 220 birds were roosting in the same area.

Australian Pelican

9 pairs with 9 large chicks were located at nests in the centre of the island, inland from the main southern beach. 8 adults were ashore at the south-west corner.

Caspian Tern

1 pair was located at a nest with egg fragments at the south-west end of the island on top of a small sand dune.

VEGETATION

The vegetation is dominated by *Poa poiformis* and *Tetragonia* sp. with *Disphyma crassifolium* and *Carpobrotus rossii* concentrated around the shoreline. There is one patch of myrtaceous scrub on the north-east shore. The *P. poiformis* is thickest at the east side of the island and *Atriplex cinerea* is found in small patches at the south end.

BIRDS

Native:

Cape Barren Goose – 1 pair with 3 fledged chicks

White-faced Heron – 2 immature birds

Black-fronted Dotterel – 12 birds

Forest Raven – 1 empty nest at the north-east corner on a rock ledge

Introduced:

Common Starling – 1 nest in a crevice 10 cm off the ground

No other fauna was recorded.

Pacific Gull and Sooty Oystercatcher midden of shellfish.

COMMENTS

There may be a low incidence of illegal harvesting of Short-tailed Shearwaters. Given that pelicans are particularly shy breeders, any disturbance on or nearby the island will disrupt their breeding activities. The high species diversity and presence of Australian Pelicans make this an important seabird island. This is the only Australian Pelican colony in western Bass Strait and one of only 3 or 4 known breeding sites in Tasmanian waters. Unlike some of the breeding sites in eastern Bass Strait, the pelicans always use this island. It is also the largest White-faced Storm-Petrel colony in western Bass Strait and the only island on which White-faced Storm-Petrels, Fairy Prions and Common Diving-Petrels all nest. Access should be by permit only.

'Curious legends have been told about these birds, and in early days it was believed that the home of the pelican was away in the land beyond the sunset, in the dead heart of Australia, in a land of plenty, beyond the burning sands of the desert…but each year a little colony is to be found in possession of the same lonely island in Bass strait, where the seasons of the dry interior affect them not at all – sea fishermen, perhaps, who scorn their land-loving relations and breed on this tiny island…They were obviously alarmed at our intrusion but at first made a brave attempt to bluff it out and assumed a show of defiance that they were obviously far from feeling…their self-assurance gave away and they fled precipitously across the island, helping themselves in their headlong flight with outstretched wings' Donald Thompson, 1930.

Bears Island

(Hunter Island Group, page 56)

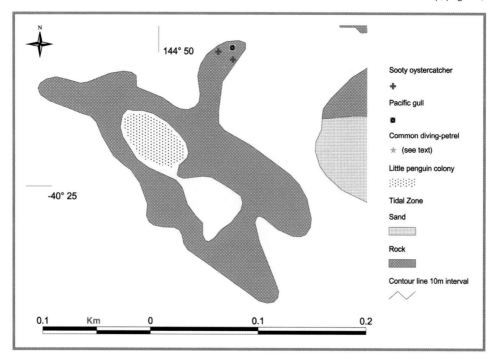

Location: 40°26'S, 144°50'E
Survey date: 14/12/82
Area: 0.34 hectares
Status: Non-allocated Crown Land

Lying off the north-west tip of Three Hummock Island, this small island is a series of three disjointed granite boulders, with patches of vegetation dominated by *Carpobrotus rossii*. The connections between the rocks are wave-washed.

BREEDING SEABIRD SPECIES
Little Penguin
20 pairs, with eggs and chicks, were located scattered under the boulders in short burrows beneath *C. rossii*.

Common Diving-Petrel
Up to 10 pairs breed in crevices and stipa. areas, where soil availability makes burrowing possible on the northernmost of the three boulders.

Pacific Gull
One nest with 1 egg was located on the northernmost boulder.

Sooty Oystercatcher
2 pairs, with eggs, were located on the northernmost boulder.

No other fauna was recorded.

COMMENTS
It is an island vulnerable to extreme weather, with low species diversity and biomass, providing an example of the species that may occupy marginal and limited habitat.

Three Hummock Island

(Hunter Island Group, page 56)

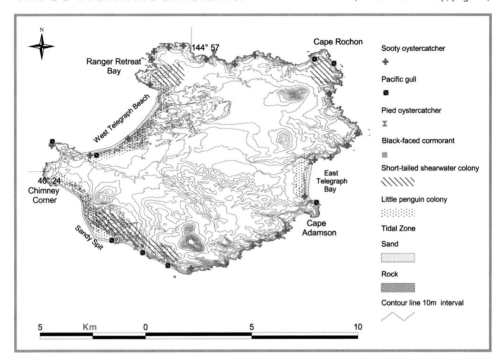

Location: 40°24'S, 144°57'E
Survey date: 24/11/99 - 29/11/99
Area: 6966.56 hectares
Status: Nature Reserve, Lease

Three Hummock Island is generally a low island with a coastline comprising 40 km of exposed granite outcrops, interspersed with 10 km of sandy beaches and dune areas. Low hills exist in a band around the east of the island from the north-east to the south-west. The three highest, after which the island is named, are North Hummock (160 m), Middle Hummock (100 m) and South Hummock (237 m). The hills border a broad, flat plain in the centre of the island. A series of swamps and lagoons have resulted from extensive dune systems blocking the drainage from the plain. Thick coastal vegetation is found around much of the coastline. Infrastructure includes a small settlement at Chimney Corner, a muttonbird hut at the north-east tip, a lighthouse at Cape Rochon, three airstrips, roads, fencing and a wharf.

BREEDING SEABIRD SPECIES
Little Penguin

An estimated 2059 pairs breed in low densities scattered around most of the coast, apart from the dune areas of south-western corner (Sandy Spit coastline), West Telegraph Beach and East Telegraph Bay. Their colonies are often interspersed with those of the Short-tailed Shearwaters. Much of the Little Penguin territory is dominated by *Poa poiformis*.

Chimney Corner, and the homestead, Three Hummock Island.

North-north-west

Short-tailed Shearwater

168,724 pairs breed in colonies, mostly along the south coast of the island and on Ranger Point and at Cape Rochon, usually co-existing with little penguins in *P. poiformis* tussock grassland. Some colonies also occcur in vegetation dominated by a succulent mat herbfield of *Carpobrotus rossii*, *Tetragonia implexicoma* and *Rhagodia candolleana*.

Pacific Gull

7 pairs were located scattered around coast often on rocky points.

Pied Oystercatcher

7 pairs were breeding on the beach and dune areas.

Sooty Oystercatcher

20 pairs were breeding around the coastline.

VEGETATION

Scrub on acid sands is the dominant vegetation community covering 54% of the island. The typical height of this scrub is between 4 and 5 metres and it comprises a dense thicket, usually dominated by *Leptospermum scoparium*, *Melaleuca ericifolia* and *Banksia marginata*. A widespread fire in the 1960s is responsible for the broad occurrence of this vegetation community. *Eucalyptus nitida* occurs extensively over the island on the acid sands and accounts for 25% of the island's vegetation, with trees growing as high as 35 metres. The major vegetation communities in the seabird rookeries are *P. poiformis* tussock grassland or *C. rossii*, *T. implexicoma* and *Rhagodia candolleana* succulent mat herbfield.

OTHER SEABIRD SPECIES
Black-faced Cormorant – 131 roosting on headlands and rocky outcrops

Great Cormorant – 1 seen on Burgess Point

Australian Pelican – 5 on Burgess Point, North West Cape and Neils Rock

MAMMALS
Introduced:
Eastern Grey (Forester) Kangaroo
Sheep
Cattle
Cat
House Mouse

BIRDS
Native:
Hooded Plover – 14 pairs breeding mainly on the west coast

Introduced:
Common Starling

REPTILES
Tiger Snake

COMMENTS
From at least the mid 1800s to the mid 1970s, the island was farmed and some of the sandy areas were cleared for stock grazing, with exotic grass species being planted, particularly around the homestead at Chimney Corner. Three feral sheep were sighted during the 1999 survey, but were not perceived to be a problem. Although part of the island is a nature reserve and should be fully protected, unrestricted access is still a problem: seasonal muttonbirding takes place between March and April; abalone poachers use the central airstrip, the Commonwealth government visits the island to maintain the lighthouse at Cape Rochon and recently expressions of interest were called for tourism operations on the island. Rob Alliston still lives in the Chimney Corner homestead.

The nature reserve status of the island should require visitors to have a permit before visiting it to protect the nesting seabirds from disturbance and to protect the island from further fire, weed invasion and seabird habitat destruction.

> '…the homestead at Chimney Corner is a loney home for weather-bound crews in times of stress. It is in fact a meeting place of skippers – a sort of informal sailors' club and weatherbound seamen put in, sure of a ready welcome. I witnessed quite a reunion when the skippers of a little craft met in the kitchen of the homestead. They smoked strong black tobacco and told tales of wild adventure and shipwreck among the islands, returning to their vessels laden with fresh vegetables from the cottage garden.' Donald Thompson, 1930 from Edgecombe J., 1986.

Albatross Island

(Hunter Island Group, page 56)

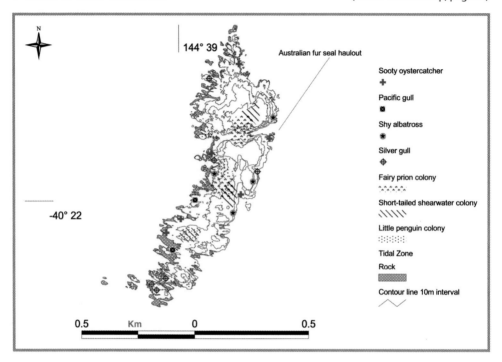

Location: 40°23'S, 144°39'E
Survey date: Yearly for the past 20 years
Area: 18 hectares
Status: Nature Reserve

It is an elongate island orientated north to south with a rocky, jagged coastline comprised of large Precambrian conglomerate boulders and gulches. The geology is unique to Tasmania's islands, having resulted from river deposition of small cobbled rocks, which have over long periods, cemented together. Large slabs have eroded and collapsed forming gulches and caves. An extensive cave complex, bisected by the major east-west gulch, includes a central depression, which is the result of the cave roof collapsing. This is known as 'The Trap' because of the albatrosses which are caught in it when unfavourable wind eddies cause them to stall and fall. The three main caves (Main, Sealers and Pauls), which probably developed 75,000 – 100,000 BP, are a geomorphic feature of outstanding State significance (Dixon, 1996).

BREEDING SEABIRD SPECIES

Little Penguin

An estimated 350 pairs occur, scattered throughout the island mainly in rock crevices and under rock ledges in Poa, *Disphyma crassifolium* and *Senecio* sp. and in the cave systems of the northern half of the island.

Short-tailed Shearwater

The main colony of 1720 pairs is above the cave in the southern part of the island, with burrows under *Senecio* sp., *D. crassifolium* and *Poa*. Another 20 to

50 pairs are scattered around the rest of the island. A fire in 1982 destroyed a second concentration of burrows in the northern section.

Fairy Prion

30 000 – 50 000 pairs are estimated to be breeding in all areas of the island apart from the albatross colonies. Large numbers occupy the cave system, with all suitable rock crevices occupied and some burrows occurring beneath *Tetragonia*.

Pacific Gull

2 pairs breed on the south-west coast each year.

Silver Gull

61 pairs were located. Most were found on the south-western coast, with a further 27 pairs located towards the northern tip. Similar numbers in the same general locality occur annually.

Sooty Oystercatcher

Breeding annually, there are usually up to 6 pairs scattered around the shoreline. Some years they are absent through the winter months, returning in late September.

Shy Albatross

Currently 5000 pairs breed on the island and the population is recovering from near extermination. This species was commonly exploited in the late 1800s for feathers. There are three main colonies on the east coast, with the largest being in the mid east and a smaller colony on the west coast.

VEGETATION

In the north of the island *D. crassifolium* is dominant (60%) with *Senecio* (40%). On the east coast, *Senecio* is dominant (60%) with *Poa poiformis* (30%) and *D. crassifolium* (10%), while on the more rugged west coast *D. crassifolium* is dominant (70%) with *Senecio* sp. (30%). *P. poiformis* is the dominant vegetation on the more exposed southern end (50%) with *D. crassifolium* (40%) and *Senecio* sp. (10%).

December carpet of pigface flowers at the north end of Albatross Island.

OTHER SEABIRD SPECIES

Chatham Island Albatross – 2 birds were regularly present with the Shy Albatrosses in the mid 1980s, but have not been seen in recent years.

Crested Tern – visit in low numbers

MAMMALS

Australian Fur Seal – regularly haul out

New Zealand Fur Seal – small numbers regularly haul out

Southern Elephant Seal – seen on one occasion

Australian Sea Lion – seen on one occasion

BIRDS

Native:

Approximately 60 species have been recorded here with the following being the most frequent:

White-bellied Sea-Eagle – breeds annually

Swamp Harrier

Brown Falcon

Peregrine Falcon

Lewin's Rail

Ruddy Turnstone

Masked Lapwing

Southern Boobook

Barn Owl

Eastern Spinebill

White-fronted Chat

Pink Robin

Grey Fantail

Black-faced Cuckoo-shrike

Forest Raven

Richard's Pipit

Welcome Swallow

Silvereye

Introduced:

Skylark

Common Starling

Common Blackbird

REPTILES

Metallic Skink

Tasmanian Tree Skink

Shy Albatross adults perform in strong winds at cliff edge.

COMMENTS

Because of their restricted occupancy and limited numbers, Shy Albatross (*Thalassarche cauta*) are listed as vulnerable under the Tasmanian *Threatened Species Protection Act, 1995* and the Commonwealth *Endangered Species Protection Act 1992*. An escaped camp fire burnt 2.5 hectares in the northern part of the island in 1982, destroying a colony of Short-tailed Shearwaters, incinerating at least 900 chicks, as well as destroying up to 100 Little Penguins and innumerable Fairy Prion chicks. It is potentially a very vulnerable island due to the combination of its relative accessibility

Fairy Prions at nest sites in Pauls cave.

Albatross adult and newly hatched chick, December

and the increased desire by people to experience wildlife in the wild. Uncontrolled access could lead to further fires or feral animal and plant introduction, which would adversely affect the seabird populations. Its nature reserve status is necessary to guarantee the unfettered survival of the seabird species, particularly the Shy Albatross, breeding there. A visitor permit system would minimise visitation and human impact on the island. Along with the projected population recovery of the Shy Albatross will be a radical alteration to the island's ecology, with possible vegetation and soil degradation, which will have a flow-on impact on the other breeding seabirds.

North end of the main Shy Albatross colony on the east cliff edge.

'This island appears to be almost white with birds, and so much excited our curiosity and hope of securing a supply of food that Mr. Bass went ashore in the boat, whilst I stood off and awaited his return. Mr. Bass returned at half past two with a boat load of seals and albatrosses. He was obliged to fight his way up the cliffs of the island with seals, and arrived at the top to make a road with his club among the Albatrosses.' Flinders 1798 from Flannery T., 2000.

Black Pyramid Rock

(Hunter Island Group, page 56)

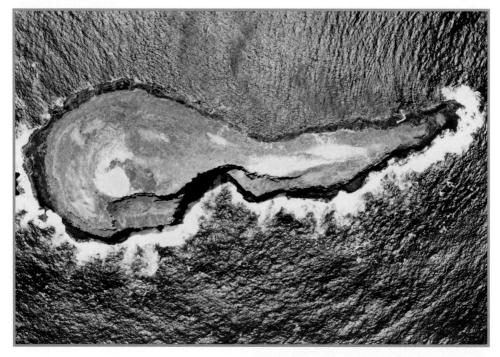

Location: 40°28'S, 144°21'E
Survey date: 2/12/83
Area: 40 hectares
Status: Nature Reserve

Lying to the west of Hunter Island, this is a spectacular tear drop shaped island rising to the summit on the northern end. Steep cliffs surround the island with gentler slopes and a rock platform on the central eastern side. It is basaltic with sedimentary upper layers.

BREEDING SEABIRD SPECIES
Little Penguin

13 pairs, with eggs to large chicks, were located. 8 pairs were nesting in the sea caves at the base of the cliff and rock platform on the east side and 5 pairs were found in crevices under boulders

Basalt and sedimentary formation on eroding west slopes.

and fissures in the rocks at the south-west end of the island.

Short-tailed Shearwater
Only 1 pair, with a fresh egg, was found on the eastern side in a burrow leading into a rock crevice lined with *Carpobrotus rossii*.

Fairy Prion
This species was nesting from very low down in cliff crevices to the summit over most of the island in average densities of 0.7/m². Their shallow burrows are mainly in *Carpobrotus rossii* over loose, light soil. In some areas where soil depth and vegetation permit, the burrows are deeper and interconnected. At the time of the survey most pairs had eggs or very small chicks.

Common Diving-Petrel
They predominantly occur throughout the Fairy Prion colonies in average densities of 0.55/m². There appeared to be a tendency for their burrows to be more abundant in the lower, steeper areas, decreasing with increasing altitude. At the time of the survey, most contained downy chicks with wings feathered.

Carefully measured pecking distance between gannet nest sites gives this regularity of nest spacing.

Pacific Gull
1 pair, with a 150 g downy runner, was located on the rock ledge at the south-east end of the island.

One of the regions most spectacular islands, Black Pyramid from the north-west with white of gannet colonies clearly visible.

Silver Gull
34 nests, 5 with 1 egg, 10 with 2 eggs, 8 with 3 eggs, 11 with no eggs, were found on the mid-eastern ledges of *C. rossii* in rock hollows.

Sooty Oystercatcher
1 pair was located 10 metres from the Pacific Gull nest.

Australasian Gannet
An estimated 12 342 pairs nest in the main gannet colony on the south-eastern side of the rock. At the time of the survey, some were still constructing nests of *C. rossii*, a third to a half of the population were on eggs and the rest were with chicks ranging from newly-hatched to fully-fledged.

No other seabirds were recorded.

MAMMALS
The island is occasionally used as a haul-out by Australian Fur Seals.

BIRDS
Native:
Swamp Harrier
Peregrine Falcon – defending site
Forest Raven – 1 pair
Welcome Swallow

Introduced:
Common Starling – breeding

REPTILES
White's Skink – 6 and 1 juvenile

COMMENTS
This is the largest Australasian Gannet colony in Tasmania. The island's nature reserve status and inaccessibility should be sufficient to ensure the sustainability of the seabird populations inhabiting it, however their viability may be threatened in the future by overfishing.

It is interesting that, at the time of this survey, only one Short-tailed Shearwater was breeding, a direct result of there being insufficient soil for a species of this size to burrow. Although the island's steep sides largely prevent the accumulation of soil, the main reason for the island's lack of soil and vegetation is believed to be due to the widespread distribution of gannets, who have destroyed it in their pursuit of nest-building material. Its volcanic and biological features make this island one of outstanding State significance.

West side looking south

South Black Rock

(Hunter Island Group, page 56)

Location: 40°34'S, 144°36E
Survey date: 10/12/82
Area: < 1 hectares
Status: Non-allocated Crown Land

It is a north to north-east orientated reef of bare rock which is often almost wave washed. From the north-north-east end, bare rock rises steeply to the highest point. A ridge slopes gradually to the west and east, then forms a sheer drop to the sea, broken by a few ledges, particularly on the eastern side.

BREEDING SEABIRD SPECIES
Fairy Prion
An estimated 100 pairs nest in burrows and crevices. The burrows occur on all the ledges of the east side, where soil is sufficient. They also nest in crevices at the north end of the main part of the rock. Of 20 burrows examined, two had unguarded chicks, 3 were with eggs and the rest had a chick and an adult.

Common Diving-Petrel
Possibly as many as 20 pairs breed here. Burrows were found at the cliff edge, but it is likely they also breed amongst rock crevices.

Pacific Gull
1 pair, with a chick, was found at the south end of the ridge.

Black-faced Cormorant

From the guano present, it is probable that this has been a regular breeding site with up to 80 pairs. Three quarter-grown chicks, which had been dead for some time, were found in the colony. 12 birds were sitting on the shoreline.

OTHER SEABIRD SPECIES

Silver Gull – 8 were present, but were not defending and no nests were found

Sooty Oystercatcher – one flew off.

BIRDS

Native:

Satin Flycatcher – 1 female was sitting on the rock ledge

Introduced:

Common Starling – 1 pair was nesting in a rock crevice.

VEGETATION

Disphyma crassifolium is the only vegetation, thinly covering the rocks a third of the way down on the western side and half way down the eastern side.

COMMENTS

South Black Rock is important as one of the most westerly land masses of Tasmania and its representative value as an island with remote, wilderness characteristics.

TREFOIL ISLAND GROUP

(Region 1, page 45)

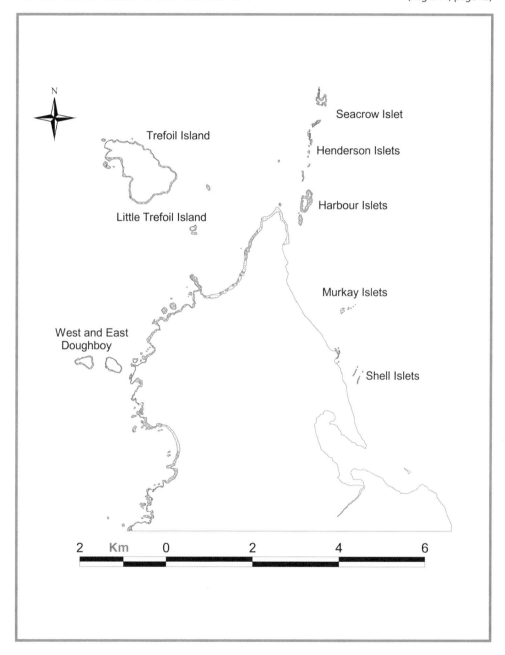

Doughboy Island East

(Trefoil Island Group, page 89)

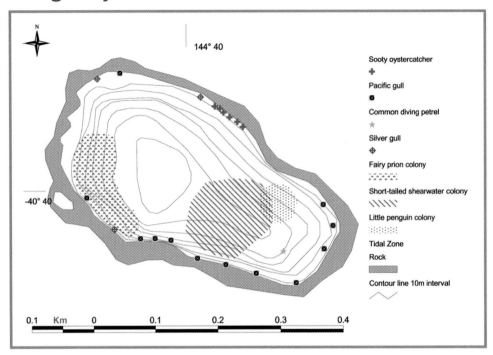

Location:	40°40'S, 144°40'E
Survey date:	11/12/82
Area:	7.01 hectares
Status:	Nature Reserve

The eastern Doughboy Island is kidney-shaped, with two summits, one at the north west and one at the south-east end, connected by a ridge with a smaller ridge running along the south-west end from the north-west summit. The island is surrounded by a 20 m wide wave platform on all sides, except the south side. This platform is wave-washed during high seas. There are two caves on the north-east side which would be awash during storms. The caves are 10 m deep with a 5 m high and 15 m wide entrance. The roof and floor of the caves converge towards the back. The island is encircled by a 20 m high cliff which grades to steep slopes. There are flat areas around the summit and along the ridges. These slopes are fragile. The south-east slopes are the most stable on the island. The remainder of the slopes are composed of a thin layer of sandy soil that is unstable.

BREEDING SEABIRD SPECIES
Little Penguin
6 pairs, with eggs, small and large chicks were found in the caves.

Short-tailed Shearwater
An estimated 19 984 pairs were located nesting under the dense mats of *Poa poiformis* on the flatter summit areas and down the slopes where

the tussocks and soil provide burrowing habitat. There is some overlap between the Common Diving-Petrel breeding habitat and that of the Short-tailed Shearwater at the top of the slopes where the tussocks replace the *Disphyma crassifolium* as the dominant vegetation.

Fairy Prion
Up to 500 pairs were estimated to be breeding all over the island on the steep slopes. The burrow density decreases where the slope eases and where the Short-tailed Shearwaters are dominant.

Common Diving-Petrel
2000 to 3000 pairs breed on the steep, lower slopes, where they form burrows under the *D. crassifolium*. The loose, shallow, stony soil restricts the construction of burrows. The colony extends into the lower patches of *P. poiformis* on the upper slopes. The majority of the birds were found on the southern end of the eastern slopes. There are none on the northern end of these slopes, probably because of the highly unstable stony soil cover.

Pacific Gull
12 pairs were nesting on the cliffs all around the island.

Silver Gull
20 pairs, at empty nests and with small chicks, were located in 3 colonies. 6 pairs were nesting on the north-east slope 2 m above the wave platform, 4 pairs on the cliff at the north-west end and 10 pairs at the south-west end of the cliff.

Sooty Oystercatcher
6 pairs were nesting on ledges on the cliffs at the north-east end of the island.

Western shore of Doughboy Island East with Cape Grim in the backgound.

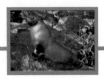

VEGETATION

The cliffs are vegetated with *D. crassifolium* and luxuriant pendulous clumps of *Sarcocornia* sp. The rest of the island is dominated by *P. poiformis*, which forms dense mats on the higher, flatter areas. On the north-east slopes, there are clumps of halophytic shrubs. As with the western Doughboy Island, there is little soil on the higher areas and the *P. poiformis* patches are sparser on the northern slopes than elsewhere. The western slopes, above the cliffs, are composed of stony, loose soil with little vegetation. There are three African boxthorn bushes on the island.

COMMENTS

The fragile nature of the slopes makes the island vulnerable to human impact. Its unique geology is of representative and outstanding significance for Tasmania (Dixon, 1996). Visits are only likely to be by researchers or managers and a few visits by determined muttonbirders. Any management or research visits should be adequately supervised on site.

No mammals or other seabirds were recorded.

BIRDS
Native:
White-faced Heron
Swamp Harrier
Nankeen Kestrel
Welcome Swallow
Forest Raven
Introduced:
Common Starling

REPTILES
Metallic Skink

It is an island with a wide variety of seabirds, appropriately conserved by its inaccessibility and its status as a nature reserve. Consideration should be given to making it mandatory for any visitors to be supervised and to have permission in the form of a permit.

'......in January 1828, another group of sealers secreted themselves in a cave at the Doughboys to ambush a group of Pennemukeer women collecting muttonbirds and shellfish. As the women swam ashore the sealers rushed out with muskets, pushed fourteen women into an angle of the cliff, bound them with cords, and carried them off to Kangaroo Island. In revenge the Pennemukeer men had later clubbed three sealers to death. Most women acquired by the sealers were taken to Kangaroo Island, Off the South Australian coast, or King Island at the western end of Bass Strait.'   L Ryan in The Aboriginal Tasmanians.

Doughboy Island West

(Trefoil Island Group, page 89)

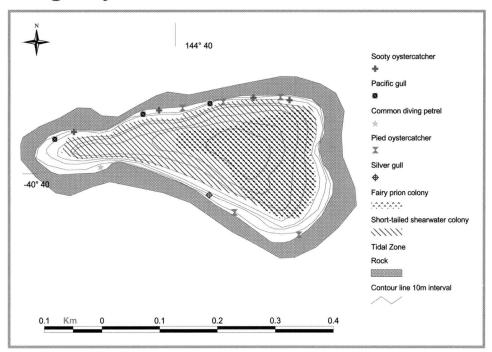

Location: 40°40'S, 144°40'E
Survey date: 11/12/82
Area: 5.4 hectares
Status: Nature Reserve

It is an oval-shaped island with two peaks at the eastern and western sides. The eastern peak is the summit of the island. The island has a wave-washed platform on the north-east side, which is 40 m wide at the widest point and averages 20 m wide. The island rises steeply from the platform and on the southern side the cliffs are vertical to overhanging, varying in height from 20 to 60 m. There is a large sea cave at the south-east tip. Many hollows and ledges on the cliffs are used by seabirds for nesting and roosting.

BREEDING SEABIRD SPECIES
Short-tailed Shearwater

An estimated 59 020 pairs nest all over the island from the central flat area to half way down the *Poa poiformis*-dominated slopes. The lowest densities were found on the flat areas at the summit and along the east to west ridge, where the birds nest under the dense *P. poiformis* cover. On the slopes, the birds burrow into loose, sandy soil, where the densities are higher. In some areas, particularly on the north slope, the burrowing activity has reduced the *P. poiformis* cover to a few sparse patches, resulting in erosion of the slope.

Impressive symmetry and rock platform of Doughboy Island. Here is an excellent example of a topographical relationship between breeding seabird and habitat distribution.

Fairy Prion

1000 to 2000 pairs nest all over the island on the steep slopes, which are covered by *Disphyma crassifolium*. The burrow density decreases as the slope eases and the vegetation changes to a *P. poiformis* dominated assemblage with Short-tailed Shearwater burrows. The highest densities of burrows were found on the eastern slopes. Burrows were also located on the western summit.

Common Diving-Petrel

An estimated 5000 pairs burrow on steep slopes, where the soil is shallow and contains a high proportion of small rocks and stones. Although a few Short-tailed Shearwaters were found in this habitat, it appears as if the shallow rocky soil limits their distribution as well as that of the Fairy Prions, whilst the Common Diving-Petrels exploit the habitat by digging short winding burrows. These birds also burrow into very shallow soil under mats of *D. crassifolium*, with burrows up to 3 m long. Burrows are concentrated from the top of the western summit along the south coast to the eastern end on the very steep cliffs.

Bird burrowing habitat on the less steep upper slopes.

Southern side.

Pacific Gull
3 pairs were found nesting on the northern cliff on ledges. Remains of Fairy Prions and Common Diving-Petrels were found in the sea cave at the south-east end, probably food remains from Pacific gulls.

Silver Gull
30 pairs were located on a steep cliff at the south-east end of the island.

Sooty Oystercatcher
4 nests were located on ledges on the cliffs along the northern and eastern sides of the island.

Pied Oystercatcher
5 pairs, at nests, were located on sandy beaches on the north and south coastlines.

VEGETATION
D. crassifolium dominates the sheer cliffs, with a few luxuriant patches of *Sarcocornia* sp. The *D. crassifolium* forms a ring around the island above which thick *P. poiformis* dominates. On the north face, *Atriplex cinerea* forms dense localised clumps. On some parts, and almost exclusively on the western ridge and summit, *P. poiformis* forms a dense ground cover.

No mammals or other seabird species were recorded.

BIRDS
Native:
White-faced Heron
Forest Raven
Little Grassbird

REPTILES
Metallic Skink

COMMENTS
The fragile nature of the slopes makes the island vulnerable to human impact. Its sloping supratidal platforms, vertical cliffs and low sea caves make it an island of geological significance at the State level (Dixon, 1996).

Like its sister island, it has a wide variety of seabirds, which are adequately conserved by its inaccessibility and status as a nature reserve. Consideration should be given to making it mandatory for any visitors to gain access permits.

Trefoil Island

(Trefoil Island Group, page 89)

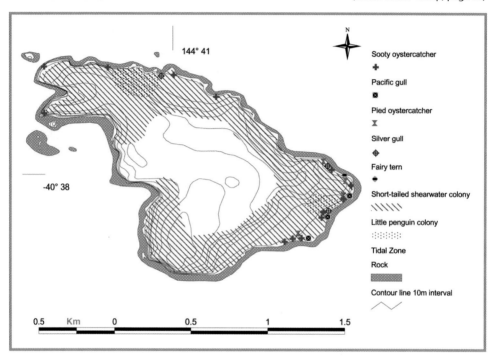

Location: 40°38'S, 144°41'E
Survey date: 15/9/80-17/9/80 plus information from the visits by G. Towney in December and February every year between 1980 and 1983.
Area: 115.79 hectares
Status: Private property – owned by the Trefoil Island Aboriginal Co-operative. Approval required prior to visit.

1.8 km long by 1.1 km at the widest point, the island rises steeply from the sea to a height of 84 metres in the south and east and 60 metres in the north and west. A layer of black, sandy soil between one and two metres thick overlies the rocky sandstone formation. The shoreline is covered with pebbly rocks. An airstrip and huts exist at several localities on the island. The steep cliff face and reefs surrounding the island make it relatively inaccessible by boat, except at the south end.

BREEDING SEABIRD SPECIES
Little Penguin

An estimated 500 pairs inhabit scattered burrows throughout the Short-tailed Shearwater colonies. Concentrations occur at Kelp Beach and Slippery Bottom.

Short-tailed Shearwater

An estimated 1.54 million pairs occupy burrows over 100 hectares of the island. Except for the

airstrip and buildings, the whole island is covered by burrows. Only scattered burrows, however, occur in the bracken and on bare ground.

Pacific Gull
40 pairs, with eggs and chicks, were located on the south-east coast and to the north and south of Kelp Beach.

Silver Gull
3 – 4 pairs were found around the island. Their distribution was not recorded.

Sooty Oystercatcher
Up to 26 pairs, with eggs or chicks, were located in the same areas as the Pacific Gulls.

Pied Oystercatcher
4 pairs, with eggs and chicks, were recorded in the same areas as the Pacific Gulls and Sooty Oystercatchers.

VEGETATION
The main vegetation is silver tussock, *Poa poiformis*, with a few small patches of bracken,

OTHER SEABIRD SPECIES
Fairy Prion
Black-faced Cormorant
Great Cormorant
Fairy Tern – 2

MAMMALS
House Mouse
Sheep – about 100 are grazed around the airstrip.

BIRDS
Native:
Cape Barren Goose – 5 pairs, breeding – introduced to the region
Pacific Black Duck
Swamp Harrier – 3 or more
Brown Falcon – 2
Nankeen Kestrel – 1
Lewin's Rail
Masked Lapwing
Pallid Cuckoo
Crescent Honeyeater – 3

White-fronted Chat – common
Willie Wagtail
Forest Raven
Richard's Pipit
Welcome Swallow – breeding, nests found in sheds

Introduced:
Common Starling – many breed in sheds
House Sparrow – abundant
Skylark – common
Turkey – 3
Red Junglefowl (Chicken) – 9
European Goldfinch – more than 16
Domestic Pigeon – near sheds

REPTILES
Metallic Skink
Tiger Snake – irresponsibly introduced

Pteridium esculentum. The only trees are six specimens of *Cupressus macrocarpa*.

COMMENTS

Short-tailed Shearwaters on Trefoil Island are harvested annually, but to a lesser degree than in the past. Kelp was harvested on the island up to 1980. Sheep are still grazed around the airstrip. Tiger Snakes are reputed to have been introduced onto the island and may have a devastating impact on the seabirds.

> 'Besides these islands and rocks, we passed another cliffy island four or five miles to the south of Steep-head, and to which I gave the name of Trefoil Island, its form appearing to be nearly that of a clover leaf; there were, also, several others of less importance, mostly lying near the barren land.' Matthew Flinders, 1798 from Flannery T., 2000.

Little Trefoil Island

(Trefoil Island Group, page 89)

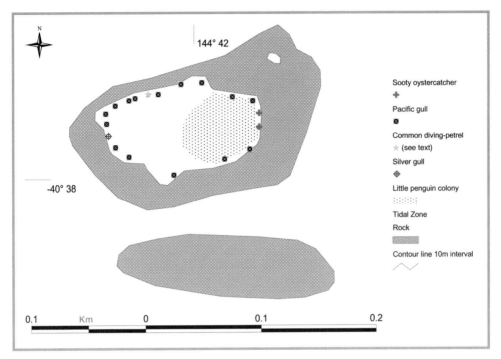

Location: 40°39'S, 144°42'E
Survey date: 10/12/82
Area: 0.64 hectares
Status: Non-allocated Crown Land ↑

Most of the island is flat or gently undulating with a rocky shoreline.

BREEDING SEABIRD SPECIES

Little Penguin

An estimated 50 -100 pairs were recorded as being scattered all over the island in burrows of less than 1 m in length below vegetation. At the time of the survey, the burrows were empty.

Common Diving-Petrel

200 were located mostly in crevices on the north west corner. 60 carcasses were found, the result of Pacific Gull predation.

Pacific Gull

15 pairs were located all around the island. Others flew in from Trefoil Island to join breeding birds here.

Silver Gull

24 pairs were located at the western end, concentrated around the stipa low down and on steep slopes.

Sooty Oystercatcher

2 pairs, with 2 eggs each, were located towards the east end of the island.

VEGETATION

The island's vegetation is dominated by *Poa poiformis* and stipa at the west end with *Tetragonia implexicoma* scattered elsewhere.

Eastern shore with Cape Grim shoreline in background.

No mammals or other birds were recorded.

REPTILES
Metallic Skink

COMMENTS
Its seabird diversity suggests an up-grading of status is warrented

Seacrow Islet

(Trefoil Island Group, page 89)

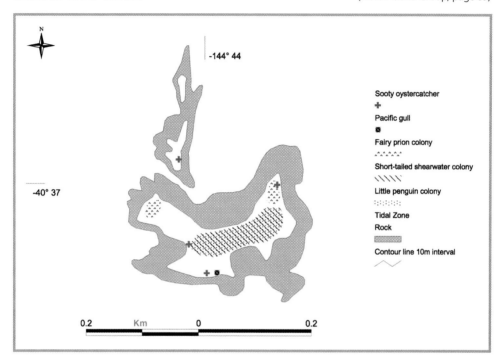

Location: 40°37'S, 144°44'E
Survey date: 10/12/82
Area: 1.56 hectares
Status: Non-allocated Crown Land

This is an irregular shaped island with a large sandy bay on the north-east coast.

BREEDING SEABIRD SPECIES
Little Penguin
An estimated 50 pairs occur in low densities amongst the Short-tailed Shearwater colony in the centre of the island.

Short-tailed Shearwater
The densest concentration of approximately 200 pairs was found in sandy, soft soil in the centre of the island. Densities averaging $0.5/m^2$ were recorded over the rest of the island, with sparse distribution.

Fairy Prion
An estimated 100 pairs were recorded mainly at the easternmost and westernmost ends. The eastern colony utilised burrows, and the western colony crevices.

Pacific Gull
A chick was located on *Poa* near to a Sooty Oystercatcher nest, but its nest was not found.

Sooty Oystercatcher
Five were counted, two under *Poa*, one on the shingle, one on *Carpobrotus rossii*, one under saltbush. Three had two eggs, one had one egg and one chick and one had a newly-hatched chick.

No mammals or reptiles were recorded.

OTHER SEABIRD SPECIES

Black-faced Cormorant – 1 flew off the rocks.

Silver Gull – 20 were recorded feeding off the west coast, but were not defensive.

Fairy Tern – 2 flew over the west coast.

BIRDS

Cape Barren Goose – a pair with 3 almost fully-fledged chicks at a nest found amongst *Poa* on the north-west tip – introduced to the region.

Sanderling – 3 on the beach

Hooded Plover

Little Grassbird – 1 pair

VEGETATION

The southern third of the island is predominantly *Poa* and scrub, the centre third predominantly *Rhagodia candolleana* and cabbage plant and the northern third predominantly *Poa* with *Rhagodia*.

COMMENTS

A status upgrade is warranted to reflect the seabird diversity.

Skipper, checking the rigging

Henderson Islets

(Trefoil Island Group, page 89)

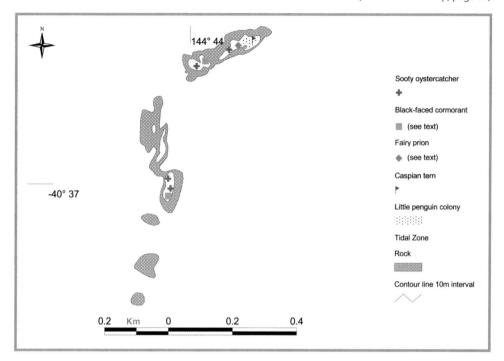

Location: 40°38'S, 144°44'E
Survey date: 10/12/82
Area: 0.41 hectares
Status: Non-allocated Crown Land

The group is made up of two small islets, which are predominantly bare rock with small patches of vegetation. They are prone to sea spray and may be wave-washed during high seas.

BREEDING SEABIRD SPECIES

Little Penguin

10 pairs in short burrows were constructed under *Poa poiformis* at the northern end of the north islet.

Fairy Prion

Up to 10 pairs were estimated to be nesting in rock crevices at the southern end of the north islet.

Black-faced Cormorant nest, now occupied by Sooty Oystercatcher

Largely bare rock of Henderson Islets.

Sooty Oystercatcher
4 pairs, all with eggs, were located. On the north islet, one pair was nesting in a Black-faced Cormorant nest and the other had constructed a nest at the northern end of the island. On the south islet, the 2 pairs were nesting in Black-faced Cormorants' nests.

Black-faced Cormorant
8 well-formed but unused nests, constructed of seaweed, were found at the south end of the north island. 34, also unused nests constructed of seaweed, were found at the south end of the south island. This, along with extensive guano deposits, suggests that the island is a sporadic breeding site.

Caspian Tern
1 pair, with a chick, was located nesting at the north end of the north islet.

VEGETATION
Patches of *Disphyma crassifolium*, *P. poiformis* and *Atriplex cinerea* occur on both islets.

No other fauna was recorded.

COMMENTS
The status of the group is appropriate for this small islet group with its small numbers of a few seabird species.

Harbour Islets

(Trefoil Island Group, page 89)

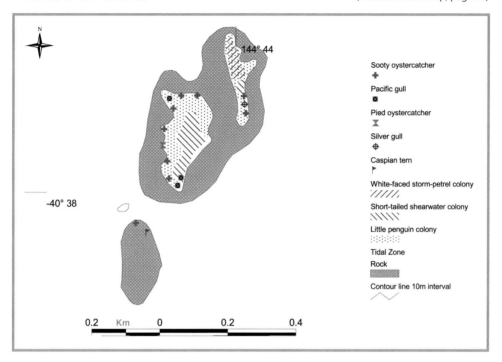

Location: 40°38'S, 144°44'E
Survey date: 15/12/82
Area: 3.13 hectares
Larger Islet (SW): 2.27 hectares
Small Islet (NE): 0.86 hectares
Status: Non-allocated Crown Land ↑

Harbour Islets is a group of two small islets that are joined at low tide. There is a smaller rock off the southern tip of the south-western islet. The coastline of the southern islet is rocky with a pebble or shell beach behind the rock band. The smaller northern islet, rising gently to a north to south running ridge, is dominated by hard rocky ground, which limits the burrowing potential.

BREEDING SEABIRD SPECIES
Little Penguin

An estimated 250 pairs were located in burrows amongst the Short-tailed Shearwater colonies on both islets with the greatest concentration of approximately 175 pairs on the larger islet. On the small islet, penguins also breed all over the island under the larger *Poa poiformis* tussocks.

Short-tailed Shearwater

An estimated 8225 pairs breed on the islets. The main colony is on the larger islet and stretches from the south-eastern shoreline two thirds of the way up the island. The best habitat for burrowing covers half of the islet, the rest is hard ground. The southern end has shrub species as the dominant

The channel, here at high tide, divides Harbour Island. (mainland Tasmania in background).

vegetation with slender thistle and *Senecio* sp also common. The soil here is sandy and soft. On the smaller islet there is a small colony of less than 50 pairs on the eastern edge.

White-faced Storm-Petrel
An estimated 50 pairs in burrows with chicks and eggs were located in a colony on the north-east end of the smaller islet. The colony is 5 m by 5 m in size on a gentle slope. The vegetation is dominated by light *P. poiformis* cover and the burrows are short and shallow because of little soil, which is hard and sandy.

Pacific Gull
2 empty nests were found at the south-east end and one with a downy chick at the north west end of the main islet. 4 pairs were seen.

Silver Gull
4 old nests were located on the smaller islet.

Sooty Oystercatcher
9 pairs, some at empty nests and others with eggs, were found on both islets around the shoreline.

Pied Oystercatcher
3 pairs were present on the main islet but only one pair displayed territorial behaviour.

Caspian Tern
1 pair, with w large chicks, was located breeding on the small rock off the southern end of the main islet.

VEGETATION
The vegetation of the south western islet is dominated by *P. poiformis* with *Senecio* sp. and other grasses also common. The southern rookery is dominated by slender thistle on bare, soft, sandy soil. stipa. forms a rim around the shoreline. The northern islet is dominated by *P. poiformis* and stipa.

No reptiles or other seabirds were recorded.

BIRDS
Native:
Cape Barren Goose
White-faced Heron
Waders (sp.)
Masked Lapwing
Brown Thornbill
White-fronted Chat

COMMENTS
The island is a roost site for waders, congregating on the mudflats, which are exposed at low tide.

The high species diversity on the islets and the presence of White-faced Storm-Petrels warrants an upgraded classification of the islets' status.

Cape Barren goslings

Murkay Islet (east)

(Trefoil Island Group, page 89)

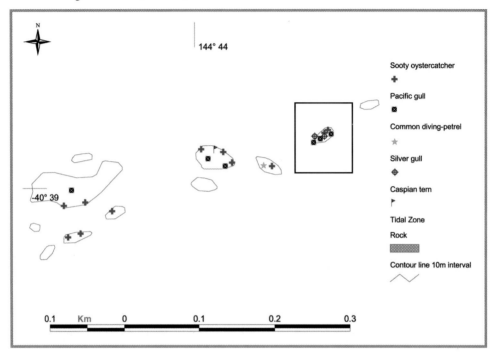

Location: 40°40'S, 140°44'E
Survey date: 15/12/82
Area: 0.03 hectares
Status: Non-allocated Crown Land

This is the smallest islet in the group. It has a rocky shoreline and is ovate in shape and orientated east to west.

BREEDING SEABIRD SPECIES
Pacific Gull
3 nests from the previous season were positioned within the stipa. clump at the eastern end of the islet.

Silver Gull
29 pairs, some at empty nests and others with eggs, were in a colony in the centre of the islet at the western end.

Sooty Oystercatcher
2 pairs, with eggs and chicks, were found in the middle of the islet on the western side.

VEGETATION
The vegetation is dominated by *Sclerostegia arbuscula* and *Sarcocornia* sp. with some patches of *Poa poiformis* and stipa.

No other fauna was recorded.

OTHER SEABIRD SPECIES
White-faced Storm-Petrel – one wing was found

Black-faced Cormorant – 150 were roosting on the islet.

Wild Wind en route

COMMENTS
Forming part of a small group of islets, this islet has low numbers of seabirds and so has appropriate conservation status.

Murkay Islets (middle)

(Trefoil Island Group, page 89)

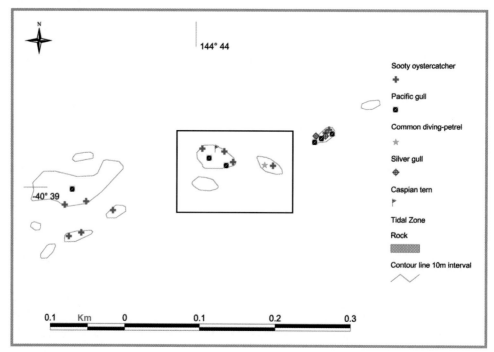

Location: 40°40'S, 140°44'E
Survey date: 15/12/82
Area: 0.3 hectares
Status: Non-allocated Crown Land

This is an irregular-shaped elongate, granite island with rocky points at opposite ends and on opposite sides of the islet. The shoreline is rocky.

BREEDING SEABIRD SPECIES
Common Diving-Petrel

20 to 30 pairs burrow in crevices in rocks that are concealed by vegetation growing over the rocks. Over 20 carcasses were also found.

Pacific Gull

8 pairs, at nests with eggs and chicks, were located in a colony on the northern side of the islet at the western end. This colony was in a patch of stipa.

Caspian Tern

1 pair, at a nest with an egg, was found in a patch of *Sarcocornia* sp. on the north-east point of the islet.

Sooty Oystercatcher

11 pairs were located amongst rocks and *Disphyma crassifolium* at either end of the islet.

VEGETATION

The vegetation is dominated by *Rhagodia candolleana* and *Poa poiformis*. *Sarcocornia* sp. and *Sclerostegia arbuscula* are also dominant.

Pacific Gull

No mammals, reptiles or other seabirds were recorded.

BIRDS
White-fronted Chat

COMMENTS
This islet, part of a small group of islets, has small populations of seabirds and is relatively secure under its current status.

Murkay Islets (west)

(Trefoil Island Group, page 89)

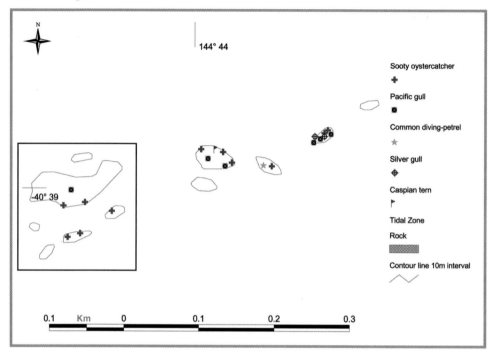

Location: 40°40'S, 140°44'E
Survey date: 15/12/82
Area: 0.13 hectares
Status: Non-allocated Crown Land

This is the largest islet in the Murkay group. The islet is roughly oval in shape with a bay on the northern side and three small subsidiary islets off the south-west end. These small islets are interconnected at low tide and are dominated by stipa and rocks.

BREEDING SEABIRD SPECIES
Pacific Gull

31 pairs, most with eggs and chicks were found all over the islet. The main colony is situated in the middle of the islet amongst *Poa poiformis* and under stipa. There were 27 pairs in this colony. Another nest was found at the western end of the islet and three more in a group at the eastern end.

Sooty Oystercatcher

5 pairs, 2 at empty nests and 3 with eggs, were found, 2 nests on the main islet and 3 on the outer subsidiary rock.

VEGETATION

The vegetation is dominated by *Rhagodia candolleana* and *P. poiformis*. *Sarcocornia* sp. and *Sclerostegia arbuscula* are also dominant.

Sooty Oystercatchers

No mammals or reptiles were recorded.

OTHER SEABIRD SPECIES
Common Diving-Petrel – remains found but no evidence of breeding

BIRDS
Cape Barren Goose - 1 old nest – introduced to the region

COMMENTS
This islet forms part of a group, with no particularly sensitive species, and is relatively secure as it is.

Shell Islets

(Trefoil Island Group, page 89)

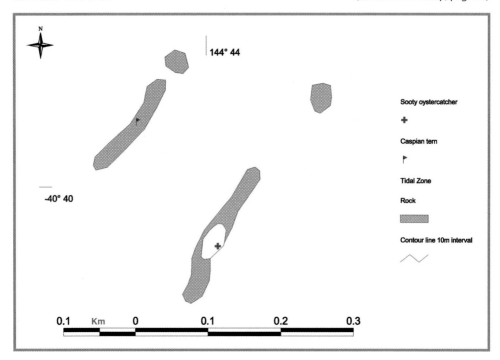

Location: 40°41'S, 144°45'E
Survey date: 15/12/82
Area: 0.082 hectares
Status: Non-allocated Crown Land

This group consists of a small islet with two subsidiary minute islets surrounded by extensive sand and mudflats at low tide. These islets are mainly comprised of exposed granite rock.

BREEDING SEABIRD SPECIES
Sooty Oystercatcher
One pair at an empty nest was located on the main islet.

Caspian Tern
1 pair, with one bird sitting on an empty nest, was recorded on the western most subsidiary islet.

No mammals, reptiles or other seabirds were recorded.

BIRDS
Waders of various species but predominantly Red-necked Stints and Sanderlings – 300 to 500 on the mudflats.

Heading home

VEGETATION
Atriplex cinerea dominates the shoreline with *Poa poiformis* dominant elsewhere.

COMMENTS
This is not such an important locality for seabirds, but is significant as a refuge for large concentrations of waders, which feed on the adjacent mudflats and in the extensive habitat within the nearby nationally-important wetland site.

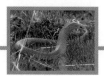

Petrel Island Group

(Region 1, page 45)

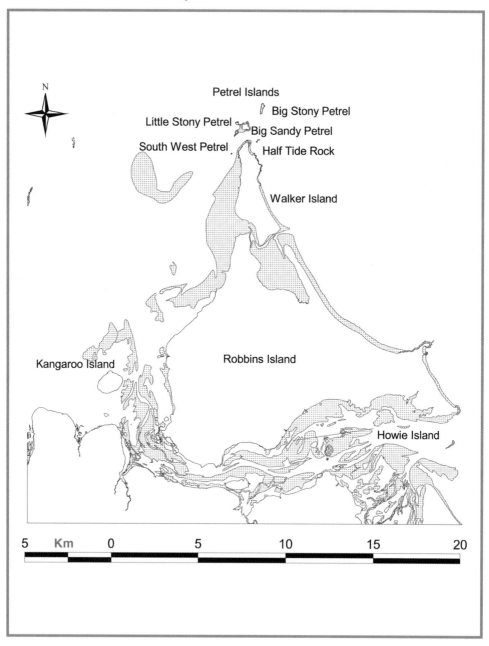

South West Petrel Island

(Petrel Island Group, page 116)

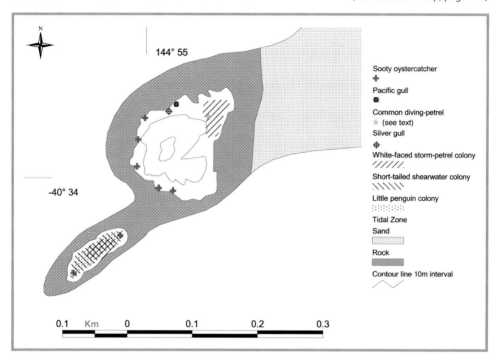

Location: 40°34'S, 144°55'E
Survey date: 11/12/82
Area: 4 hectares
Status: Game Reserve ↑

This is an elongate island, which is dome-shaped at the northern end where the summit is surrounded by steep slopes. The island slopes off towards the southern point. The shoreline is rocky with a couple of sandy and pebble beaches at the east south-east side. The southern extension of the island is separated from the main body of the island at high tide.

BREEDING SEABIRD SPECIES
Little Penguin

70 pairs, some with empty nests and others with eggs and chicks, are concentrated in burrows on the east-south-east side. A few birds were found breeding under *Poa poiformis* tussocks and rocks on the west side. There is very little suitable habitat for burrowing. 20 pairs were also found breeding at the southern end of the island.

Short-tailed Shearwater

50 pairs were located breeding in the same habitat as Little Penguins on the east south-east side of the island with a few pairs on the southern tip.

Common Diving-Petrel

100 – 200 pairs were located in similar habitat to the storm-petrels. However none were recorded at the southern side, though a few scattered pairs may burrow there.

Aerial view.

White-faced Storm-Petrel

An estimated 250 – 500 burrows occur on the island. 150 to 300 pairs were located breeding all over the steep slopes on the northern end of the island in places between the rocks where soil had accumulated and under tussocks where the soil is shallow. At the southern end, a colony of 100 to 200 pairs in burrows was located amongst stipa and *Disphyma crassifolium*.

Pacific Gull

19 pairs, with chicks, were located nesting on the north-west side of the summit. The colony extended from the summit to part of the way down the steep slope.

Silver Gull

5 pairs, at empty nests, were located in a colony south of the Pacific Gull colony on a steep slope. 4 adults were seen and the nests were empty.

Sooty Oystercatcher

7 pairs, on empty nests and with eggs, were scattered all around the shoreline with 2 pairs on the southern end on the eastern shoreline of this portion of the island.

No mammals or reptiles were recorded.

OTHER SEABIRD SPECIES

Caspian Tern – 1 flew by
Crested Tern – 4 flew by

BIRDS

Cape Barren Goose – 1 pair with 2 fledged chicks
Forest Raven – 1

VEGETATION

The vegetation is dominated by *P. poiformis*, *D. crassifolium* and stipa between boulders. There is straggly scrub amongst the boulders and on the summit.

COMMENTS

The island's status as a game reserve, which implies harvesting of wildlife, is unwarranted as there are very few Short-tailed Shearwaters. The island should, however, be granted conservation area status and access discouraged to protect the small burrow nesting species from further disturbance to their fragile habitat.

Little Stony Petrel Island

(Petrel Island Group, page 116)

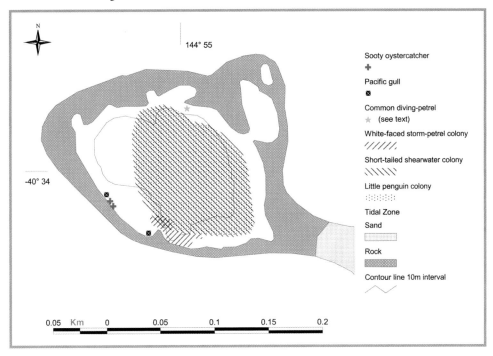

Location: 40°34'S, 144°55'E
Survey date: 11/12/82
Area: 4 hectares
Status: Game Reserve ↑

The island is dome-shaped with rocky outcrops. Bare rock constitutes a quarter of its area.

BREEDING SEABIRD SPECIES
Little Penguin

A quarter of the island is unoccupied because it is composed of bare rock. An estimated 150 pairs breed in burrows and under tussocks over the rest of the island in association with Short-tailed Shearwaters.

Short-tailed Shearwater

An estimated 10 500 pairs inhabit the island in a similar distribution pattern to the penguins, with the highest densities of $1/m^2$ over a quarter of the island and lower densities of $0.2/m^2$ over half of the island. The colonies are covered by thick *Poa poiformis* tussocks and to a lesser degree by

Aerial view.

Senecio sp and *Disphyma crassifolium*. There is a mixture of burrows and vegetation-covered nests.

Common Diving-Petrel
An estimated 100 pairs occur in a colony at the northern end of the island.

White-faced Storm-Petrel
An estimated 100 pairs breed amongst stipa in pockets between exposed rocks. They are preyed on by Pacific Gulls.

Pacific Gull
2 pairs, at nests with chicks, were found nesting on the south-west side of the island and at the southern tip on the edge of stipa and *D. crassifolium*.

Sooty Oystercatcher
2 pairs were located on the south-west side of the island. One pair was nesting within 2 metres of a Pacific Gull nest on the edge of stipa and *D. crassifolium*.

VEGETATION
The vegetation is dominated by *P. poiformis* and *Senecio* sp. with *D. crassifolium* also common. Scrub is confined to the rocky outcrops and the summit.

No mammals or reptiles were recorded.

OTHER SEABIRD SPECIES
Black-faced Cormorant – 4 were roosting at the northern end.

Pacific Gull – 35 adults flew over from the nearby islands.

BIRDS
Cape Barren Goose – 1 pair

White-faced Heron

COMMENTS
Excessive harvesting of Short-tailed Shearwaters may not only adversely affect their distribution on the island but also that of the other burrowing petrels. Visits to engage in this activity are likely to be detrimental to the fragile nesting habitat of the burrowing petrels, especially the minor species, as they are restricted to marginal habitat, whilst the Short-tailed Shearwaters occupy the more suitable burrowing habitat. The high diversity of breeding seabirds warrants the upgrading of the island's status to nature reserve.

Burrowing habitat of Storm-Petrels and Shearwaters.

Big Stony Petrel Island

(Petrel Island Group, page 116)

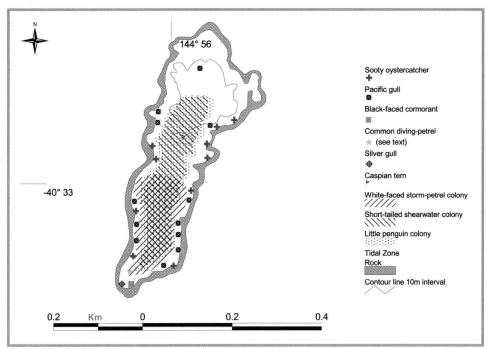

Location: 40°34'S, 144°56'E
Survey date: 14/12/82
Area: 7.2 hectares
Status: Game Reserve ↑

It is an elongate island that is gently undulating and orientated with the long axis north to south. The shoreline is very rocky on the east side, while the western shoreline is more extensive and flatter, though still rocky. There are no beaches.

BREEDING SEABIRD SPECIES
Little Penguin
300 to 500 pairs were scattered in burrows over most of the island amongst Short-tailed Shearwaters and in rock crevices where there is insufficient soil to burrow. The overall density is light and restricted to approximately two-thirds of the island.

Short-tailed Shearwater
An estimated 11 148 pairs inhabit a colony which extends over approximately two-thirds of the island in sandy soft soil. Several colony areas are bare but the majority have a good cover of *Poa poiformis*.

Common Diving-Petrel
100 pairs, with chicks, were located in a small colony at the southern end of the island.

White-faced Storm-Petrel
100 pairs, most with eggs, were located in burrows in sandy soil on the southern half of the island. The burrows are shallow, winding and

Looking towards Little Stony Petrel.

interconnected. The colony area is vegetated with stipa It is possible that there are other small pockets of birds in similar habitat on the island.

Pacific Gull
12 pairs, mostly at nests with eggs and chicks, were located scattered around the island. A colony of 5 pairs was nesting at the south-east end of the island. Just to the north, a solitary pair was nesting. In the middle of the island and on the eastern shore another 3 pairs were nesting together. A lone pair had a nest on the north of the centre ridge of the island. A further 2 pairs had nests on the west side.

Silver Gull
One colony, with 11 pairs mostly at empty nests, and another, with 20 pairs at old nests, were both located at the south-west end of the island.

Sooty Oystercatcher
10 pairs, 8 on empty nests, one with an egg and another with a chick, were evenly scattered around the coast.

Black-faced Cormorant
55 pairs, at new nests without eggs or chicks, were located at the southern tip of the island. 23 nests were positioned on bare rock ledges 15 m above the sea. Another colony of 32 pairs was located on the western side of the island.

Caspian Tern
One pair, with a chick, was located on the central rock ridge in association with *P. poiformis* and *Disphyma crassifolium*.

VEGETATION

The dominant vegetation is stipa and *Poa* with patches of *D. crassifolium*.

No mammals, reptiles or other seabird species were recorded.

BIRDS

Native:

Cape Barren Goose – 1 pair with 4 flightless goslings – introduced to the region

White-fronted Chat

Welcome Swallow

Introduced:

Common Starling – breeding.

COMMENTS

The island is a game reserve, and is subjected to the harvesting of Short-tailed Shearwaters. The high species diversity suggests an upgrading of the status to preclude harvesting and the unavoidable destruction to the fragile nesting habitat of White-faced Storm-Petrel occurring during visits. The Black-faced Cormorant colony is also vulnerable to disturbance and should be protected.

Big Sandy Petrel Island

(Petrel Island Group, page 116)

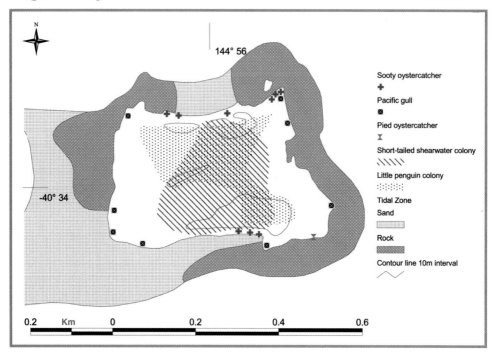

Location: 40°34'S, 144°56'E
Survey date: 14/12/82
Area: 15.2 hectares
Status: Game Reserve ⇑

This is a crescent-shaped island with extensive sandy beaches and flat, jagged boulders on the shoreline. The interior of the island is dominated by undulating, vegetated dunes that are orientated north-east to north-west. There are deep gullies between the dunes supporting the major concentrations of burrowing birds. The dunes at the north and south-west of the island rise steeply from the beaches.

BREEDING SEABIRD SPECIES

Little Penguin

400 pairs were located, scattered in burrows over most of the island, amongst Short-tailed Shearwaters and under dense vegetation.

Short-tailed Shearwater

An estimated 16 380 pairs breed all over the island in very sandy soil. The burrows are deep and long. The densest colony is located opposite Little Stony Petrel in a gully formed by two parallel dunes. This area is 90 m by 90 m and is dominated by *P. poiformis*. Birds are present in lower densities over the remainder of the island.

Vegetated dunes looking towards Little Stony Petrel.

Pacific Gull
8 pairs, at nests with eggs and small chicks, were nesting on the shoreline all around the island. There was no colony formation, with only two pairs nesting close to each other, otherwise the birds were solitary. All nests were in association with stipa.

Sooty Oystercatcher
9 pairs, at new empty nests, one with an egg, were located on sandy patches all around the shoreline. On the rocky sections of the coast, the birds still utilise sandy patches for their nests.

Pied Oystercatcher
1 pair was located on a sandy beach at the south-east end of the island.

VEGETATION
The vegetation is dominated by *P. poiformis* with *Myoporum* sp. also common in some areas. The coastal strips are vegetated by *Atriplex cinerea*.

No mammals or other seabird species were recorded.

BIRDS
Brown Quail

Cape Barren Goose – 2 pairs with 3 fledged chicks – introduced to the region

Swamp Harrier

Hooded Plover – 1 pair

Silvereye

Little Grassbird

REPTILES
Tiger Snake

Skink – 1 seen, not identified

COMMENTS
There is a high incidence of harvesting of Short-tailed Shearwaters. This is the only island in the Petrel Island group with reptiles on it. The sandy soils make burrows susceptible to destruction by birders. Status should be upgraded.

Half Tide Rock

(Petrel Island Group, page 116)

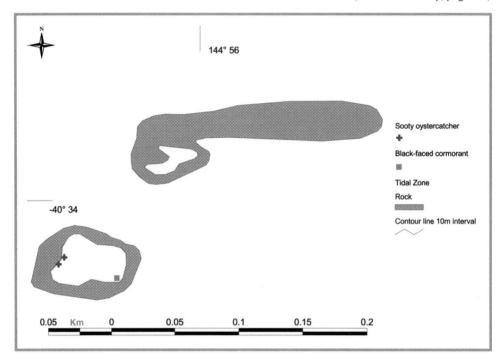

Location: 40°35'S, 144°56'E
Survey date: 14/12/82
Area: 0.13 hectares
Status: Non-allocated Crown Land

Half Tide Rock is an irregular-shaped, jagged and steep-sided islet. It consists mainly of bare rock with some vegetation in the cracks and on the ledges.

BREEDING SEABIRD SPECIES
Sooty Oystercatcher

2 pairs, one at an empty nest and one with an egg, were located in a crevice vegetated by *Disphyma crassifolium* on the western side of the rock.

Black-faced Cormorant

39 pairs, at newly-constructed nests, were found in a colony on rock ledges at the south-east corner of the island from 5 to 20 m above water level.

VEGETATION

D. crassifolium covers over half of the area and *Senecio* sp. grows on the sheltered ledges.

COMMENTS

The presence of a Black-faced Cormorant colony on this small rock is of significance because of the few islands where these birds breed. It appears as if this colony may transfer to Big Sandy Petrel Island in some years.

Kangaroo Island

(Petrel Island Group, page 116)

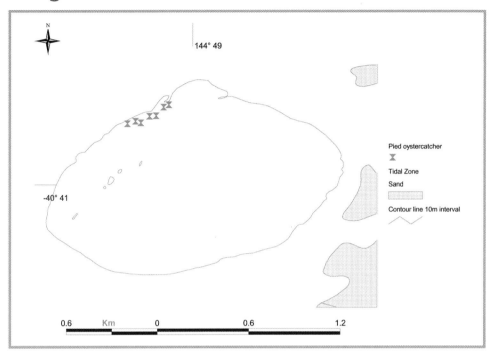

Location: 40°44'S, 144°58'E
Survey date: 8/12/89
Area: 124.55 hectares
Status: Non-allocated Crown Land

Lying directly to the west of Robbins Island this is a flat, oval shaped island with mudflats surrounding it, particularly extensive on the north to north west coasts. The interior of the island comprises rolling dunes.

BREEDING SEABIRD SPECIES
Pied Oystercatcher

7 pairs were located on the north west coast amongst *Sarcocornia* sp.

VEGETATION

S. quinqueflora dominates the coastline and bracken and *Leptospermum* sp. are found on the dunes in the interior.

No reptiles were recorded.

OTHER SEABIRD SPECIES
Sooty Oystercatcher – 15 individuals

Fairy Tern – 46 fishing offshore

MAMMALS
Cattle

Tasmanian Pademelon

BIRDS
Waders, mostly stints – thousands in *Sarcocornia* sp.

Eastern Curlews – 150

COMMENTS
Grazing cattle have caused extensive damage to the *Sarcocornia*. This island is part of the significant Boullanger Bay/Robbins Island wetland site.

Howie Island

(Petrel Island Group, page 116)

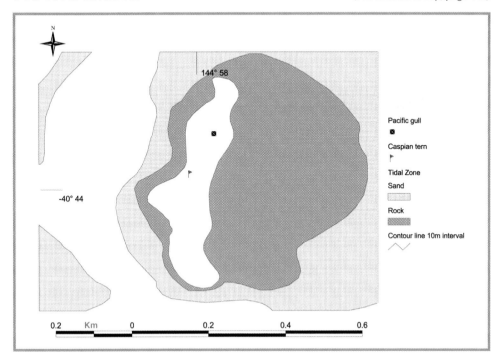

Location: 40°44'S, 144°58'E.
Survey date: 8/12/89
Area: 4.1 hectares
Status: Non-allocated Crown Land

This is a long, narrow island orientated north to south. The shoreline is rocky, spasmodically interrupted by sandy beaches, which occur on all coastlines. There is extensive low tide mudflat exposure around the island.

BREEDING SEABIRD SPECIES
Pacific Gull
120 pairs, most with 2 to 3 eggs, some with one or none and a few with newly-hatched young still in nests, were recorded. The relatively dense colony was situated at the extreme northern end of the island in amongst the *Poa*, stipa and rocks.

Caspian Tern
1 pair, with an egg in its nest was located at the northern end amidst the Pacific Gull colony.

OTHER SEABIRD SPECIES
Pied Oystercatcher – several pairs on the sandy beaches, but no sign of breeding

Sooty Oystercatcher – several pairs on the sandy beaches, but no sign of breeding

MAMMALS
Rabbit – a lot of digging throughout the island, but damage so far minimal

VEGETATION

The southern end is densely covered in *Leptospermum* sp., with moss-covered soil or rock underneath. The northern end is dominated by *Poa*, stipa and *Acaena*.

COMMENTS

Howie Island forms part of the Boullanger Bay/Robbins Island significant wetland site.

Taking in the beautiful island scenery.

North Coast Islands – Region 2

(Regions, page 44)

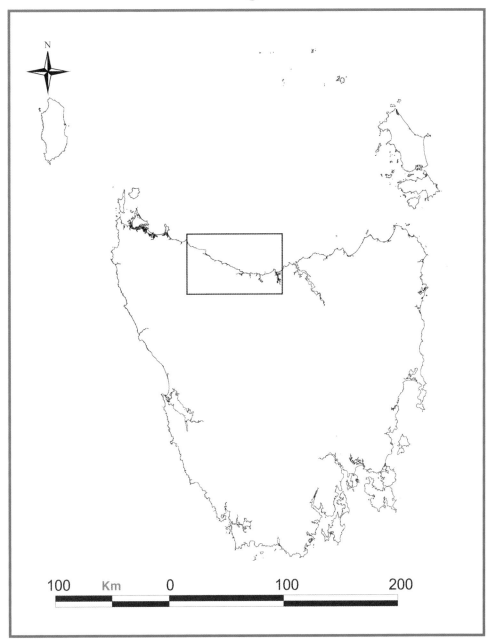

North Coast Group

(Region 2, page 131)

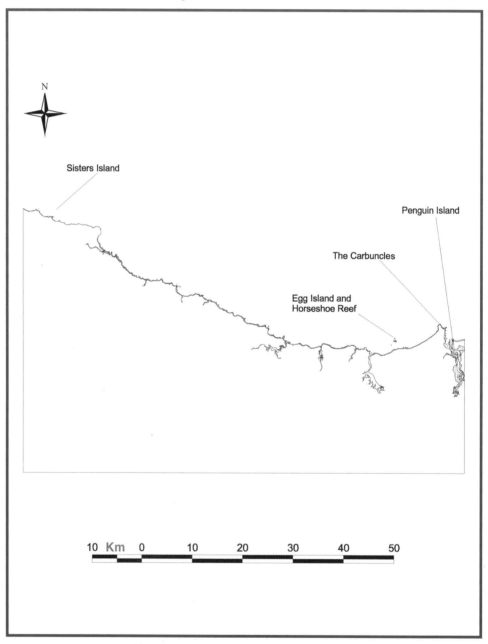

Sisters Island

(North Coast Group, page 132)

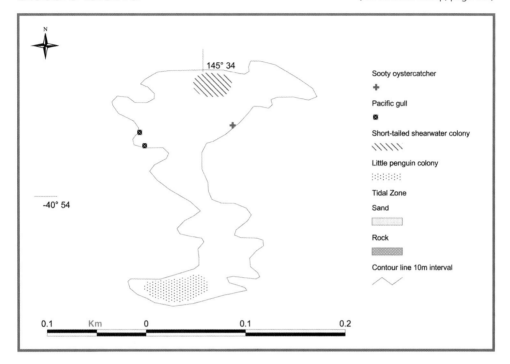

Location: 40°54'S, 145°34'E
Survey date: 28/11/87
Area: 1.43 hectares
Status: Non-allocated Crown Land

Running in a north-north-west to south-south-easterly direction, it is a long, narrow islet. It has jagged rock formations with almost vertical spikes, interspersed with dense *Coprosma repens* of up to 2 metres high, which forms a canopy over the entire island. Its shoreline is rocky.

BREEDING SEABIRD SPECIES
Little Penguin
60 pairs were located nesting extensively under *Coprosma*. All were found in surface nests, which were protected by the cover formed by overhanging rocky outcrops and the surrounding *Coprosma*.

Short-tailed Shearwater
A small colony of short burrows, harbouring about 20 pairs, was found in patches of *Poa poiformis* towards the northern end of the island.

Pacific Gull
2 pairs, one with 2 eggs and one with runners, were located on the north west coast.

Sooty Oystercatcher
One pair with 2 eggs was located on the east coast pebble beach.

VEGETATION
C. repens forms a dense canopy over much of the island. Other vegetation species are scattered and sparse.

No mammals, reptiles or other seabirds were recorded.

BIRDS
Common Starling – a very dense roost site was located in the *Coprosma*.

COMMENTS
A native vegetation rehabilitation program needs to be undertaken, introducing suitable seabird nesting habitat before instigating weed eradication.

Bass Strait Shearwater

Egg Island, Horseshoe Reef

(North Coast Group, page 132)

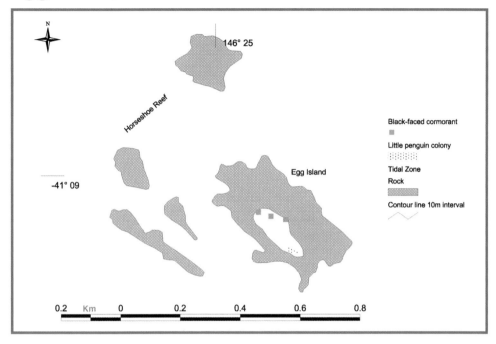

Location: 41°09'S, 146°25'E
Survey date: 28/11/87 and 22/12/87
Area: 0.97 hectares
Status: Conservation Area

Forming a part of Horseshoe Reef, just off the coast from the Devonport airport, Egg Island is a small, flat, rounded pebble islet with a rocky outer shore.

BREEDING SEABIRD SPECIES
Little Penguin
3 pairs were found breeding under the old hut at the southern end of the island.

Black-faced Cormorant
542 pairs were located breeding in three distinct colonies on the rocky shores, mainly in the north of the island. Many were also roosting.

VEGETATION
The vegetation is a mixture of Australian mallow *Lavatera plebeia*, African boxthorn *Lycium ferocissimum* and a few bushes of *Myoporum* sp.

OTHER SEABIRD SPECIES
Sooty Oystercatcher – 2
Pacific Gull – 2
Silver Gulls – Not nesting. 60+ dead
Caspian Tern – 1 pair flew by. In 1970, they were recorded breeding on the island.
Crested Tern – recorded breeding in 1969/70.

No other fauna was recorded.

COMMENTS
This is a significant island because of its large Black-faced Cormorant population. Crested Terns have been recorded breeding here in the past. Silver Gulls have bred here in large numbers but their apparent conflict with the adjacent Devonport Airport precipitated population control through poisoning programs.

Penguin Island

(North Coast Group, page 132)

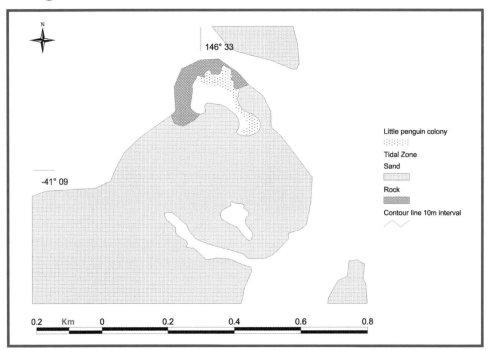

Location: 41°09'S, 146°33'E
Survey date: 28/12/87
Area: 2.73 hectares
Status: National Park

Part of the Narawntapu National Park, Penguin Island comprises 3 islets, all low and flat, joined at low tide by sandbars and extensive rocky shorelines. The northernmost islet is the largest and measures 80 metres (east to west) by 50 metres (north to south)

BREEDING SEABIRD SPECIES
Little Penguin

An estimated 100 pairs are concentrated in sandy soil at the south east end, while a few are lightly scattered in the north.

VEGETATION

The vegetation is dominated by *Senecio* sp., *Tetragonia implexicoma* and *Rhagodia candolleana*, with stipa just before the rocky shoreline. Introduced grasses are also prevalent.

No other fauna was recorded.

COMMENTS

A sign warning people not to disturb the breeding penguins had been erected on the northernmost islet.

The Carbuncles

(North Coast Group, page 132)

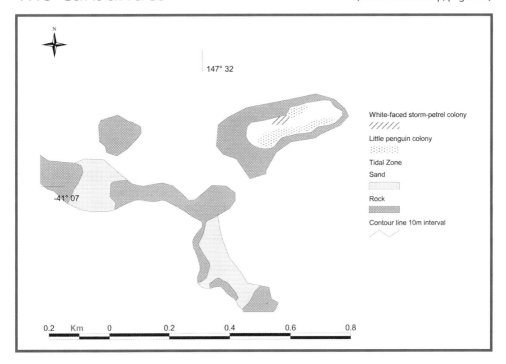

Location: 41°07'S, 147°32'E
Survey date: 27/12/87
Area: 1.91 hectares
Status: National Park

Part of the Narawntapu National Park, the Carbuncles is in two sections, with access from mainland Tasmania via a 50 metre channel at low tide. It has a 10 – 20 metre wide rocky shoreline.

BREEDING SEABIRD SPECIES
Little Penguin

An estimated 30 pairs mostly nest under rocks overgrown with creeping *Tetragonia* in the northern section. About 80 pairs occur in the southern section throughout all habitats in short, shallow burrows and also under rocks and scrub with matted *Tetragonia* growing over it.

White-faced Storm-Petrel

10 burrows were found in the central north of the islet in a patch of mixed stipa and *Poa*. None of the burrows were occupied and 2 freshly-eaten adults were found just beneath the stipa, near the burrows. Faeces would indicate that rats are responsible for the decimation of the colony. No burrows were found in the south.

VEGETATION

The north is dominated by *Carpobrotus rossii*, *Tetragonia* and *Rhagodia* with scattered stipa and *Poa* clumps and patches of *Correa alba*. Two-thirds of the southern section is a mixture of *Poa*, *C. rossii* and *Tetragonia*, with the north east of that section dominated by large *Coprosma* shrubs and patchy *Correa*.

MAMMALS

Rats

No reptiles or other birds were recorded.

COMMENTS

A sign warning people not to disturb the breeding penguins had been erected.

White-faced Storm-Petrel

North Bass Strait Islands – Region 3
(Regions, page 44)

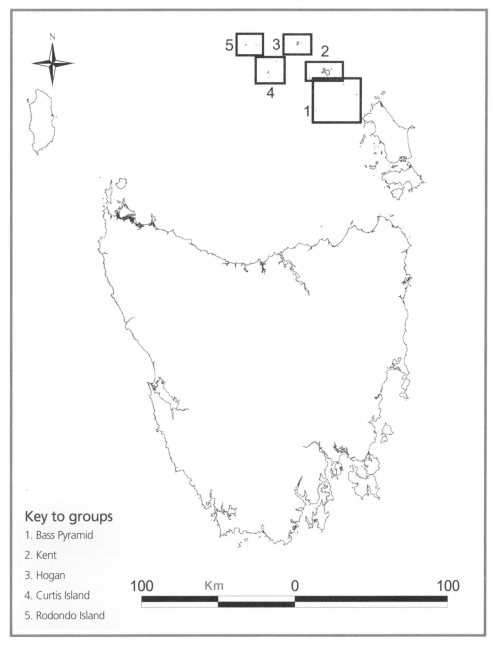

Key to groups
1. Bass Pyramid
2. Kent
3. Hogan
4. Curtis Island
5. Rodondo Island

Kent Group

(Region 3, page 139)

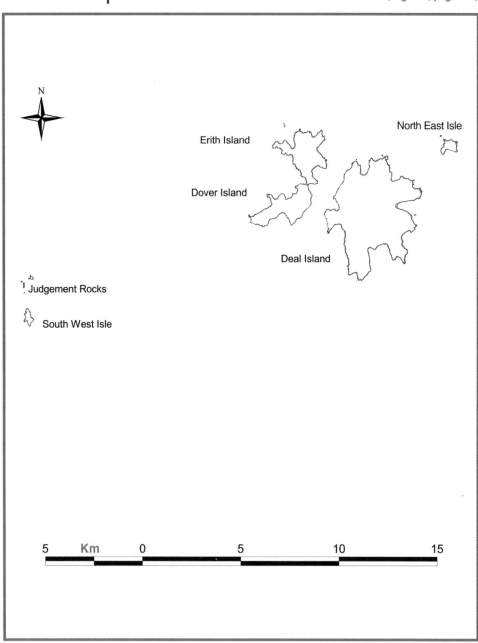

Judgement Rocks

(Kent Group, page 140)

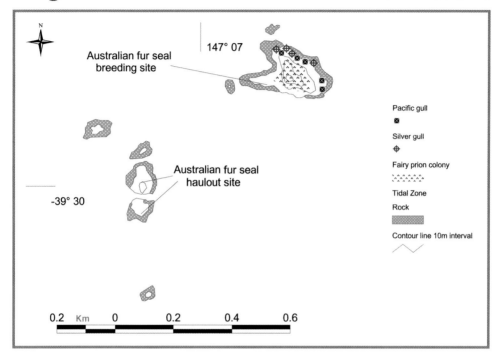

Location: 39°30'S, 147°07'E
Survey date: 11/11/84
Area: 0.39 hectares
Status: Nature Reserve

An impressive granite island with a dome-like base, it rises to cliffs topped by a plateau in the north-west. There are numerous small caves in the north and gulches in the west. Large boulders create caves in the east. All around the island are steep slopes and ledges, which provide the only shelter for vegetation, small patches of *Tetragonia implexicoma* on the western-facing slopes. The rest of the rock is bare with impressive quartz veins running through it. The west side has a steep slope known as 'the ladder', which seals climb up every morning and evening to feed their pups.

North-west end section of Australian Fur Seal colony.

BREEDING SEABIRD SPECIES
Fairy Prion
50 pairs, with eggs, were located in crevices on the south-east, south and west sides of the top plateau.

Pacific Gull
5 pairs were located on the north-east ridge slopes alongside the Silver Gulls.

Silver Gull
Up to 25 pairs at new nests were located on the north-east ridge slopes, nesting in rock crevices above seal colonies.

Sooty Oystercatcher
1 pair was sighted but its nest was not located.

VEGETATION
Only small patches of succulents survive on the ledges.

Australian Fur Seal pup, January, now several weeks old.

OTHER SEABIRD SPECIES
Australasian Gannet – 3 were seen flying low over the rock.

Black-faced Cormorant – 8 were roosting on western point.

Crested Tern – hundreds roosting

MAMMALS
Australian Fur Seals – 2434 pups were counted in the 1999-2000 census, making it Tasmania's largest breeding colony. Seals also haul out on the three offshore rock stacks.

Killer Whales visit to feed on seals.

BIRDS
Peregrine Falcon

Forest Raven

REPTILES
Metallic Skink

COMMENTS
It is an important Australian Fur Seal breeding site, the largest of only 5 sites in Tasmanian waters. It is especially significant because unlike other sites, it is secure from the high seas when pups are young and vulnerable. Under the *Small Bass Strait Island Reserves Draft Management Plan October 2000*, by the Department of Primary Industry, Water and Environment, access without permission is prohibited. Under this plan near-shore boating or fishing activity is also discouraged to minimise disruption.

Australian Fur Seals stand out as black dots against the bare granite rock of Judgement Rocks.

'Seeing a cruel sailor armed with a heavy stick run sometimes for fun through the midst of these marine herds, killing as many seals as he hits and soon surrounding himself with their corpses, one cannot help bemoaning the kind of lack of foresight or the cruelty of nature which seems only to have created such strong, gentle and unfortunate beings in order to deliver them to all the blows of their enemies.' Francois Peron, official historian to the French explorer Nicholas Baudin (January 6, 1803) from Cornell 1974.

South West Isle

(Kent Group, page 140)

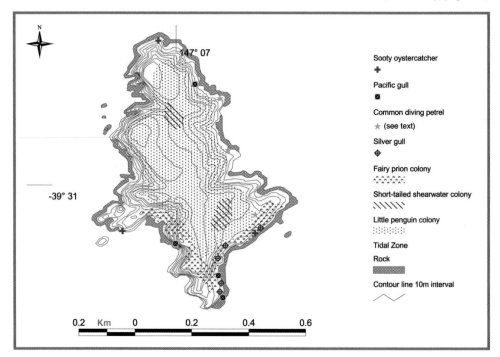

Location: 39°31'S, 145°07'E
Survey date: 11/11/84
Area: 19.09 hectares
Status: Non-allocated Crown Land ↑

The island is elongate, running north to south rising to a 120 metre central summit. It is composed of granite and an estimated one quarter of it is unsuitable for burrowing birds. To the south of the summit is a small lagoon. There are also several freshwater soaks evident around the island. The west coast has steep cliffs, but there are several inlets which provide safe landing sites. It has a navigation light, a helipad and two rock shelters, one in the north-east and one in the south-west.

BREEDING SEABIRD SPECIES
Little Penguin

150 pairs, some with eggs, but most in empty burrows, were located nesting over most of the island in low densities, mainly under rocks with a small minority in burrows. Penguins appeared to come ashore only in one area – the inlet on the southern shore.

Short-tailed Shearwater

An estimated 9500 pairs are in two main concentrations both in depressions that slope gently to the east from the summit before the eastern cliff. Burrows are particularly dense just north of the southern landing site in co-dominant *Poa* and *Senecio* sp., intermingled with rock.

East side looking northward.

Fairy Prion
Up to 5000 pairs, most on fresh eggs, were located with the diving-petrels on the lower steeper slopes, with some also occupying crevices beneath rocks on the top eastern slopes of the western cliffs.

Common Diving-Petrel
3000 – 5000 pairs, with three-quarter grown chicks, some already fledged, were primarily on the lower steeper slopes, with the highest densities occurring in *Poa* and *Disphyma crassifolium* on the eastern side.

Pacific Gull
4 pairs, 2 with eggs, were located around the coastline with 2 pairs nesting near the Sooty Oystercatcher and gulls on the south-eastern coast.

Silver Gull
5 pairs, none with eggs, were nesting together near the Pacific Gulls and Sooty Oystercatchers on the south-eastern coast.

Sooty Oystercatcher
One pair was located on the northern tip, one on the south-western top, and a pair with 2 eggs, was found on the south-eastern side.

VEGETATION
D. crassifolium is the dominant vegetation especially on the east coast lower slopes and is co-dominant with *Poa* and *Senecio* on the other slopes.

OTHER SEABIRD SPECIES
Caspian Tern – 1 dead

No mammals were recorded.

BIRDS
Native:
Brown Quail

Cape Barren Goose – 3 pairs with 5 large runners

Peregrine Falcon – a pair was defending aggressively.

Welcome Swallow

Little Grassbird

Introduced:
Common Blackbird

REPTILES
Metallic Skink

Bougainvilles Skink

White's Skink

COMMENTS
This is undoubtedly a regionally important island for Common Diving-Petrels and Fairy Prions along with its diversity of reptiles. To reflect this, its status should be upgraded to nature reserve.

Construction of helipad and navigation beacon, incorporating artificial nest burrows for displaced seabirds – Australian Maritime Safety Authority

Erith Island

(Kent Group, page 140)

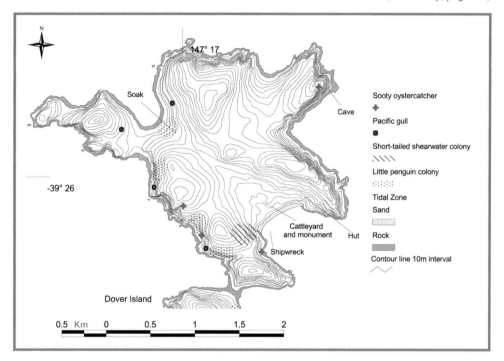

Location:	39°27'S, 147°17'E
Survey:	8/1/84
Area:	323.19 hectares
Status:	Non-allocated Crown Land, Lease, Australian Bush Heritage Fund

Erith Island has been highly modified in the process of improvement for cattle grazing and has in fact, degenerated to a paddock with sparse natural vegetation and patches of soil erosion. There are sandy beaches on the east coast with extensive rugged and cliffy coastline elsewhere. A large raised sea cave in granite at the base of the sheer cliffs is an oustanding geological feature. The cave is 23 metres deep and is now 20 metres above sea level. Deposits in the cave floor contain archaeological material indicating occasional occupation between 13 000 and 8000 before present (B.P.). A house is situated on the northern end of the east beach with a grave site and monument at the southern end. Other human-made features include a dam, fences, cattle yards and numerous well developed campsites.

BREEDING SEABIRD SPECIES
Little Penguin
300 pairs were found inland from all major access points on west and southern coasts. None were on the east coast.

Short-tailed Shearwater
Up to 20 pairs were recorded in a small colony in the mid southern section of the island.

Pacific Gull
4 pairs were located scattered around the west coast.

Sooty Oystercatcher
4 pairs were recorded, one on north-east coast and 3 on the south-east coast,

VEGETATION
Apart from a few patches of native tussock, the natural vegetation has been converted into pasture.

Commemorative cross on the east shoreline.

OTHER SEABIRD SPECIES
Black-faced Cormorant – 6 were near the hut on the rocks.

MAMMALS
Native:
Southern Brown Bandicoot

Long-nosed Potoroo

Brushtail Possum

Introduced:
Cattle – 40+ were there during the survey, but not under the current lease arrangement.

BIRDS
Native:
Brown Quail

Brown Falcon – pair

Fan-tailed Cuckoo

Tasmanian Thornbill

Crescent Honeyeater

Olive Whistler

Forest Raven

Beautiful Firetail

Silvereye

Introduced:
Skylark

Common Blackbird

REPTILES
Metallic Skink

Three-lined Skink

White's Skink

White-lipped Whip Snake

COMMENTS
Grazing has now ceased with the island being leased to the Australian Bush Heritage Fund in 1997. Native vegetation should recover in time but, with easy access and frequent visitation, the risk of fire is of concern. The diversity of mammals, reptiles and birds along with the delicate geomorphological features should be protected.

Dover Island

(Kent Group, page 140)

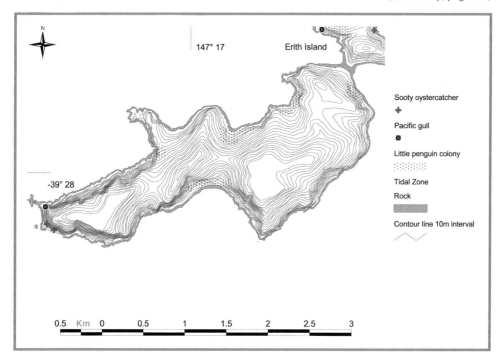

Location: 39°28'S, 147°17'E
Survey date: 9 & 10/11/84
Area: 295.02 hectares
Status: Non-allocated Crown Land

Connected to Erith Island via a tidal gravel bank, it is a densely vegetated granite island with steep slopes to its shoreline.

BREEDING SEABIRD SPECIES
Little Penguin
250 pairs were found inland from major access points, primarily on the west and south coasts. Ten pairs inhabit a cave on the north-west coast.

Pacific Gull
One pair was located on a small offshore rock off the south-west end.

Sooty Oystercatcher
2 pairs were found on the south coast.

VEGETATION
Because Dover Island has not been grazed by stock or cleared, it is covered by low forest of which is dominated by *Allocasuarina verticillata*, closed scrub and heath. It is almost devoid of exotic species.

No other fauna was recorded.

BIRDS
Native:
Brown Quail
White-bellied Sea-Eagle
Brown Falcon – pair
Fan-tailed Cuckoo
Tasmanian Thornbill
Crescent Honeyeater
Olive Whistler
Forest Raven
Beautiful Firetail
Silvereye
Introduced:
Skylark
Common Blackbird

COMMENTS
Further surveys are likely to reveal terrestrial mammals and an extensive botanical and avian inventory.

Deal Island
(Kent Group, page 140)

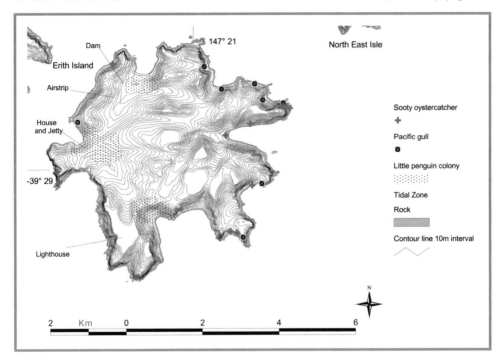

Location: 39°28'S, 147°21'E
Survey: 8/11/84
Area: 1576.75 hectares
Status: Conservation Area

Because of its highly modified landscape with shallow coarse soil, Deal Island has very few seabirds. It has a lighthouse, airstrip, roads, jetty, four houses, assorted sheds, dam and museum. Its coastline features a number of beautiful coves with sandy beaches. On its west and south-west coasts, it has an extensive granite rock or cliff shoreline.

BREEDING SEABIRD SPECIES
Little Penguin
500 pairs, all on fresh eggs, were found around the coast near suitable landing sites. Nests are generally under rocks draped with *Tetragonia* or under rock slabs and boulders to about 130 metres, as there is very little soil for burrowing.

Lower eastern slopes where penguins nest.

Pacific Gull
8 pairs were recorded, mainly on the east coast.

Sooty Oystercatcher
11 pairs were recorded all around the coast.

OTHER SEABIRD SPECIES
Black-faced Cormorant – 23 were roosting on the central eastern coast and 4 on north-west coast.

Silver Gull – in low numbers

Caspian Tern – 1 pair was seen feeding all day in the East Cove but not breeding. One dead Caspian Tern was found on the Garden Cove Beach.

MAMMALS
Native:
Red-necked (Bennett's) Wallaby - 36 counted in pasture in one spot (introduced to the island)

Brushtail Possum

Southern Brown Bandicoot

Swamp Rat – possibly

Introduced:
Rabbit

Rat

Sheep

Cow

Cat

BIRDS
Native:
Cape Barren Goose – at houses

Straw-necked Ibis – feeding on pasture near houses

White-bellied Sea-Eagle – 1

Brown Falcon – 1

Nankeen Kestrel – 1

VEGETATION
Although the island's vegetation is dominated by introduced species, there are small important communities of *Apium insulare* and *Ixiolaena supina*.

Masked Lapwing – 6 at East Cove (breeding)

Green Rosella – 1 dead

Blue-winged Parrot – 2

Pallid Cuckoo

Tasmanian Thornbill

Crescent Honeyeater

Scarlet Robin

Flame Robin

Olive Whistler

Forest Raven – 4+

Beautiful Firetail – common

Welcome Swallow – nesting under jetty

Silvereye

Introduced:
Skylark

European Goldfinch

Common Blackbird

Common Starling

REPTILES
Metallic Skink

Bougainvilles Skink

White's Skink

White-lipped Whip Snake

OTHERS
Introduced land snail

Garden Cove with airstrip to settlement on distant western skyline.

COMMENTS

This is a highly modified island, with very little of its natural environment intact due to fire, land clearing and grazing. Its tenure change to conservation area may result in recovery. Its diversity of reptiles and terrestrial birds is significant. The feral plant *Euphorbia* (sea spurge) is having an adverse impact on the coastline.

From Deal Island looking across Mercury Passage to Erith Island

> *Deal Island lighthouse was built in 1847 with immense labour using bullock teams and convict workers. In the nineteenth century, it became home of families like the Baudinets and Browns, large families brought up in this remote environment. If the three monthly supply vessel from Hobart Town was a month or two overdue, the lusty boys of the family would think little of pulling in the whaleboat five or six miles offshore to hail a becalmed ship for news and perhaps some badly needed supplies. Murray-Smith, in 'Bass Strait Australia's Last Frontier', 1969*

North East Isle

(Kent Group, page 140)

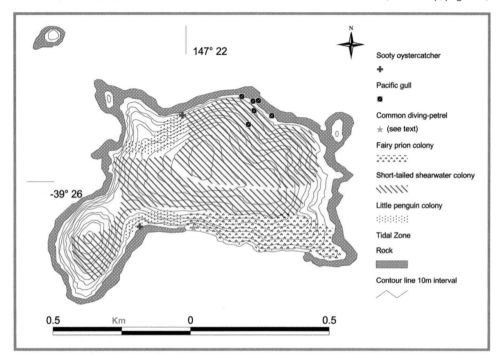

Location: 39°27'S, 147°22'E
Survey date: 16/11/84 and 22/2/87
Area: 32.62 hectares
Status: Non-allocated Crown Land ⇑

This is an irregular-shaped, topographically striking island, whose steep coastline is interrupted by gulches. The southern knob of 105 metres is joined to the northern summit of 125 metres by a saddle. Jumbled granite boulders dominate the shoreline. An estimated one third of the island has either very shallow soil or bare granite slabs, making it unsuitable for burrowing birds.

BREEDING SEABIRD SPECIES
Little Penguin

An estimated 140 pairs come ashore in the south-east and the north-west gulches, where they also tend to congregate. They mainly nest under boulders, with a few in burrows amongst *Poa* on steep slopes. Some birds manage to reach burrows near the highest peak.

Short-tailed Shearwater

An estimated 197 276 pairs, most with chicks (22/2/87), nest all over the island with the highest concentrations in the north in areas where *Poa* is particularly thick, generally above 80 metres. Here burrows are under matted vegetation. Elsewhere burrows are short and shallow on gently undulating slopes. Burrow density is consistently greater on the south slopes from the summit. On 22/2/87, burrow occupancy was reported as being low, thought to be the result of predation by

Northern sector of North-East Isle, looking east.

ravens and gulls who take advantage of the short, sometimes near-surface, burrows.

Fairy Prion
1000 – 2000 pairs, most with eggs, generally inhabit similar areas to the diving-petrels, but tend to be in greater numbers in the sites dominated by rock crevices.

Common Diving-Petrel
2000 – 4000 pairs, most with chicks, are abundant throughout the island, mainly below 80 metres on low slopes of thick *Poa* and on cliff ledges where dense *Poa* also grows. Some nest with Fairy Prions in patches of small jumbled boulders.

Pacific Gull
26 pairs were counted, the largest colony of 20 pairs nesting in a narrow strip of *Poa* winding through crevices in flat, gently north-sloping granite slabs. Nests were as close as 3 metres apart, but generally more than 5 metres apart. Most nests contained eggs. 6 pairs were defending the south area.

Sooty Oystercatcher
One pair was defending in the vicinity of the south-east landing and one in the northern gulch area.

VEGETATION
The vegetation is dominated by *Poa*, with *Senecio* sp. On the western cliffs there are various shrub species growing, as well as stipa and *Rhagodia* sp.

No mammals were recorded.

OTHER SEABIRD SPECIES
Black-faced Cormorant – 30 were roosting in a regular spot off the south-east to east side

Silver Gull – 2 flew by

MAMMALS
White-footed Dunnart

BIRDS
Native:
Cape Barren Goose

White-bellied Sea-Eagle – flew by

Forest Raven – 6+

Welcome Swallow – 5

REPTILES
White's Skink

Looking towards Deal and Erith Islands.

COMMENTS
This is a pristine island, with a diverse seabird presence, making it significant for the Bass Strait region. Its status should be upgraded.

Far south section toward north end summit.

Hogan Group

(Region 3, page 139)

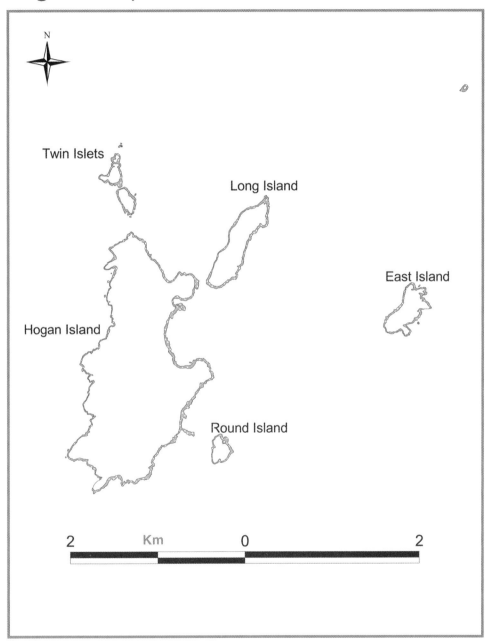

Hogan Island

(Hogan Group, page 157)

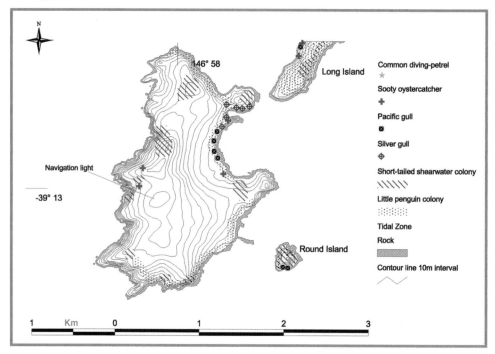

Location: 39°14'S, 146°59'E.
Survey date: 14/11/84
Area: 232.27 hectares
Status: Non-allocated Crown Land

The largest of the group, it is an elongate island running north to south with a highly modified natural environment. 80% of the island is grassland with little soil and a lot of bare granite slabs. Scrub is largely confined to the west coast slopes amongst boulders. The shoreline of the east side gently slopes up from rock slabs and boulders. The west side is steep. Calcarenite (lime-rich sand) veneers the granite on parts of the island with springs at the base on the east coast. Peat of up to 0.5 metres deep has formed around the springs, overlying clays derived from granitic gravels. Soils elsewhere on the island are shallow and sandy. There is a navigation beacon on the summit, with disused cattle watering troughs fed by natural springs.

BREEDING SEABIRD SPECIES
Little Penguin

800 pairs, ranging from single birds, birds with empty nests and some with eggs, nest in dense colonies on the steep north to north-east slopes up to 30 metres in crevices, mainly under rocks. Some are also lightly scattered at the north end and down the eastern and southern slopes, both in crevices and under vegetation up to 80 metres.

Short-tailed Shearwater

In average densities of $0.46/m^2$, an estimated 83 825 pairs were scattered all over the island.

Colonies at the south end were in association with *Senecio* sp. and *Poa poiformis*. At the northern end and on the north-east knob they were also in a mixture of *Senecio* and *P. poiformis* while the eastern colony directly in from the eastern offshore island has 100% *Poa*.

Pacific Gull

4 pairs were located south of beach on the east coast, with nests built of *Poa*.

Silver Gull

6 pairs, 3 with empty nests, 2 with eggs and 1 with chicks were located at the north-east point of the main beach, nesting in rock crevices and ledges 5 metres up by *Poa*.

Sooty Oystercatcher

4 pairs were breeding on the island. One nest contained shell fragments, one contained an addled egg, one had eggs and one had chicks. One nest was located on the north-east point of the main beach, one on the west end of main beach, one 50 metres further west and one centre west, just north of the navigation light.

VEGETATION

Introduced pasture species dominate the island.

COMMENTS

The island's natural vegetation and springs were damaged by uncontrolled livestock grazing, which has now ceased. The correlation between degraded islands and high reptile diversity, as exemplified by this island, needs greater investigation. The island's peat soils, which have formed around the springs, are considered to be of outstanding geomorphological significance for the State (Dixon, 1996).

OTHER SEABIRD SPECIES

Cape Barren Goose – 25+

Common Diving-Petrel – 2 dead birds were found on central east coast. There were no signs of breeding.

Fairy Prion – suspected to be breeding in west coast crevices

MAMMALS

Cattle – about 60

Rat – remains were found, suspected to have been the prey of Barn Owls.

BIRDS

Native:

Brown Quail – 3

White-faced Heron

White-bellied Sea-Eagle – on the cliff on the northern tip

Swamp Harrier

Brown Falcon

Peregrine Falcon

Nankeen Kestrel

Masked Lapwing

Barn Owl – in the cave on the south side of the north-east point

Welcome Swallow

Silvereye

Little Grassbird – few

Introduced:

Skylark – 1 nest with 2 eggs and 1 chick. Abundant at the north tip

European Goldfinch

Common Blackbird

Common Starling

REPTILES

Blue-tongue Lizard

Metallic Skink

Three-lined Skink

Bougainvilles Skink

White's Skink

OTHERS

Land Snail

Burrowing Crayfish – in water soak on the western side of the island, of unknown species, but, according to Pierre Horowitz (pers.comm.), it could have been endemic. It was not found in a 1993 survey of the island and is now considered to be extinct, destroyed by the cattle using the water soak in which it lived.

Twin Islets (north)

(Hogan Group, page 157)

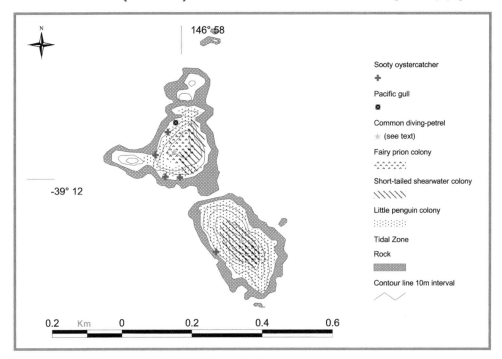

Location: 39°12'S, 146°58'E
Survey date: 15/11/84
Area: 2.6 hectares
Status: Non-allocated Crown Land ↑

Part of the Hogan group, this is a steep sided granite island, teardrop in shape, with three small rock stacks off the pointed northern end. The northern end is composed of unvegetated granite boulders while the other sides are steep granite slabs or cliffs.

BREEDING SEABIRD SPECIES
Little Penguin
20 pairs, on eggs, were recorded on the west side of the island under rocks and in crevices.

Short-tailed Shearwater
20 to 30 burrows were located. They are restricted by shallow soils and, as a result, burrows are short and shallow. The burrowing habitat was located amongst *Senecio* sp., *Poa poiformis* and *Disphyma crassifolium*. Fairy Prions and Common Diving-Petrels burrow in association with the Short-tailed Shearwaters.

Fairy Prion
150 burrows were located along the central portion of the island in association with Common Diving-Petrels and Short-tailed Shearwaters.

Common Diving-Petrel
150 burrows were located along the central portion of the island.

Pacific Gull
1 pair was nesting on the north-west end of the island.

From Hogan Island looking north to Twin Islets.

Sooty Oystercatcher
4 pairs, on nests with eggs, were found on the western and southern sides of the island.

VEGETATION
The vegetation is dominated by *D. crassifolium*. 60% of the area is unvegetated. There are small patches of *P. poiformis*..

No mammals or reptiles were recorded.

BIRDS
Cape Barren Goose – 2 adults were seen.

COMMENTS
This is an undisturbed island, whose status should be upgraded.

Twin Islets (south)

(Hogan Group, page 157)

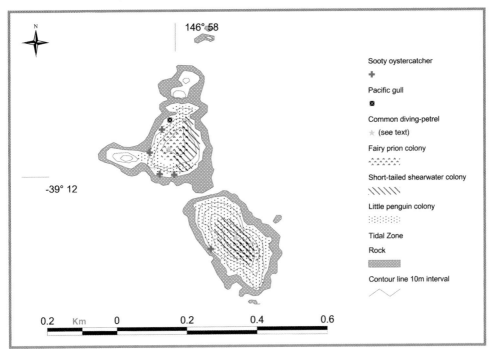

Location: 39°20'S, 146°58'E
Survey date: 15/11/84
Area: 3.01 hectares
Status: Non-allocated Crown Land ↑

The southern Twin Islets have steep massive granite slabs on the western side and granite boulders and slabs on the eastern side. Approximately 60% of the island is granite with insufficient soil for birds to burrow in, whilst there is shallow soil on the summit area between the granite boulders and *Disphyma crassifolium* on the eastern slopes.

BREEDING SEABIRD SPECIES
Little Penguin

50 to 80 pairs inhabit the island. The only access for these birds is on the eastern side of the island amongst granite rocks and boulders. They breed all over the island in rocks and crevices, but mainly near the water at the landing.

Short-tailed Shearwater

30 pairs were recorded breeding in short, shallow burrows on the summit area.

Fairy Prion

50 pairs were found in crevices, under boulders and in burrows where there is sufficient soil. Few nests were found on the western side of the island, and of these, three contained abandoned eggs. It is likely that this area is too exposed to the prevailing weather conditions and high seas.

Common Diving-Petrel

50 pairs were found breeding where soil is sufficient to support burrows.

Sooty Oystercatcher

One was located on the western side of the island on granite gravel amid *D. crassifolium*.

VEGETATION

The vegetation is dominated by *Poa poiformis*, *D. crassifolium* and *Senecio* sp. These species alternate in their dominance. Small patches of *P. poiformis* have been burnt.

No other seabirds, mammals or reptiles were recorded.

BIRDS
Introduced:
Common Blackbird

COMMENTS

A relatively undisturbed island, however the evidence of fire and a rock shelter which contains plastic and tin pieces on the eastern side, suggests that shipwrecked fishers have spent some time here. The status of the two islets should be upgraded to reflect their relatively pristine condition and high seabird diversity.

Long Island

(Hogan Group, page 157)

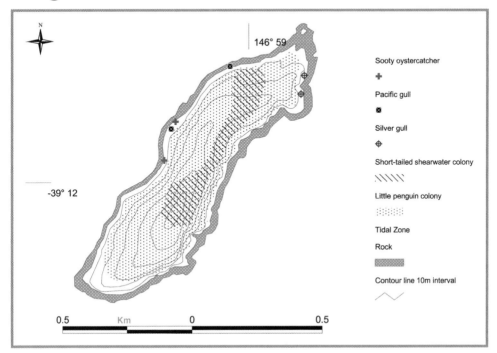

Location: 39°12'S, 146°59'E
Survey date: 14/11/84
Area: 22.85 hectares
Status: Non-allocated Crown Land

Part of the Hogan group, it is a long, narrow island running north-east to south-west with a central ridge, which is largely bare rock. The island rises to its maximum height of 80 metres in the north-east. The soil is black, soft and sandy and is very sparse. Apart from the south-east end which has slab and boulder cliffs, the rest of the coast slopes away gradually.

BREEDING SEABIRD SPECIES
Little Penguin
200 pairs are lightly scattered all over the island, mainly under rocks and boulders with a few in burrows and crevices. They mostly land on the west side.

Short-tailed Shearwater
An estimated 3000 pairs breed mainly along the eastern slopes of the central ridge in *Senecio* sp, with the highest concentrations at the northern end. A few patches extend 20 metres down the western slope. Burrows are short and shallow due to the lack of soil depth.

Pacific Gull
2 pairs, one with an empty nest and one with 2 eggs were recorded nesting on the west coast.

Silver Gull
2 pairs were seen defending, but their nests were not located.

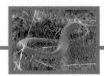

Sooty Oystercatcher
2 pairs, one with an empty nest and one with 2 eggs, were located on the west coast.

VEGETATION
Senecio lautus is the dominant species.

Australian Fur Seal haul-out on bare granite slopes to the north.

No mammals were recorded.

OTHER SEABIRD SPECIES
Crested Tern – 3 were perched on rocks at the south end

Black-faced Cormorant – 10 were sitting on rocks on the north-east end

BIRDS
Native:
Brown Quail - 3

Cape Barren Goose – 6 seen

Swamp Harrier

Introduced:
Common Blackbird - fledgling and adults

Common Starling

REPTILES
White's Skink

COMMENTS
This is an undisturbed island with low seabird species diversity due to shallow soil. The black, soft sand is worthy of further investigation.

Round Island

(Hogan Group, page 157)

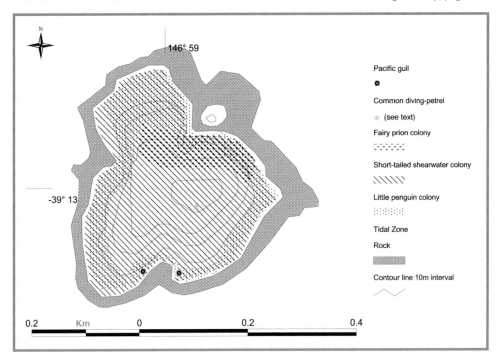

Location: 39°14'S, 146°60'E
Survey date: 15/11/84
Area: 3.95 hectares
Status: Non-allocated Crown Land ⇑

Part of the Hogan group, it is a round island with slabs of granite forming cliffs which slope from the water on the eastern and northern sides. The other sides are more gentle with the north broken cliff providing a gentle slope to the summit.

BREEDING SEABIRD SPECIES
Little Penguin

150 pairs, some with eggs, the majority still with empty nests, were located all over the island, mainly in crevices and under rocks. They come ashore anywhere on the south-east or north-west coasts.

Short-tailed Shearwater

200 pairs were located around the island, with burrows amongst *Senecio* and *Poa*. Their distribution is patchy due to the majority of the island being devoid of soil or having insufficient soil for burrowing.

Fairy Prion

200 – 300 pairs were located in rock crevices, under boulders and in burrows beneath *Disphyma crassifolium* all around the island. They are in the same habitat as the Common Diving-Petrels, but are not using the crevices to their full extent, possibly only a quarter to a half of their potential.

Common Diving-Petrel
Up to 100 pairs are estimated to be very lightly scattered around the island on the lower vegetated slopes beneath *Poa* and *Disphyma crassifolium*.

Pacific Gull
2 pairs, at nests with chicks, were located beside *Poa* at the southern tip of island

VEGETATION
Vegetation is dominated by *Senecio* sp. and *Poa poiformis* with *Disphyma crassifolium* becoming dominant on the eastern and southern slopes.

COMMENTS
A small pristine island with a representative diversity of seabirds and skinks, it is a particularly important refuge in the Hogan group, considering the environmental degradation of the larger islands, particularly Hogan Island itself.

No mammals were recorded.

OTHER SEABIRD SPECIES
Black-faced Cormorant – 29 sitting on regular roost site on western side

BIRDS
Native:
Cape Barren Goose – 5 ashore

Introduced:
Skylark – 1 nest in *Poa* with 3 eggs

Common Starling

Common Blackbird

REPTILES
White's Skink

Metallic Skink

East Island

(Hogan Group, page 157)

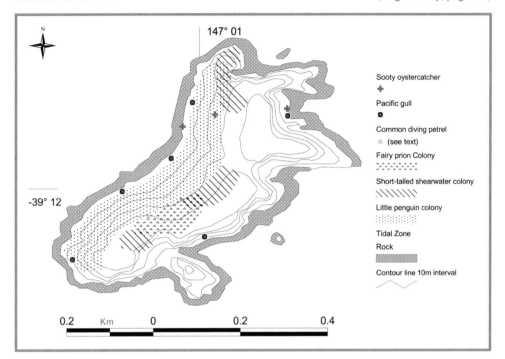

Location: 39°13S, 147°12'E
Survey date: 15/11/84
Area: 12.42 hectares.
Status: Non-allocated Crown Land ↑

Part of the Hogan group, this is a small, elongate island running north to south. Half of the island is bare granite with insufficient soil for any burrowing birds. The lower slopes of the west and south sides are bare with no *Poa* cover. The northern end, except for the very northern tip, has sparse soil with *Poa poiformis* the dominant vegetation.

BREEDING SEABIRD SPECIES
Little Penguin

50 pairs with empty nests and eggs were sparsely distributed under rocks all over the western side.

Short-tailed Shearwater

An estimated 1878 pairs were found along the main ridge, where there is sufficient soil They were in three small concentrated colonies one in the north, one in the centre and one in the south.

Fairy Prion

Although difficult to detect, an estimated 2000 – 3000 pairs burrow sparsely amongst Common Diving-Petrels. Although more noticeable in the west coast cliff crevices, they were still not using them to their full extent.

Common Diving-Petrel

An estimated 5000 – 10 000 pairs were located along the ridge, where there is enough soil to

construct burrows, and on the flat upper part of island. They are less abundant, where the soil is deep enough for Short-tailed Shearwaters.

Pacific Gull

6 pairs with eggs were located primarily on the west coast.

Sooty Oystercatcher

One nest with 2 eggs was located on the north-east coast and 2 pairs were on the west coast.

COMMENTS

Its relatively high seabird species diversity combined with its pristine natural environment makes it worthy of a status upgrade to conservation area.

No mammals were recorded.

OTHER SEABIRD SPECIES

Black-faced Cormorant – 12 were roosting on the north-west tip

Silver Gull – 1

BIRDS
Native:
Brown Quail

Cape Barren Goose – 6

Swamp Harrier

Peregrine Falcon – one was displaying territorial behaviour in the north-east corner

Forest Raven

Introduced:
Common Blackbird

Common Starling

REPTILES

White's Skink

Metallic Skink

Curtis Island Group

(Region 3, page 139)

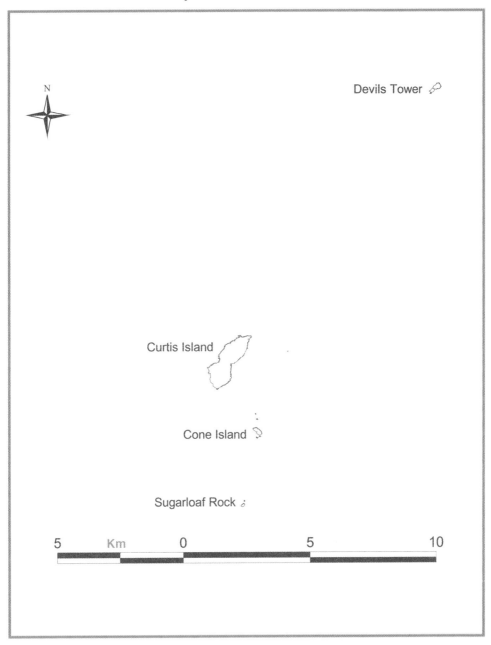

Curtis Island

(Curtis Island Group, page 171)

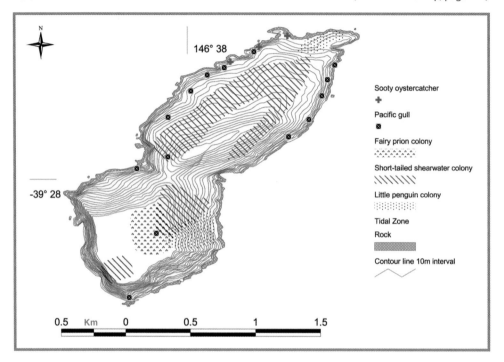

Legend:
- Sooty oystercatcher
- Pacific gull
- Fairy prion colony
- Short-tailed shearwater colony
- Little penguin colony
- Tidal Zone
- Rock
- Contour line 10m interval

Location: 39°28'S, 146°39'E
Survey date: 12/11/84
Area: 149.21 hectares
Status: Nature Reserve

Curtis Island is a granite island, orientated north-east to south-west with steep cliffs dominating most of the coastline. Its ridge, 1.78 m long, rises steadily from north to south but is broken into two peaks by a saddle 50 m deep. The bare rounded northern peak reaches 224.3 metres, while the square-capped and vegetated summit in the south rises to 335.3 metres and ends in a precipitous cliff 250 – 300 metres high. The soils are granite-derived, consisting of a mixture of granitic gravels and bird-derived organic matter.

BREEDING SEABIRD SPECIES

Little Penguin

1000 pairs, many with eggs, were primarily around their two main access routes, one at the east coast bay on steep *Poa*-covered slopes and one in the north-west corner nesting up to saddle height.

Short-tailed Shearwater

In average densities of $0.78/m^2$, an estimated 514 298 pairs inhabit the large colonies both on the south summit slopes and on the north-west summit slopes in dense *Poa* in short burrows. They have extensively burrowed in the granite-derived sandy soils, wherever soil depth is greater than 25 metres. Sometimes they are also found under granite slabs.

Fairy Prion
70 pairs were located in rock crevices mainly around southern peak slopes with no evidence to suggest that this species breeds elsewhere on the island. However, the east to south-east facing slopes of the north section are potentially most suitable and were not thoroughly surveyed.

Pacific Gull
22 pairs were located all around the coast of island with one pair on the summit ridge.

Sooty Oystercatcher
3 pairs were found on the north-west coast.

VEGETATION
Poa poiformis is the dominant vegetation with scattered areas of *Melaleuca armillaris* on the ridge slopes. Swards of the succulent *Disphyma crassifolium* occur in crevices on the steepest slopes and the areas most exposed to salt-bearing wind and *Carpobrotus rossii* prevalent on the deeper soils in the coastal strip. Stipa occupies areas with slightly deeper soils and less exposure than the *D. crassifolium* and *C. rossii*.

COMMENTS
The shearwaters' digging activities in the granite-derived lithosols have contributed to soil development and the occurrence of partially-decomposed organic matter throughout the deeper soils. This is considered of representative and outstanding geoconservation significance for Tasmania (Dixon, 1996). It is a significant island because of its soil type, species diversity and pristine state.

No mammals or other seabirds were recorded.

BIRDS
Native:
Cape Barren Goose – 2 pairs with 2 goslings each were located at the northern end
White-bellied Sea-Eagle – a nest was found
Swamp Harrier
Brown Falcon
Little eagle
Olive Whistler
Forest Raven
Richard's Pipit
Welcome Swallow
Silvereye

Introduced:
Common Starling – 2
Common Blackbird
European Goldfinch

REPTILES
Metallic Skink
Bougainvilles Skink
White's Skink
White-lipped Whip Snake

Cone Island

(Curtis Island Group, page 171)

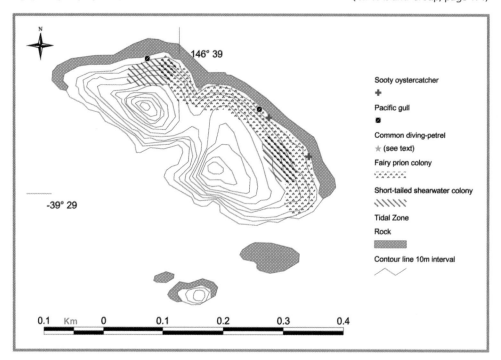

Location: 39°30'S, 146°40'E
Survey date: 12/11/84
Area: 4.82 hectares
Status: Non-allocated Crown Land

This is a small, granite island, oval in shape. The east side of the island is dominated by cliffs and jumbled boulders, 20 to 30 m above water level, which extend around to the northern end. The east side has 50 m high cliffs which slope gradually to the southern peak (110 metres) and steeply to the northern peak (80 metres). The peaks are separated by a saddle. The slope from the saddle to the highest peak is composed of unstable boulders. At least half of the island is bare rock and jumbled boulders.

BREEDING SEABIRD SPECIES
Short-tailed Shearwater

Approximately 150 pairs, in two colonies, were located at either end of the saddle. The burrows are in soil under boulders and loose rocks. The northernmost colony has 20 pairs and the southern colony 130 pairs. There are also several pairs scattered close to these two concentrations.

Fairy Prion

An estimated 2000 to 3000 pairs were located mainly in crevices (80%) and under boulders, largely confined to the edge of the island.

Common Diving-Petrel
300 to 500 pairs were located in burrows under rocks and boulders, mainly around the edges of the island. They were also occasionally in crevices away from edges of the cliffs which surround the island.

Pacific Gull
2 pairs with eggs were located. One pair was nesting on the northern summit and the other midway along the island on the eastern side.

Sooty Oystercatcher
2 pairs, with eggs, were nesting 20 to 30 metres above water level on the eastern side on a ledge with *Disphyma crassifolium*.

VEGETATION
The vegetation is dominated by *D. crassifolium* growing on steep granite slabs and boulders. On the east side where soil and grit are sufficient for burrowing birds, there are patches of stipa. At the time of the survey, large areas of *D. crassifolium* were being killed by browsing black and brown caterpillars.

No mammals were recorded.

OTHER SEABIRD SPECIES
Crested Tern – one was seen flying over the island.

BIRDS
Native:
Orange-bellied Parrot – 1 flew off giving an alarm call
Little Raven –1 seen
Grey Fantail –1 seen
Silvereye –2 seen

Introduced:
Common Blackbird – 1 male seen

REPTILES
Metallic Skink

COMMENTS
This is a remote island, free of human disturbance with a high seabird species diversity, which provides a good example of a natural island for comparison with those in the region that have been disturbed.

Sugarloaf Rock

(Curtis Island Group, page 171)

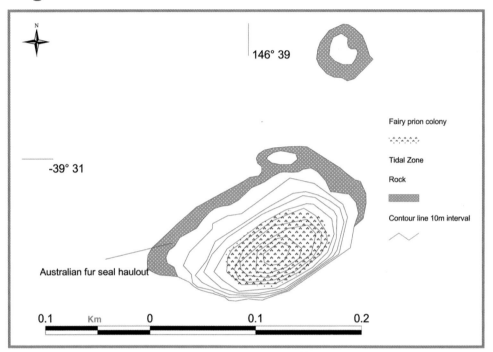

Location: 39°31'S, 146°39'E
Survey date: 20/11/84
Area: 1.07 hectares
Status: Non-allocated Crown Land ↑

This is a small island with near vertical granite cliffs surrounding it. There is scant vegetation found in cracks in the rock. Only a small area from the sea to approximately half way up could be surveyed from land, the rest was checked by boat.

BREEDING SEABIRD SPECIES
Fairy Prion
An estimated 500 pairs were found almost exclusively in rock crevices around the stack.

The steep slopes of Sugarloaf Rock, from Cone Island.

East side.

Common Diving-Petrel
This species is likely to be breeding as the habitat present is similar to other islands in the area on which they breed.

OTHER SEABIRD SPECIES
Black faced cormorant – 1 seen on shoreline rocks.

MAMMALS
Australian Fur Seals – 20 were seen on a boulder ledge at the south-west tip.

COMMENTS
The unique shape of the island is a landmark to fishermen and sailors and of considerable aesthetic value. This, along with the presence of Fairy Prions and Australian Fur Seals, is sufficient to warrant increased protection, and a status upgrade.

Devils Tower

(Curtis Island Group, page 171)

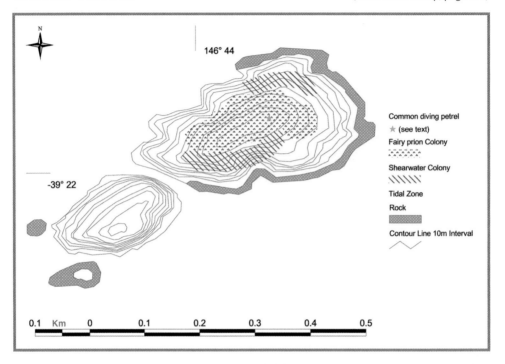

Location: 39°22'S, 146°44'E
Survey date: 20/11/84
Area: 4.77 hectares
Status: Nature Reserve

There are two islands orientated north to south and elongate in shape. They are both surrounded by near vertical cliffs. Land surveys were only carried out on the northern island, as the other is almost impossible to access, as vertical cliffs rise directly from the sea. Half the island is bare granite boulders and slabs sloping to about 60 m from the sea where most of the cliffs are bare granite. There are shrubs around the summit boulders.

North side of western sector.

BREEDING SEABIRD SPECIES
Short-tailed Shearwater
There are two colonies with an estimated total of 5213 pairs, one on the north-west side and the other on the south-east side of the island. The density is higher where there is *Poa poiformis*. The colonies' soils are sandy with granite gravel.

Fairy Prion
2000 to 3000 pairs nest in crevices and burrows under boulders from the summit down to 50 m above the sea.

Common Diving-Petrel
2000 to 3000 pairs inhabit the same area as the Fairy Prions on the flatter areas of the south-east slope between the summit and the east cliff.

OTHER SEABIRD SPECIES
Black-faced Cormorant – 1 was seen on the shoreline rocks.

Silver Gull – 1 was seen being eaten by a female Peregrine Falcon.

MAMMALS
Australian Fur Seals use this island as a regular haul-out.

BIRDS
Native:
Peregrine Falcon

Little Raven

REPTILES
Metallic Skink

VEGETATION
Where there is sufficient soil, *Disphyma crassifolium* and stipa. are the dominant species.

COMMENTS
This island is an authorised target for military practice. The unique shape of the island is a landmark to fishermen and sailors and of considerable aesthetic value. This, along with the presence of breeding seabirds, is sufficient to warrant protection via statutory regulations to prevent the island from ever being used for military practice.

The second unvisited islet is comparable in area and habitat to the one surveyed and so is expected to harbour a similar variety and abundance of species.

Devils Tower to Rodondo in distance.

Rodondo Island Group

(Region 3, page 139)

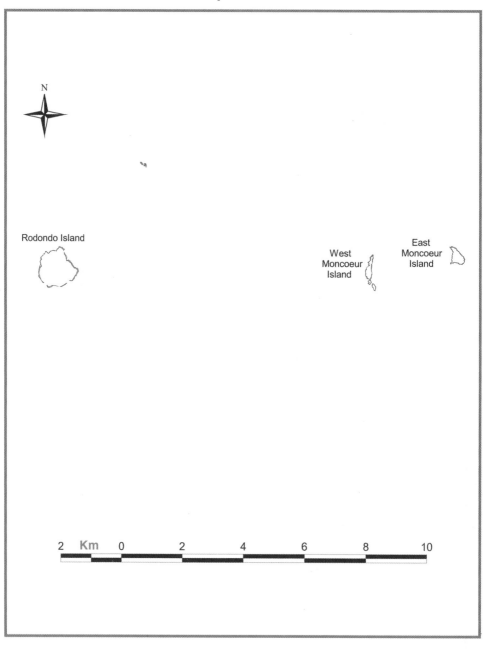

Rodondo Island

(Rodondo Island Group, page 180)

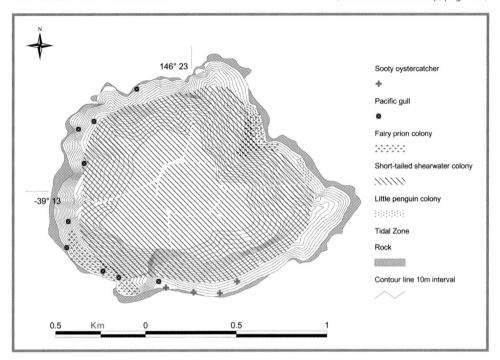

Location: 39°14'S, 146°23'E
Survey date: 13/11/84
Area: 106.15 hectares
Status: Nature Reserve

Situated only 10 kilometres from Wilsons Promontory in Victoria, Rodondo Island is a spectacular granite island ringed by steep cliffs varying from 70 to 200 metres. The central summit rises to approximately 350 metres, giving it a distinctive profile from afar. It is well-vegetated, with open herbfields, grasslands and forests.

BREEDING SEABIRD SPECIES
Little Penguin
50 - 100 pairs were located nesting on the east side landing site up to a maximum altitude of 80 metres.

Short-tailed Shearwater
In average densities of $0.85/m^2$ over the island, an estimated 1 135 948 Short-tailed Shearwaters are concentrated in the mid north and south-east of the island under *Melaleuca*, where there is *Poa poiformis* even up to the summit. Colonies, however, are densest in clearings with few scattered shrubs on steep slopes, where *Poa* is dominant. Little burrowing occurs on the north ridge of the summit due to inadequate soil.

Fairy Prion
They were predominantly found nesting in burrows on the north-east ridge from 50 to 200 metres in altitude. They were also found on the south-west side to a maximum height of 280 metres.

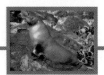

Pacific Gull

9 pairs, 7 with eggs and 2 with new nests, were located on the west and south-west ridges.

Sooty Oystercatcher

4 pairs, on nests and with eggs, were located predominantly on the southern side.

VEGETATION

The major vegetation communities are:

1. *Disphyma crassifolium* herbfield with *Lepidium foliosum*, *Sarcocornia quinqueflora*, and *Asplenium obtusatum*. The latter two are confined to the cliffs. *Pelargonium australe*, *Senecio lautus*, *Bulbine bulbosa*, *Tetragonia implexicoma* and *Rhagodia candolleana* are also relatively common both within and outside this community.

2. Stipa tussock grassland which occupies the area with slightly deeper soils and less exposure than the herbfield.

3. *Poa poiformis* tussock grassland, which includes *P. poiformis*-dominated closed tussock and tussock grassland and smaller areas where *P. poiformis* is subdominant to *S. lautus*, *Bracteantha bracteata* and *Lavatera plebeia*.

4. *Melaleuca armillaris* low closed forest which varies from an open scrub to a closed forest and is the most common community on Rodondo Island. Plants found in this community include *Olearia phlogopappa*, *Acacia stricta*, *R. candolleanaa*, *T. implexicoma*, *Wahlenbergia gracilis* and *Correa backhouseanna*.

5. *Allocasuarina verticillata* low open forest is limited to the western slopes of Rodondo and consists of *A. verticillata* and a generally bare understorey.

6. Shrubland is found only on the upper cliffs on the north-east of Rodondo, with the main shrub present being *Paraserianthes lophantha* with *C. alba* and *Myoporum insulare* also being reasonably common.

7. *Eucalyptus* aff. *globulus* open forest, is confined to the area around the summit and parts of the western slopes of the island. The eucalypts that characterise this community are typically 10 – 15 metres tall. *M. armillaris* forms a ten metre tall understorey in some parts. Elsewhere there is a generally scattered shrub layer including *C. backhousiana*, *Pimelea linifolia* *Leucopogon parviflorus* and *Cyathodes juniperina*. The ground layer is dominated by *Poa poiformis* with occasional plants of, *S. lautus*, *B. bracteata* and *Pteridium esculentum*.

A short distance from the Tasmania-Victoria border, this island is the northernmost extremity of Tasmania. A summit at around 350m often attracts a halo of cloud. This view is of the south-east coast.

OTHER SEABIRD SPECIES

Black-faced Cormorant – 13 roosting on the north-west cliff

Australasian Gannet – 2 flew by

Silver Gull – 3

No mammals were recorded.

BIRDS

Native:

White-bellied Sea-Eagle – a pair was breeding east-north-east of the central gully

Wedge-tailed Eagle – 1

Peregrine Falcon – 2 were defending

Fan-tailed Cuckoo

Tasmanian Scrubwren

Yellow-faced Honeyeater

Olive Whistler – common

Forest Raven – 3 pairs

Silvereye – 1 pair nesting

Introduced:

Common Blackbird – nesting

REPTILES

Metallic Skink

White's Skink

Southern Water Skink

Pristine vegetation. This island is especially unique for having escaped distruction by people.

COMMENTS

Due to its climax *Eucalyptus globulus* community, which has evolved due to the absence of fire, its large Short-tailed Shearwater population and it being the only known locality in Tasmania where the Southern Water Skink is found, Rodondo Island is considered to be of outstanding State significance. A permit is required to gain access to the island.

West Moncoeur Island

(Rodondo Island Group, page 180)

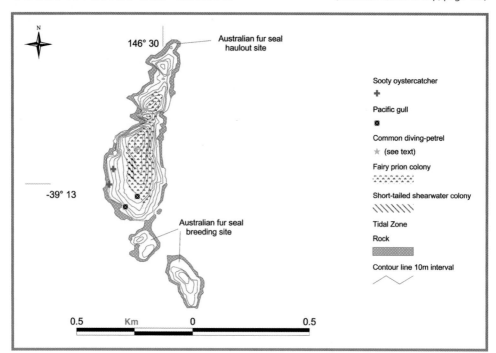

Location: 39°14'S, 146°30'E
Survey date: 2/12/81
Area: 9.18 hectares
Status: Nature Reserve

It is a slab-sided, granite island with very steep to sheer cliffs on all sides sloping down at each end. Seventy percent of the eastern sector is bare rock. The west side is characterised by patches of soil with *Poa* and *Carpobrotus rossii* interspersed with bare rock and boulders. A central ridge runs along the lower part of the island. The northern end is dominated by massive boulders and ravines, which make access difficult. There are two small islands off the southern end.

BREEDING SEABIRD SPECIES
Short-tailed Shearwater
Up to 100 pairs were found in burrows on the western side and along the ridge where soil patches were suitable and *Poa poiformis* and *C. rossii* dominant.

Fairy Prion
3000 – 5000 pairs, with eggs, have made full use of all rock crevices, boulders and any other suitable habitat all over the island.

Common Diving-Petrel
200 - 400 pairs in burrows were found in similar habitat to the shearwater colonies.

Pacific Gull
2 pairs were located on the south western ridge slopes.

Sooty Oystercatcher
2 pairs, with eggs, were recorded on the central western coast.

OTHER SEABIRD SPECIES
Black-faced Cormorant – 23 were roosting on the west side. They may also nest there.
Silver Gull – 8 sitting about.

MAMMALS
Australian Fur Seals – about 450 were seen at the south end mostly on the flat, sloping granite slabs. Each year approximately 250 pups are born here.

BIRDS
Native:
Cape Barren Goose
Peregrine Falcon – 1 defending aggressively
Introduced:
Common Blackbird – 2

REPTILES
Metallic Skink

East slopes.

VEGETATION
Poa and *C. rossii* are the co-dominant species on the island.

COMMENTS
This is an important Australian Fur Seal breeding colony, well-protected from storms with 252 new pups in the 1999 – 2000 census. The population is stable with room for expansion on the main island, however it is vulnerable to disturbance from trans-Tasman ferries travelling close to the island. Joint management with Victoria should be discussed. Two smaller islands off the southern end also have breeding populations of Australian Fur Seals, with the seals occupying low-roofed caves on the southernmost island. Accumulations of organic deposit – scats and dead seals – are found in these caves, a unique phenomenon to Tasmanian islands.

East Moncoeur Island

(Rodondo Island Group, page 180)

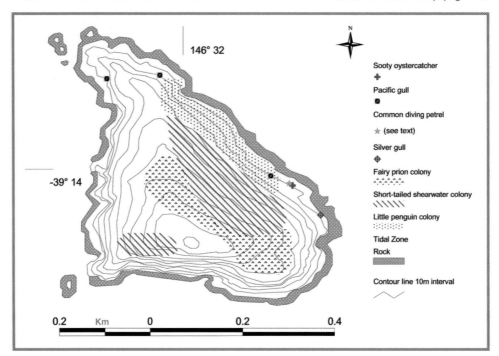

Location: 41°00'S, 148°21'E
Survey date: 13/11/84
Area: 14 hectares
Status: Non-allocated Crown Land ↑

It is an elongate, granite island with slopes all around. One quarter of the island is bare rock and boulders dominate the west side slopes.

BREEDING SEABIRD SPECIES
Little Penguin

200 pairs, most with eggs, were recorded during this survey. They come ashore on the north-east coast and are concentrated in this area, breeding almost to the summit in *Poa poiformis*, with one nest recorded at 70 metres. There are also many breeding under granite slabs and boulders, interspersed with shearwaters.

Short-tailed Shearwater

An estimated 34 320 pairs breed on the island. There are two major colonies, the largest in the centre of the island where *P. poiformis* is densest. The other is in the central southern area of the island, where the birds burrow in gravelly, dry soil or under slabs of granite. Their burrows are mostly short.

Fairy Prion

An estimated 5000 pairs occur all over the island beneath boulders. They are also on the lower east slopes in crevices and burrows.

Common Diving-Petrel

Up to 300 pairs breed only on the lower eastern slopes in *Disphyma crassifolium* and also patchy *Poa* amongst boulders.

Pacific Gull
3 pairs, at nests with eggs, were located on the northern and eastern coasts.

Silver Gull
50 pairs, most with empty new nests, some with eggs, were located in a colony on the south-east coast, with nests often tucked in under boulders.

Sooty Oystercatcher
1 pair, with 2 eggs, was located on the eastern shore near the Pacific Gull nest.

COMMENTS
Due to the island's high diversity of seabirds and pristine nature, a status upgrade to nature reserve is desirable.

No mammals or other seabirds were recorded.

BIRDS
Native:
Cape Barren Goose – 1 pair

Wedge-tailed Eagle

Hooded plover – dead in rock crevice on lower east side.

Forest Raven

Introduced:
Common Blackbird

REPTILES
Metallic Skink

Bougainvilles Skink

White's Skink

Bass Pyramid Group

(Region 3, page 139)

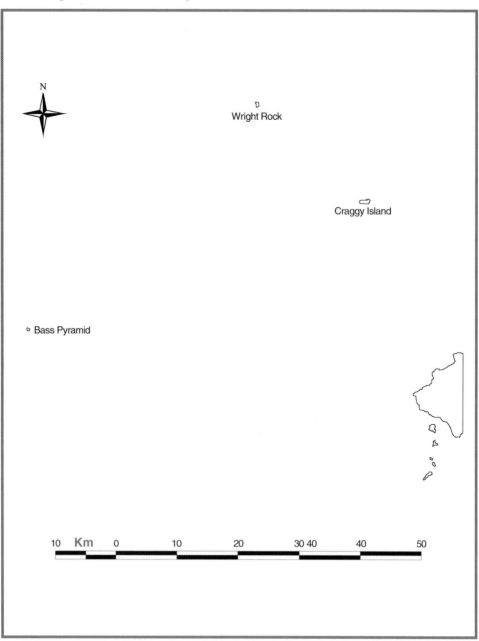

Craggy Island

(Bass Pyramid Group, page 188)

Location: 39°41′S, 147°41′E
Survey date: 17/12/86
Area: 38.88 hectares
Status: Non-allocated Crown Land ⇧

An elongate island orientated east-west with a lobed eastern side and gently curving northern side, it has a pronounced point on the mid southern side. It is composed of granite with a central jagged ridge rising steeply on all sides from a low flat coastal area. The shoreline is rocky and rugged with smooth sloping slabs on the southern side. Between the coastal flats and the central ridge there are gentle slopes which grade as they get higher. At least 60% of the island is bare rock. The rest of the island is covered by sparse soils.

BREEDING SEABIRD SPECIES
Little Penguin

1000 pairs were distributed all over the island with highest densities on the northern side on the lower slopes, though nests were found all the way to the top of the jagged ridge. Most nests were in rock crevices as there is a scarcity of deep soil on the island.

Short-tailed Shearwater

500 pairs were found nesting in two main colonies on the southern and western lower flats. There was also an area of scattered burrows on the southern shore. Some birds were even found nesting in rock crevices in cliff-like habitat. The burrows on the lower flats were vulnerable to heavy rains.

Fairy Prion

15 000 pairs were found nesting all over the island. They are concentrated on the lower slopes in habitat dominated by *Rhagodia* at the north-west end and elsewhere sparse *Senecio* and patches of very low matted *Poa poiformis* especially at the southern side. They are also found on the lower slopes, but in lower numbers than the common diving petrels. Prions were also found in rock crevices all over the island, up the higher slopes and in the cliff habitat.

Common Diving-Petrel

An estimated 10 000 pairs were found nesting all over the island. They tend to be concentrated on the lower slopes in habitat dominated by *Rhagodia* at the north-west end and elsewhere by sparse *Senecio* sp. and patches of very low, matted *P. poiformis*, especially at the southern side. They are on the lower slopes in higher numbers than the Fairy Prions. The burrows on the higher slopes are restricted in depth by the soils, whereas on the lower flats they are long and winding.

Spectacular rugged terrain.

Pacific Gull

14 nests were located and 37 adults were seen. There was a colony of 6 pairs on the east side of the south shore. Solitary nests were concentrated on the north shore and at the east end.

Sooty Oystercatcher

9 pairs were located on all shores apart from the south.

No mammals or other seabird species were recorded on the island.

BIRDS

Native:
Brown Quail
Cape Barren Goose
White-bellied Sea-Eagle
Peregrine Falcon – defending
Forest Raven
Little Grassbird
Silvereye
Thornbill

Introduced:
Common Blackbird

REPTILES

Metallic Skink

VEGETATION

The dominant vegetation is *Senecio* sp., *Rhagodia* sp., and *Atriplex cinerea*, in patches.

COMMENTS

The high species diversity justifies the island's status being upgraded.

Wright Rock

(Bass Pyramid Group, page 188)

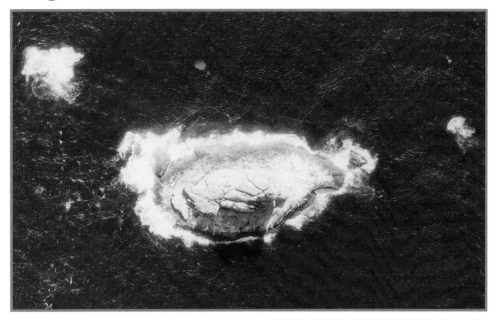

Location: 39°36'S, 147°33'E
Survey date: 21/2/87
Area: <1 hectare
Status: Nature Reserve

This is a steep-sided rock composed entirely of granite. On the lower half of the slopes a big smooth slab of granite juts out from the water then rises to steep to vertical cliffs. There is no soil for burrowing birds. The rock has a north to south orientated central ridge.

BREEDING SEABIRD SPECIES
Pacific Gull
1 pair, with a fully-fledged chick and a 2 – 3 year old bird, were located just south of the summit on a patch of *Disphyma crassifolium*.

Silver Gull
2 pairs, with fully-fledged chicks, were located in the crevice on the south-east corner on bare rock. 20 adults were sitting in the water throughout the visit.

OTHER SEABIRD SPECIES
Black-faced Cormorant – about 10 were sitting about on a regular roost site.

MAMMALS
Australian Fur Seal – they use all lower areas as haul-out sites, but are concentrated along the east side and at the south-west end.

No reptiles or other birds were recorded.

COMMENTS
Occasionally seal pups are born, the maximum recorded being four. It is a major haul-out site, with numbers ranging from 50 to 1000 and is appropriately protected.

Bass Pyramid

(Bass Pyramid Group, page 188)

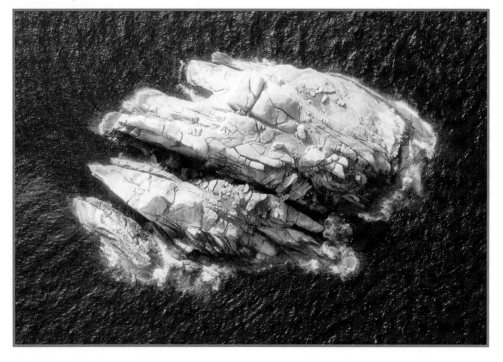

Location: 39°49'S, 147°15'E
Survey date: 8/10/84 and 7/11/84
Area: >1 hectare
Status: Nature Reserve

Bass Pyramid is a spectacular, two-sectioned, oval island with steep granite rock cliffs (90°) rising to a more gently sloping, sparsely vegetated area to a flat summit. The two sections are connected by a rock bridge. The main east to west gulch is cluttered with the remains of artillery shells and missiles. There are steeply sloping wave platforms in the north and south, on which seals haul out.

BREEDING SEABIRD SPECIES
Fairy Prion
Up to 50 pairs were on eggs principally in rock crevices with a few in burrows in the same area as the diving-petrels.

Common Diving-Petrel
Up to 100 pairs, with big chicks, were located on ledges above 40 metres on the south side (7/11/84).

Western side.

Pacific Gull
1 pair, with 3 eggs, was located on the north-east ledge 5 metres from the Sooty Oystercatcher nest.

Silver Gull
8 pairs, 5 with no eggs, 3 with 2 eggs, were located on the north-east lower ledge above the sheer cliff.

Sooty Oystercatcher
1 pair, with 2 eggs, was located on the north-east side 5 metres from the Pacific Gull nest in *Rhagodia* and *Carpobrotus rossii*.

Ballistics damage to granite.

Australian Fur Seal haul out on lower eastern platform.

OTHER SEABIRD SPECIES

Australasian Gannet – roosting on the ledge at the south-west corner, plus 2 flew by

Black-faced Cormorant – 2 were sitting at the southern end

Caspian Tern – 2 flew by

Crested Tern – 1 was sitting at the southern end

MAMMALS

Australian Fur Seals – 102 were counted at a haul-out on the east north-east end and 124 were counted at the southern end haul-out.

BIRDS
Native:
Azure Kingfisher

Flame Robin – dead

Richard's Pipit

Silvereye – dead

Introduced:
Domestic Pigeon – dead

No reptiles were recorded.

COMMENTS

Various craters and pits in its cliffs, caused by the impact of explosions from bombs and shells, are the result of this specatacular rocky island being used intermittently from the 1940s to 1988, as an Australian airforce and naval bombing target. There are also explosion-shattered boulders near the island's summit.

Few localities are used by Australasian Gannets as resting sites and any that are, may, in time, also become breeding sites. Given the increased abundance of this species in the south-east Australian region, Bass Pyramid should be monitored for signs of breeding.

Upper east slope of fractured granite boulders caused by artillery fire.

Furneaux Islands – Region 4

(Regions, page 44)

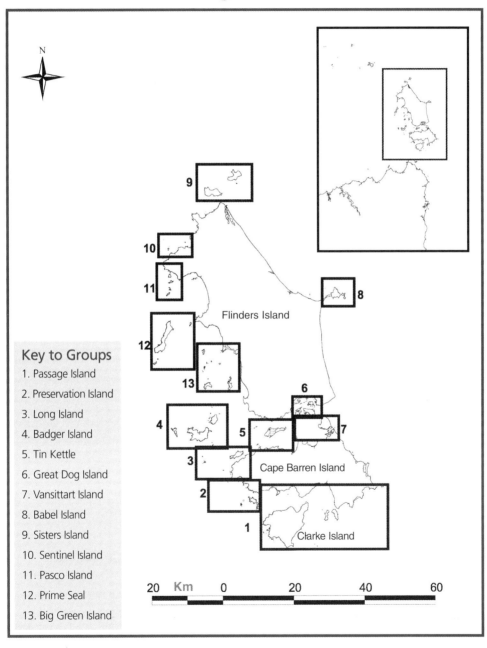

Key to Groups
1. Passage Island
2. Preservation Island
3. Long Island
4. Badger Island
5. Tin Kettle
6. Great Dog Island
7. Vansittart Island
8. Babel Island
9. Sisters Island
10. Sentinel Island
11. Pasco Island
12. Prime Seal
13. Big Green Island

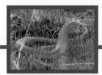

Passage Island Group

(Region 4, page 195)

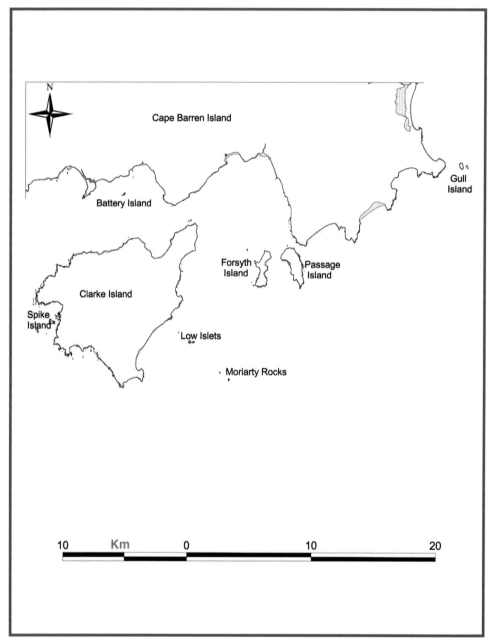

Gull Island

(Passage Island Group, page 196)

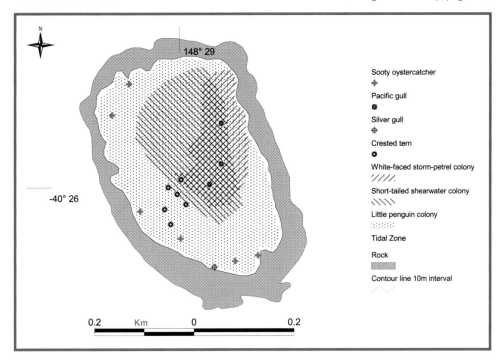

Location: 40°26'S, 148°30'E
Survey date: 24/11/86 and 2/8/95
Area: 8.49 hectares
Status: Conservation Area

Gull Island is an oval shaped granite island with sandy soil, barely deep enough for burrowing birds. The shoreline is composed of gently sloping granite slabs extending 20 metres from the sea (30 metres on the north-west to north-east end). Many inland areas also have extensive bare rock slabs or boulder rocks protruding from among the vegetation. There is less soil towards the south end, where rock slabs are dominant.

BREEDING SEABIRD SPECIES

Little Penguin

An estimated 11 500 pairs, either attending small chicks or with eggs, nest over the entire island, mainly under rocks covered with *Rhagodia* or *Carpobrotus rossii*. Few occur in real burrows. They land all around the shoreline.

Short-tailed Shearwater

Up to 50 pairs are estimated to be breeding on the island, generally in short burrows due to poor soil depth.

White-faced Storm-Petrel
An estimated 4320 pairs breed extensively through the centre of the island, extending on all sides in narrow bands towards the sea following wherever soil is sufficient between bare granite. They also breed in the north-west under *Poa* sp. and *C. rossii*.

Pacific Gull
3 pairs and 2 juveniles were located in the central area south of the White-faced Storm-Petrel colony.

Silver Gull
48 pairs, 15 with no eggs and the rest with 1 to 3 eggs, were breeding on the southern coast. 81 adults were counted.

Sooty Oystercatcher
6 pairs, one with 2 eggs, 2 with 1 egg and 3 with no eggs, were located scattered around the island with nests usually at the vegetation/soil interface.

Crested Tern
6 pairs, with one egg each, were located on the south-west coast just west of Silver Gull colony. 50 individuals were sitting on the rocky reef east of the island.

Little Penquin and Storm-Petrel nesting habitat.

VEGETATION
The dominant species are *Poa poiformis*, *Rhagodia candolleana* and *C. rossii*, which grow in the rock crevices.

COMMENTS
This is a significant seabird island, which has had no human interference.

OTHER SEABIRD SPECIES
Black-browed Albatross – flew by (2/8/95)

Shy Albatross – flew by (2/8/95)

Australasian Gannet – foraging offshore (2/8/95)

Black-faced Cormorant – 118 were on the reef to the east and many were foraging.

No mammals or reptiles were recorded on the island.

BIRDS
Native:
Brown Quail

Cape Barren Goose

INTRODUCED:
Common Starling

Skylark

Passage Island
(Passage Island Group, page 196)

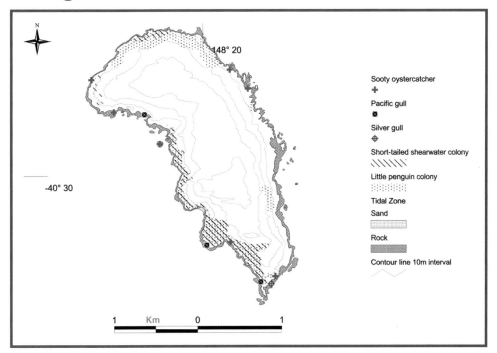

Location: 40°30'S, 148°39'E
Survey date: 25/11/86 and 21-23/8/00
Area: 253.36 hectares
Status: Non-allocated Crown Land, Lease

It is an elongate island, which has been extensively grazed especially in the north. Seabird colonies exist in the sandy soils and low vegetation almost all the way around the island. A dolerite band, broken by granite slabs, stretches along the east side of the island. Infrastructure includes a hut, shed and an air strip.

BREEDING SEABIRD SPECIES
Little Penguin
An estimated 1500 pairs were scattered all the way around the island in sandy soil, often

Central east with Cape Barren Island in distance.

Central west, to north-west Forsyth Island.

extending into thick scrub. Most are interspersed with Short-tailed Shearwaters.

Short-tailed Shearwater
An estimated 30 149 pairs inhabit colonies, which extend from the north to the south along the western coast, with *Senecio* sp. and *Disphyma crassifolium* dominant. Other colonies are found in bare soil patches surrounded by *Poa* sp.

Pacific Gull
3 pairs, one nest with 3 eggs, were located on the west coast and on the southern tip.

Silver Gull
3 pairs were located on the southern tip.

Sooty Oystercatcher
10 pairs, one pair with 2 eggs and one with one chick, were located aound the coast.

VEGETATION
Dense scrub on the granite soils includes *Callitris rhomboidea*, *Allocasuarina verticillata*, *Leptospermum laevigatum* and *Melaleuca ericifolia*. In the south on the dunes, the scrub is dominated by *Acacia sophorae*. Other vegetation includes tussock grasslands on the sandy soils.

OTHER SEABIRD SPECIES

Black-faced Cormorant – 200+ were roosting on large boulder slab on the southern end of island. There was no sign of breeding.

Australian Pelican – 1 flew by

Crested Tern – 1 flew by

MAMMALS

Introduced:

Rat

Rabbit – numerous

Cattle – about 30

House Mouse

BIRDS

Native:

Brown Quail

Cape Barren Goose – breeding with goslings at various stages

White-bellied Sea-Eagle

Swamp Harrier

Masked Lapwing – breeding in open paddocks at northern end

Yellow-tailed Black-Cockatoo – 3

Brown Thornbill

Crescent Honeyeater

New Holland Honeyeater

White-fronted Chat

Flame Robin

Olive Whistler

Grey Fantail

Forest Raven – nesting

Welcome Swallow

Silvereye

Introduced:

Common Blackbird

Common Starling – nesting

Skylark

European Greenfinch

REPTILES

White's Skink

Metallic Skink

COMMENTS

The grazing of cattle and the proliferation of rabbits has had a marked effect on the vegetation and seabird habitat. Rabbit control is now taking place as a joint venture between the lessee and the Department of Primary Industries, Water and Environment. A comparison between this and nearby Forsyth Island gives an indication of the effects of grazing on seabird habitation, although the geomorphology varies. The southern end is especially sensitive to grazing and disturbance because of the sandy soils, sand blow and shearwater colonies. Cattle should be excluded from this area.

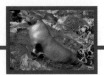

Forsyth Island

(Passage Island Group, page 196)

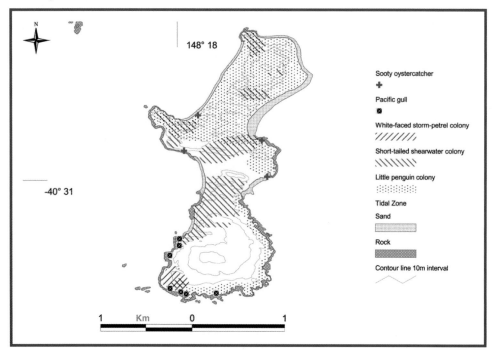

Location: 40°30'S, 148°19'E
Survey date: 23/11/86
Area: 166.85 hectares
Status: Non-allocated Crown Land ⇑

It is an elongate island orientated north to south, scrub-covered with numerous sandy beaches abutting granite slabs and rocks. There is a central ridge and gently undulating vegetated dunes running east to west across the island. Generally two larger dunes lie parallel to each other with a smaller dune in the valley between them. These dunes do not all reach both the east and west coast, and most tend to break up on the east side. The soil is very sandy. There is a sandblow in the central region.

BREEDING SEABIRD SPECIES
Little Penguin

An estimated 147 318 pairs nest densely on slopes at the back of or near the granite slabs that form the coastline away from sandy beaches. The birds tend to avoid the areas behind the sandy beaches. The birds nest within the Short-tailed Shearwater colonies and in a 30 m wide strip along the granite shoreline. They are also scattered in lower densities along the side of and the tops of the east-west running dunes across the northern half of the island. There is a dense coastal strip where the birds extend back 60 m north of the sandblow.

North end, west side.

Short-tailed Shearwater

An estimated 157 451 pairs were located at 6 colony sites on the west central point, the north-west point, north central region, near the blowout and at the south-west end of the island. The colony at the south-west end has the highest densities located in *Senecio* sp. distributed amongst granite slabs. The western colonies are dominated by *Poa poiformis* where burrows are densest and co-dominant with *Senecio* sp. elsewhere. The dense colony near the sand blowout is dominated by *Poa*, *Tetragonia* sp. and *Senecio* sp.

White-faced Storm-Petrel

Up to 100 pairs were located nesting in small scattered groups at the edges of the Short-tailed Shearwater colony at the south-west end of the island. The burrows are located in areas dominated by stipa and *Carpobrotus rossii*.

Pacific Gull

7 pairs, most with eggs, were concentrated to the south-west end of the island. The nests were scattered, except for 2 pairs nesting adjacent to each other in vegetated (mainly stipa) cracks between granite slabs.

Sooty Oystercatcher

4 pairs, at nests with eggs, were located scattered from the central region down either shoreline.

South-east corner.

VEGETATION

On the east side Aizoaceae succulents in sandy soil dominate, whilst the west side and south central region have denser scrub cover. *P. poiformis* is co-dominant with *Senecio* sp. and *Tetragonia* sp., stipa and *C. rossii*. The shrubby vegetation is dominated by *Olearia axillaris* and *Ozothamnus turbinatus* with patches of *Leptospermum laevigatum* and *Myoporum insulare* up to 5 metres in height.

COMMENTS

The Short-tailed Shearwater colonies attract birders. The rodents are introduced. There is a high diversity of both passerines and non-passerines, possibly a consequence of the dense scrub. Fires would affect this assemblage and probably the stability of the dunes which many of the birds, including the seabirds, rely on. The dunes should be protected, not only because of their habitat value, but also because of their geoconservation value. Further investigations should be undertaken to ascertain the identification of the small mouse and whether the rodents are affecting the distribution and abundance of the white-faced storm petrels. This is a seabird island of significance, whose status should be upgraded to nature reserve to reflect this.

OTHER SEABIRD SPECIES

Black faced cormorant – 1 flew by

Australian Pelican – 9 seen gliding around. Suitable nesting localities occur but no nests were found.

Crested Tern – 1 flew by

MAMMALS

A small mouse, the size of a House Mouse, was seen near small burrows at the north-east end.

Rat – trapped

Swamp Rat – burrows were located at the south end

BIRDS

Native:

Brown Quail

Cape Barren Goose – 3 adults and 4 downy chicks

White-bellied Sea-Eagle – 1 adult

Swamp Harrier

Brown Falcon

Peregrine Falcon

Hooded Plover

Blue-winged Parrot

Crescent Honeyeater

Brown Thornbill

White-fronted Chat

Olive Whistler

Grey Fantail –1 seen

Forest Raven

Tree Martin

Little Grassbird

Silvereye

Introduced:

Common Starling – 2 seen

European Goldfinch

REPTILES

Blue-tongue Lizard – common

Tiger Snake – common

Low Islets (east island)

(Passage Island Group, page 196)

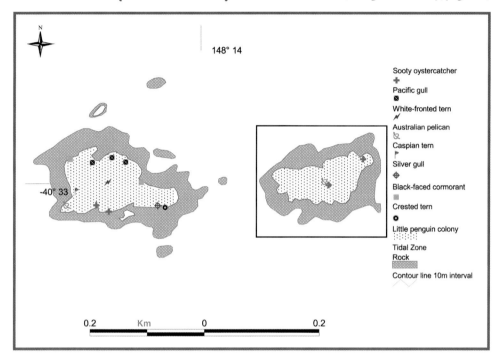

Location: 40°34'S, 148°14'E
Survey date: 24/11/86
Area: 1.28 hectares
Status: Nature Reserve

The east islet is circular in shape with a low, flat profile and considerable areas of exposed granite.

BREEDING SEABIRD SPECIES

Little Penguin
Up to 20 pairs were nesting under boulders covered by *Tetragonia* around the island.

Sooty Oystercatcher
2 pairs, at nests with eggs, were located near the pelican colony.

White-fronted Tern nest.

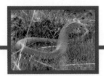

Sparse stipa amongst granite boulders and slate.

Australian Pelican
29 pairs, with chicks in all nests, were located in a colony in the centre of the island.

No mammals or other birds were recorded on the island.

REPTILES
Metallic Skink

VEGETATION
The vegetation is dominated by stipa, *Poa poiformis*, *Tetragonia* sp. and other Aizoaceae succulents.

COMMENTS
See comments for Low Islets (west island).

Low Islets (west island)

(Passage Island Group, page 196)

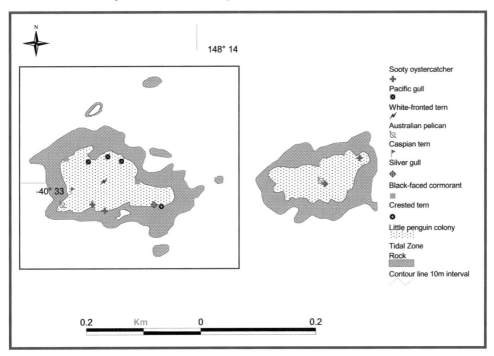

Location: 40°34'S, 148°14'E
Survey date: 24/11/86
Area: 8 hectares
Status: Nature Reserve

It is a triangular-shaped island, rising gently to a low central area.

BREEDING SEABIRD SPECIES

Little Penguin
Up to 20 pairs were nesting in burrows under *Tetragonia implexicoma* and under boulders.

Pacific Gull
9 pairs, several with eggs and runners, were nesting on the north end of the island in association with stipa.

Silver Gull
55 pairs, most with empty nests and some with chicks, were located nesting at the point on the eastern end of the island on a patch of stipa adjacent to the beach where the crested terns nest.

Sooty Oystercatcher
2 pairs, with eggs, were located on the rocky coastline on the south-west end of the island.

Black-faced Cormorant
76 old nests were located on bare granite 3 m above water level.

Australian Pelican

At the west end of the island there was an old colony with old nests and 5 dead chicks. This is used as an alternative breeding site to the eastern island.

Caspian Tern

1 pair, with eggs, was located nesting near the centre of the island at the west end on bare granite slab.

Crested Tern

49 pairs, most with eggs, were nesting at the point on the east end of the island. The colony is on the sand and pebble beach below the Silver Gull colony and is vulnerable to high seas.

White-fronted tern

1 pair, with eggs, was located in the centre of the island.

VEGETATION

The island is dominated by patches of *Poa poiformis*, *T. implexicoma* and *Carpobrotus rossii* and slabs of bare granite.

No mammals, reptiles or other seabirds were recorded.

BIRDS

Native:
Ruddy Turnstone – 3
Forest Raven – 1

COMMENTS

The Australian Pelican colonies are vulnerable to disturbance by human activity, both onshore and close offshore. The Crested Tern colony is vulnerable to heavy seas.

This island group is an important seabird breeding site, with a high species diversity, including crested and White-Fronted Terns and is one of only three sites where Australian Pelicans breed in Tasmania. Consideration should be given to restricting access to the islands and surrounding waters during the breeding seasons of the vulnerable species.

Moriarty Rocks

(Passage Island Group, page 196)

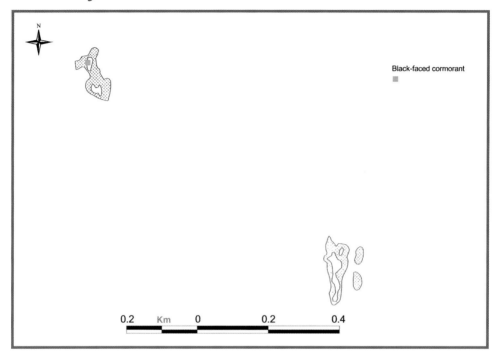

Location: 40°35'S, 148°016'E
Survey date: 1/8/95 and 3/8/95
Area: 2.46 hectares (total)
Status: Nature Reserve

Moriarty Rocks comprises two major granite rocks and several smaller ones in a reef formation.

BREEDING SEABIRD SPECIES
Black-faced Cormorant

50 occupied nests, 10 with chicks, others with eggs were located on the westernmost rock. Some of the chicks were well-advanced.

VEGETATION

There is very little vegetation on Moriarty Rocks due to their being constantly wave-washed.

Nearly at sea level, Australian Fur Seal pups born here are vulnerable to high seas

OTHER SEABIRD SPECIES

Sooty Oystercatcher

Kelp gull

Pacific Gull

Silver Gull

Crested Tern

White-fronted Tern

MAMMALS

Australian Fur Seal – this is an important breeding colony with between 97 and 1035 pups being born here annually during the past ten years. The numbers are so variable due to the rocks being wave-washed and subject to storms.

COMMENTS

The vulnerability of the rocks to being wave-washed causes a high variability in the population of Fur Seals. Monitoring should continue and disturbance must be kept to a minimum.

There are many dramatic stories of human suffering in quest of seal skins. 'Parish went to the reef (here referring to Moriarty Rocks) and found three men alive: Robert Drew, Edward Tomlins and Sydney Masell. They had come from the Hunters. Their boat had been wrecked on the rocks of the reef. The five men reached the rock but had no sustenance and had to kill seal and drink the blood. John Williams and John Brown, finding no help came from Penguin Island, built a canoe of seal skins by sewing them together. The rocks were about five miles from Clarke's Island, and there is an exceedingly bad tide rip. They put to sea, Williams saying it was the last drop of seals's blood that he would drink on that island. They were not seen after they enterred the rip.'   from Fowler 1980.

Spike Island

(Passage Island Group, page 196)

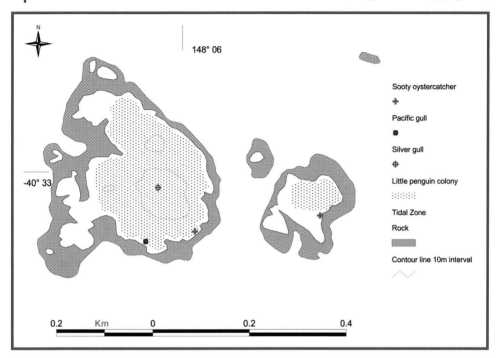

Location: 40°33'S, 148°07'E
Survey date: 26/11/86
Area: 6 hectares
Status: Non-allocated Crown Land

There are two islands, outer and inner. Outer Spike Island, the larger island, is broken by protruding granite boulders, massive bare boulders and granite slabs. Fifty per cent of the island is bare rock. Inner Spike Island is flatter, particularly in the north, with bare rock boulders and slabs to the south side.

BREEDING SEABIRD SPECIES
Little Penguin
250 pairs breed on the outer island and 50 pairs on the inner island. Burrows are all around the outer island in thick stipa and on the inner island the penguins nest under rocks and *Tetragonia* as there is no burrowing habitat.

Pacific Gull
1 pair was breeding on the outer island at the south end.

Silver Gull
3 pairs were breeding on the outer island alongside stipa just below the summit.

Sooty Oystercatcher
2 pairs were located, one on the east-south-east side of the outer island and one on the inner island.

OTHER SEABIRD SPECIES
Caspian Tern – 2 pairs flew by

BIRDS
Native:
Cape Barren Goose – 2 pairs
White-fronted Chat
Introduced:
Common Starling

No mammals or reptiles were recorded on the island.

VEGETATION
Vegetation is dominated by stipa and *Tetragonia* which forms a dense mat in places, covering rocks. *Poa poiformis* dominates the north of Inner Spike Island.

COMMENTS
This is an unaltered island of limited significance as a seabird breeding area.

Silver Gulls

Battery Island

(Passage Island Group, page 196)

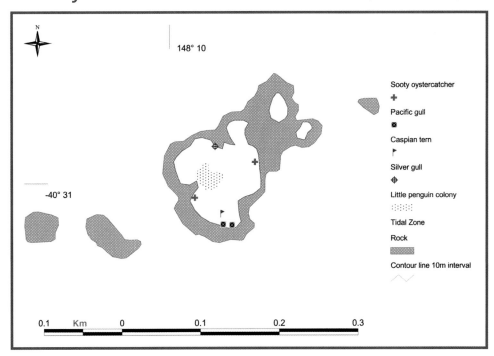

Location: 40°28'S, 148°11'E
Survey date: 25/11/86 and 2/8/95
Area: 0.68 hectares
Status: Non-allocated Crown Land

It is a low, gently sloping islet of rounded granite with boulders protruding at the northern end. Two rock islets off the north-east side are connected by a low sand bar, just above high water mark. 30% of the island is bare rock.

BREEDING SEABIRD SPECIES
Little Penguin
Up to 50 pairs, many with eggs or small chicks, were located in the main colony in the centre of the island. The penguins were primarily in burrows amongst stipa that is overgrown with *Tetragonia*. A few were also under rocks.

Pacific Gull
12 pairs, 6 with empty nests, 2 with 1 egg, 1 with 2 eggs, 2 with tiny chicks and one with a small chick, were located amongst stipa patches in the central southern area.

Silver Gull
34 pairs, 5 nests with no eggs, 2 with 1, 15 with 2, 10 with 3, one with one egg and one chick and one with 2 dead chicks and one dead chick runner, were located. The colony was situated on the northern coast amongst boulders with stipa and some were on lower nests on the bare sand of the beach.

Sooty Oystercatcher
2 pairs, with one small chick each, were located on the central east and south-west coasts.

View over Battery Island toward Clarke Island to the south.

Caspian Tern
1 pair, with a quarter-grown chick, was located alongside the stipa with *Disphyma crassifolium* in south-central area.

VEGETATION
Stipa and *Tetragonia* dominate the coastline.

COMMENT
This is an island of no specific significance for seabirds.

OTHER SEABIRD SPECIES
Common Diving-Petrel – dead

Black-faced Cormorant – 100+ were sitting on the shore

Australian Pelican – 1 on the sand bar

BIRDS
Cape Barren Goose – 1 pair ashore

No mammals or reptiles were recorded.

Preservation Island Group

(Region 4, page 195)

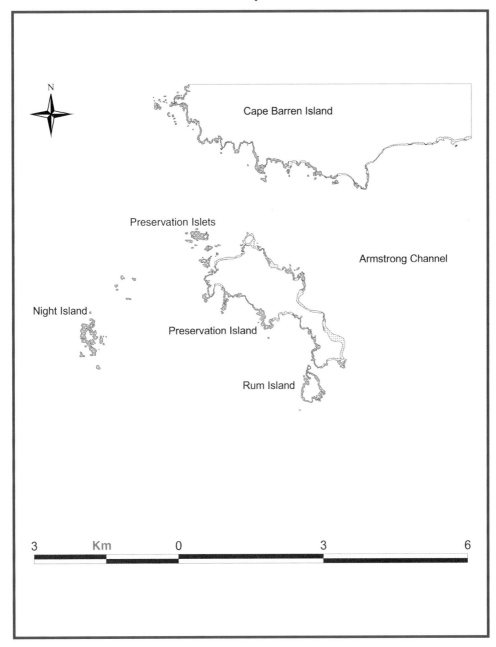

Night Island

(Preservation Island Group, page 215)

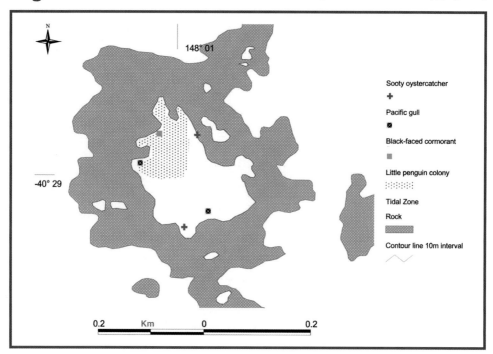

Location: 40°29'S, 148°01'E
Survey date: 26/11/86 and 3/8/95
Area: 2.59 hectares
Status: Conservation Area

It is a low islet, the east-south-east half of which is dominated by dunes covered with creeping grass. Much of the island is bare rock with big boulder points. The north-east beach is broken up by a number of protruding granite boulders.

BREEDING SEABIRD SPECIES
Little Penguin
An estimated 100 pairs were located. All checked were on eggs. The highest densities were at the north-west end of the island decreasing eastward. Burrows are either under *Tetragonia*-covered rocks or in very sandy soil beneath *Tetragonia* and stipa

Pacific Gull
2 pairs, one with runners and one dead juvenile, were located on the north-west coast and one pair was located on the south-east summit slope.

Sooty Oystercatcher
2 pairs were located, one on the north-east coast and one near the Pacific Gull nest on the south-east slopes of the summit.

Black-faced Cormorant
A colony of 138 pairs with nestlings, 26 of which were dead, was situated in a granite gully 10 metres wide running north-west to south-east. Most nests, built from seaweed, were on bare granite but extended back into the *Disphyma crassifolium*. Many juveniles were near the colony.

OTHER SEABIRD SPECIES

Short-tailed Shearwater – 2 dead alongside a dead cormorant

Australian Pelican – one flew off the south-east beach

Silver Gull – 14 ashore at north-east beach. There was no sign of breeding.

Crested Tern

No mammals or reptiles were recorded on the island.

BIRDS
Native:
Cape Barren Goose – 1 pair with 2 goslings

Ruddy Turnstone

Hooded Plover – 2

White-fronted Chat

VEGETATION
Stipa is dominant at the west-north-west end with *Tetragonia* and *Rhagodia* co-dominant.

COMMENTS
It is an important seabird island due to the presence of the Black-faced Cormorant colony. Its status is appropriate to protect its values.

Protruding granite boulders.

Preservation Island

(Preservation Island Group, page 215)

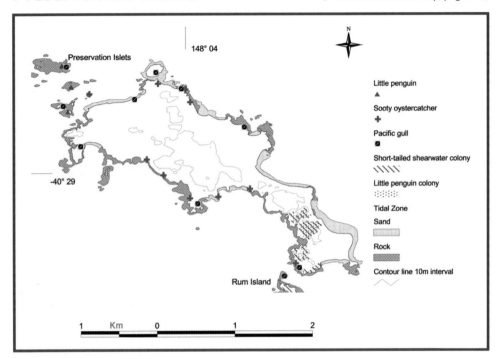

Location: 40°29'S, 148°04'E
Survey date: 27/11/86 and 24-26/3/99
Area: 207.45 hectares
Status: Multiple Land Tenure: Lease, Private property, Non-allocated Crown Land, Historic Site

The island is low and undulating, rising to only about 26 metres above sea level. It is, therefore, very exposed to both strong westerly and easterly winds throughout the year. It is formed of medium to coarse-grained Devonian granite with some remnant calcarenite exposed on the sheltered parts of the shoreline. A sandblow covers the south-eastern end of the island. Wetlands occur in the eastern section of the island. The coastline is dominated by sandy beaches, particularly on the north-eastern sides. Preservation Island is especially significant, historically, being the site of Australia's first wreck of a commercial ship and Tasmania's first commercial sealing venture. Over a hundred thousand seal skins were obtained in the area between 1800 and 1806. (Ryan 1996) There are many historical remains on the island including huts, walls, grave sites and shipwrecks off the coast. There is also a new dwelling and airstrip.

BREEDING SEABIRD SPECIES
Little Penguin

An estimated 2100 burrows were found in low densities interspersed with shearwater colonies in the south-eastern end of the island.

Saltbush at south-east end.

Short-tailed Shearwater
41 821 pairs are estimated to be breeding on the island. Burrows are confined to the south-eastern end of the island in average density. East of the sandblow, the colony follows the flat ground around bare boulders. The main colonies are in *Tetragonia*, *Senecio* and *Carpobrotus rossii*. In two places the colony extends to within 40 metres of north shore.

Pacific Gull
9 pairs, one with 3 eggs, one with 2 eggs and 7 empty nests, were scattered around the coast.

Sooty Oystercatcher
10 pairs, 3 with 2 eggs each, the remainder with no eggs, were located predominantly on the southern side.

VEGETATION
The vegetation has been highly modified by fire and the grazing of goats and cattle for almost two hundred years. The spread of introduced species has been occurring ever since the wreck of the merchant ship *Sydney Cove* in 1797 provided a catalyst for ensuing visits and inhabitation. The northern section, particularly, is infested with African boxthorn *(Lycium ferocissimum)* which are now being eradicated. There are 12 vegetation communities on the island, *Poa poiformis* tussock grassland being the largest, covering 71 hectares or approximately 39% of the island. Stipa covers 10% of the island, predominantly at the western extremity of the island, where there is full exposure to onshore westerly winds.

OTHER SEABIRD SPECIES
Black-faced Cormorant – 12 roosting on the eastern shoreline

Australian Pelican – 1 ashore on the north-east end beach

Pied Oystercatcher – 1 pair on the beach outside the hut, visiting, not breeding

Silver Gull – a few about

Caspian Tern – 1 flew by

Fairy Tern – 2 off west end feeding and off north-west tip fishing in shallow channels

White-fronted Tern – 8 feeding off south-east

MAMMALS
Cattle

BIRDS
Native:
Brown Quail – many and one with 10 eggs

Cape Barren Goose

Grey Teal

Chestnut Teal – 2

Brown Falcon – 2

Hooded Plover – off north-west tip

Masked Lapwing

White-fronted Chat – many and one nest with 3 eggs

Olive Whistler

Grey Fantail

Forest Raven – 2

Welcome Swallow

Little Grassbird – in the saltbush at the southern end.

Silvereye

Introduced:
Skylark – many and 2 eggs in one nest

European Greenfinch

Common Blackbird

Common Starling

REPTILES
Blue-tongue Lizard

Tiger Snake

White-lipped Whip Snake

COMMENTS
Early land use led to the depletion of soils through burning, stock trampling and subsequent wind erosion which resulted in a decline in the population of burrowing seabirds. The Short-tailed Shearwater colony, which once covered almost the whole island, is now confined to small areas in the south-east in *Atriplex cinerea* shrubland.

The reduction in the number of grazing cattle and the provision of adequate fencing, especially in the eastern section of the sand dune area, would help to reduce further

Coastal vegetation.

adverse impact on the soil structure and vegetation and help to restore the Short-tailed Shearwater burrowing habitat. The eradication or control of boxthorn, sea spurge and thistles is necessary to help reduce the source of propagules for further invasion and restore the island's native vegetation communities. The control of mutton-birding and quail hunting and the implementation of a fire management strategy for the island would also help to restore its ecological integrity. The south-eastern peninsular has been fenced off from cattle to protect the remnant vegetation, dunes, wetlands and shearwater colonies. The native vegetation is rapidly recolonising and colonies are expanding. (Karen Ziegler pers. comm.) Historical remains should be properly documented and protected. This is a site of national historical significance and should be managed as such.

The Hobart Town Gazette of 8 April, 1826, reported:

'Preservation Island is between two and three miles in circumference, bounded at one end by Cape Barren and on the other by Clark's Island. It affords a commodious harbour and is frequently visited by vessels passing between the two colonies and their respective settlements. In December 1825 there were no less than five at anchor there at one time. The soil is light and sandy, of which not more than a hundred acres are capable of tillage. Munro raises considerable quantities of vegetables, which he barters with the sealers. He has also several black women, who labour for him and procure the mutton birds, which also afford him an article of traffic. We learn that there are about twenty children, the offspring of such connection among the various islands of the Straits. Many advantages would arise if this island, so conveniently situated, were put under the charge of some responsible person, vested with authority to check the illicit conduct of the characters who have of late resorted thither. Among others the fishery of the seal would be protected from that indiscriminate destruction, which, if persisted in, must soon annihilate this valuable article of commerce.'

Preservation Islets

(Preservation Island Group, page 215)

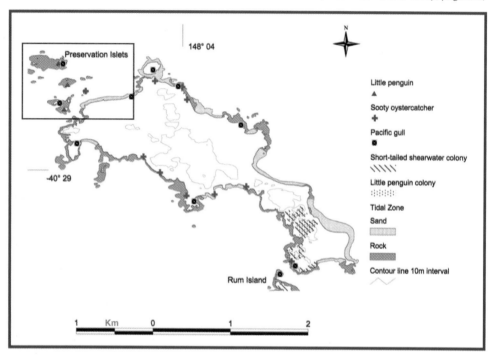

Location: 40°28'S, 148°03'E
Survey date: 28/11/86.
Area: 0.93 hectares in total
Status: Non-allocated Crown Land, Lease

These islets are a compilation of three round granite islets joined at low tide and a fourth low, round, mainly bare rock reef.

BREEDING SEABIRD SPECIES
Little Penguin

An estimated 36 pairs breed on the islets, 8 on islet 2, 7 on islet 3 and 21 on islet 4. The birds had burrows under rocks covered by *Tetragonia* and under the mats of *Tetragonia* hanging off the sides of the rocks.

Pacific Gull

2 pairs with runners were nesting on islet 2 and on islet 4. At both sites they were nesting amongst stipa

Sooty Oystercatcher

3 pairs were located. One pair was defending a nest site on islet 1 and 2 pairs with eggs were nesting on a sand gravel beach.

VEGETATION

The islets are dominated by following vegetation; Islet 1 – *Disphyma crassifolium*, Islet 2 – stipa, Islet 3 – *D. crassifolium* and stipa, Islet 4 – *Atriplex cinerea*, *Tetragonia* sp. and *Rhagodia candolleana*.

OTHER SEABIRD SPECIES
White-fronted Tern – flew over

No mammals or reptiles were recorded.

BIRDS
Native:
Cape Barren Goose – 2 pairs with goslings

White-fronted Chat

COMMENTS
The islets have no importance in terms of species diversity or numbers, but they provide an example of the types of seabirds that exploit the most marginal habitat – with no soil, limited vegetation and vulnerability to high seas.

Looking from Preservation Island to islets to the north.

Rum Island

(Preservation Island Group, page 215)

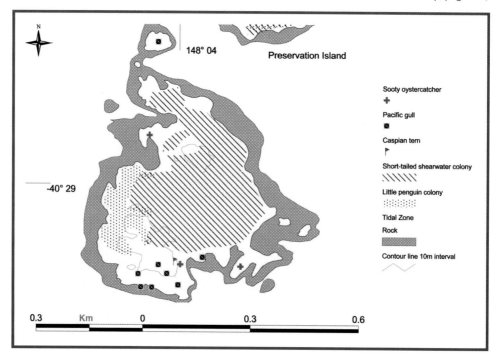

Location: 40°30'S, 148°04'E
Survey date: 27/11/86
Area: 13.46 hectares
Status: Historic Site (part of the Sydney Cove historic site)

It is a circular shaped island with an irregular granite coastline. There is a large beach on the north-east side. There is sandy soil in places, bare rock and patches of vegetation.

BREEDING SEABIRD SPECIES
Little Penguin
100 to 150 pairs were located, scattered in low densities all over the island with the highest numbers found at the western end amongst the Short-tailed Shearwaters.

Short-tailed Shearwater
An estimated 66 537 pairs occur across the island, restricted only by the lack of suitable soil for burrows. The burrow density is lowest in saltbush at the western side, where the soil is very sandy. In this area, burrows are fairly short and shallow and collapse easily. Densities range from $0.8/m^2$ in 100% thick saltbush to $0.15/m^2$ in 80% *Tetragonia*, 20% *Atriplex* co-dominant with *Rhagodia*.

Pacific Gull
8 pairs, at nests with eggs and small chicks, were concentrated at the south-west end of the island where there was a colony of 6 pairs located on an inland patch of *Disphyma crassifolium* with small clumps of *Poa poiformis* scattered around bare

rock patches. There was another nest on the shoreline east of this colony and one on the small islet off the north-east point of the island.

Sooty Oystercatcher
3 pairs, at nests with eggs, were located on the southern and northern shorelines.

Caspian Tern
One nest was located at the south-west end adjacent to the Pacific Gull colony, in a depression in *D. crassifolium* surrounded by bare granite slabs.

VEGETATION
The vegetation is dominated by *Poa poiformis*, *Tetragonia* sp., *Rhagodia candolleana* and *Atriplex cinerea*.

COMMENTS
There may be a low incidence of harvesting of Short-tailed Shearwaters. The low species diversity and the commonality of the species present makes the status of the island adequate.

OTHER SEABIRD SPECIES
Australian Pelican – 7 on the beach

No mammals were recorded on the island.

BIRDS
Native:
Cape Barren Goose –5 pairs with goslings – introduced to region

Swamp Harrier

White-fronted Chat

Little Grassbird

Silvereye

Introduced:
Common Starling

REPTILES
Metallic Skink

Central area.

Long Island Group

(Region 4, page 195)

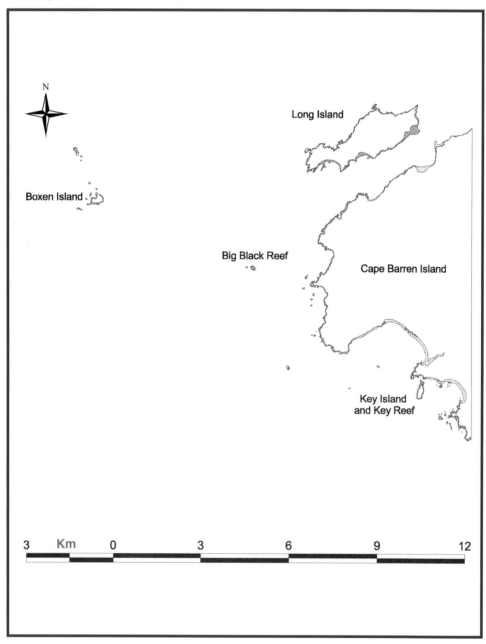

Long Island

(Long Island Group, page 226)

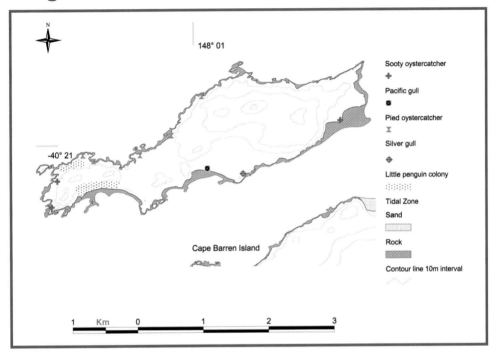

Location: 40°21'S, 148°00'E
Survey date: 1/12/86
Area: 312.66 hectares
Status: Multiple Land Tenure:
Crown Land (16 hectares),
Lease (296 hectares)

Long Island has an irregular coastline and lagoons. It is largely composed of granite and has two dolerite bands running east to west on the north coast. The island is actively farmed and grazed by sheep. There are two sheds and a house. The island is composed of largely shallow soil with good patches of *Melaleuca* scrub. There are sandy beaches and dunes which in some cases are vegetated. The island still has a diverse flora despite the farming activity. There is a freshwater soak on the island.

BREEDING SEABIRD SPECIES
Little Penguin

An estimated 56 pairs inhabit the island with 6 burrows located at the north-west end under *Melaleuca* sp. scrub. The major colony is located at the south-west end where the birds nest under rocks and in burrows in an old vegetated dune. There are also burrows behind a 15 m high dune.

Pacific Gull

One pair, with 2 fledglings, was located on the central southern shore.

Silver Gull

9 nests on bare granite ledges with stipa were located on the south-east end of the island. 19 adults were defending.

Sooty Oystercatcher

3 pairs, one with 2 eggs, were counted around the coast.

Pied Oystercatcher

5 pairs were breeding on the northern coastline, with the nests positioned in piles of seagrass on the beach. There were 6 individuals on the south coast, but they were not defensive.

VEGETATION

There are several large patches of relatively intact *Melaleuca* scrub at the western end of the island. Coastal vegetation is dominated by stipa on granite. The southernmost community of *Melaleuca armillaris* in Australia exists here. Gorse is invading the island.

COMMENTS

The island has been heavily grazed and burnt but stock levels are now low. In its present state, it is an unimportant seabird island. Feral goats have been successfully eradicated from the island and there are currently no cattle on it. If the sheep were removed, burrowing seabirds may return. The island's size, diverse vegetation and freshwater result in a high bird diversity and there is a possibility of small mammals having survived on the island. This should be investigated.

OTHER SEABIRD SPECIES

Australian Pelican
Caspian Tern
Fairy Tern

MAMMALS

Native:
Tasmanian Pademelon

Introduced:
Sheep

BIRDS

Native:
Brown Quail
Black Swan – breeding
Cape Barren Goose – breeding
Australian Shelduck
Pacific Black Duck
Grey Teal – breeding on lagoons
Hooded Plover – 4 pairs breeding
Masked Lapwing
New Holland Honeyeater
Flame Robin
Grey Fantail
Olive Whistler
Welcome Swallow
Silvereye

Introduced:
Common Blackbird
Common Starling
European greenfinch
Skylark

REPTILES

Metallic Skink
Ocellated Skink
Three-lined Skink
Copperhead Snake

Big Black Reef   (Long Island Group, page 226)

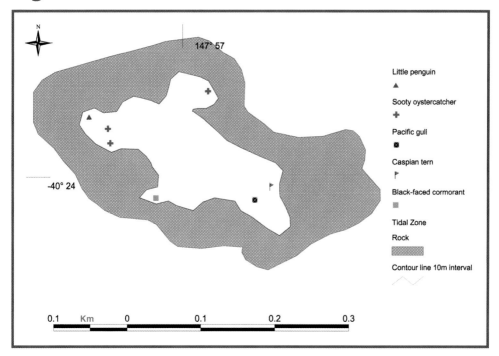

Location: 40°24'S, 147°57'E
Survey date: 28/11/86
Area: 0.54 hectares
Status: Non-allocated Crown Land ⇑

It is a small, triangular, flat, dolerite islet with no burrowing habitat. Beneath the vegetation, the islet is composed of round dolerite pebbles and rocks.

BREEDING SEABIRD SPECIES
Little Penguin
1 pair, with 2 fresh eggs, was located beneath the saltbush in the west of the island.

Pacific Gull
1 pair, with one chick, was located in *Tetragonia* at the east end of the island near the Caspian Tern's nest.

Sooty Oystercatcher
2 pairs were located to the west of the saltbush on the western coast of the island and one pair was on the north-eastern coast.

Black-faced Cormorant
40 pairs were nesting and 123 individuals roosting. Nests constructed of *Tetragonia* and seaweed were located a few metres above the sea on bare rock ledges on the south-western coast.

Caspian Tern
1 pair, with 2 eggs, was located in *Tetragonia* at the east end of the island near the Pacific Gull's nest. The Caspian Tern behaved aggressively towards the Pacific Gull.

OTHER SEABIRD SPECIES

Black-faced Cormorant – 123 roosting on eastern end

Australian Pelican – one ashore

Silver Gull – 10 at the western end

Crested Tern – 34 sitting on rocks at the western end

No mammals, reptiles or other birds were recorded.

VEGETATION

The vegetation is dominated by *Tetragonia* and *Bulbine semibarbata* co-dominant with a patch of dense *Atriplex cinerea* at the western end.

COMMENTS

This is a significant seabird island due to the existence of the Black-faced Cormorant colony. Its status should be upgraded to conservation area.

Key Island

(Long Island Group, page 226)

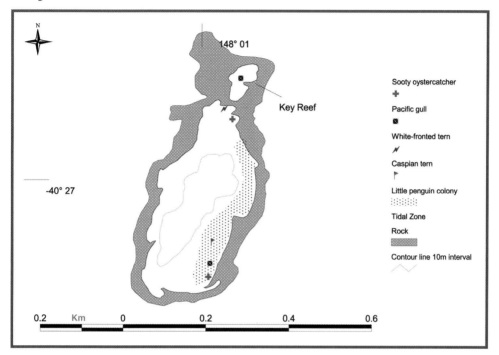

Location: 40°26'S, 148°01'E
Survey date: 28/11/86
Area: 5.9 hectares
Status: Non-allocated Crown Land ⇑

It is an elongate island orientated north to south. It is basically a massive granite slab, with the western half and up to within 3 metres of the summit devoid of vegetation.

BREEDING SEABIRD SPECIES
Little Penguin
Up to 100 pairs were located nesting under mats of stipa and *Tetragonia* sp. on the eastern side of the island.

Pacific Gull
34 pairs with eggs and chicks were nesting in a colony on the eastern side of the island on an open flat area with *Disphyma crassifolium* dominant with *Poa poiformis* and stipa.

Sooty Oystercatcher
2 pairs, one pair with eggs, was located on the eastern side.

Caspian Tern
1 pair with 2 eggs was located at the eastern side alongside the Pacific Gull colony.

White-fronted Tern
8 pairs, at nests with eggs, and 20 adults were recorded. The colony was located on a point towards the north-east tip of the island. The nests were in rock cracks located on granite slabs that sloped gently 3 m down to the water. The cracks had scant clumps of *Coprosma* sp. and

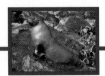

East side.

D. *crassifolium* but were mainly bare. The terns were behaving aggressively towards the Pacific Gulls nearby.

VEGETATION

The western half is devoid of vegetation and the eastern half is dominated by stipa with D. *crassifolium*, *Tetragonia* sp., *Senecio* and *P. poiformis* also present.

OTHER SEABIRD SPECIES

Black-faced Cormorant – 2 sitting on the shoreline

Silver Gull – 6 on the island

No mammals or other birds were recorded.

REPTILES

Metallic Skink

COMMENTS

The presence of White-fronted Terns breeding on the island suggests a revision of the status of the island is needed.

White-fronted Tern nests in vegetated rock crevice.

Key Reef

(Long Island Group, page 226)

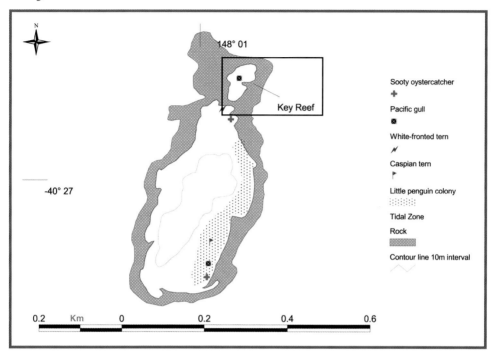

Location: 40° 26'S, 148° 02'E
Survey date: 28/11/86
Area: 0.25 hectares
Status: Non-allocated Crown Land

Key Reef is a small granite islet lying alongside Cape Barren Island.

BREEDING SEABIRD SPECIES
Pacific Gull
1 pair is nesting on the granite shoreline.

VEGETATION
The small vegetation patches are dominated by *Disphyma crassifolium*.

BIRDS
Cape Barren Goose – 1 pair

No other species were recorded.

COMMENTS
Although it is not so important as a seabird island, the Nomenclature Board of Tasmania should be notified of its existence.

Boxen Island

(Long Island Group, page 226)

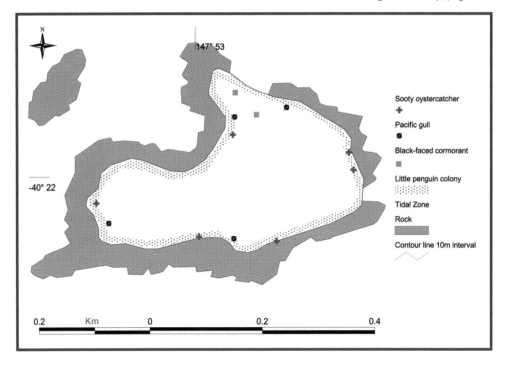

Location: 40°23'S, 147°54'E
Survey date: 28/11/86
Area: 6.74 hectares
Status: Non-allocated Crown Land ⇑

It is a cuboid-shaped, flat dolerite island with a series of gentle depressions and rises surrounded by a 30+ m wide rocky shoreline at low tide. There is a round pebble beach before the start of vegetation. There are some sedimentary strata exposed on the western side and a water-filled depression in the centre of the island. There are no burrowing birds due to the lack of soil.

BREEDING SEABIRD SPECIES
Little Penguin

975 pairs of penguins nest under thick vegetation all over the island. The densest colonies were

South-west side.

located under a strip of *Rhagodia candolleana* and *Senecio* sp. with occasional clumps of stipa at the north-east side of the island. Another dense colony was located nesting under *Tetragonia* sp. on the western side at the head of the bay. Another smaller colony was located at the south-west end.

Pacific Gull

4 pairs were nesting on the shoreline all around the island.

Sooty Oystercatcher

6 pairs were nesting on the shoreline all around the island. One pair was nesting in an old cormorant nest.

Black-faced Cormorant

208 pairs, with newly-constructed nests, were located in two colonies at the north-east side of the island. The smallest had 35 nests and the largest 173 nests. The colonies were located on bare rock ledges about 1 m above the sea. The nests were constructed of seaweed and *Tetragonia* sp. 22 birds were roosting nearby.

VEGETATION

The vegetation is dominated by *Poa poiformis* inland and Aizoaceae succulents around the coast.

COMMENTS

The presence of a Black-faced Cormorant colony makes the island an important seabird island. Along with the significance of the island as a seabird colony, it also supports a large number of other avian species. This is probably a reflection of the relatively undisturbed dense vegetation, the presence of fresh water and the large intertidal area supporting waders. This, in conjunction with the presence of a reptile, suggests that the status of the island should be upgraded.

OTHER SEABIRD SPECIES

Giant Petrel sp. – offshore
Australian Pelican
Silver Gull –2 pairs
Crested Tern – 95 on rocks
White-fronted Tern

No mammals were recorded on the island.

BIRDS
Native:
Brown Quail
Cape Barren Goose – 1 pair with 2 goslings

Grey Teal – nest under *P. poiformis* with 6 eggs
Chestnut Teal
White-faced Heron
Eastern Curlew – 1
Ruddy Turnstone – 100+
White-fronted Chat
Forest Raven – 2
Little Grassbird

Introduced:
Skylark

REPTILES
Metallic Skink

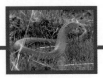

Badger Island Group

(Region 4, page 195)

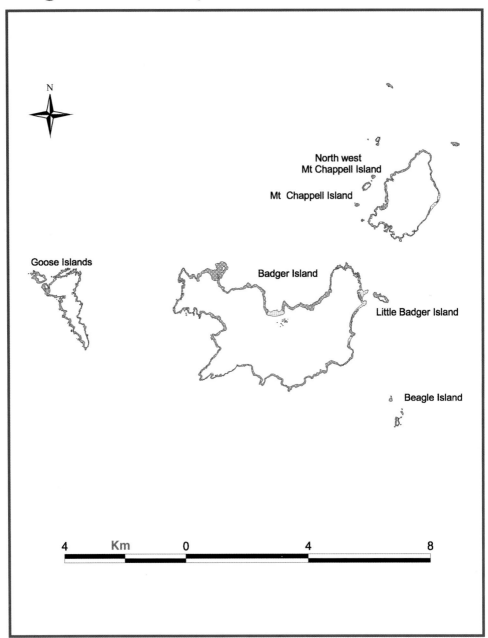

Badger Island

(Badger Island Group, page 236)

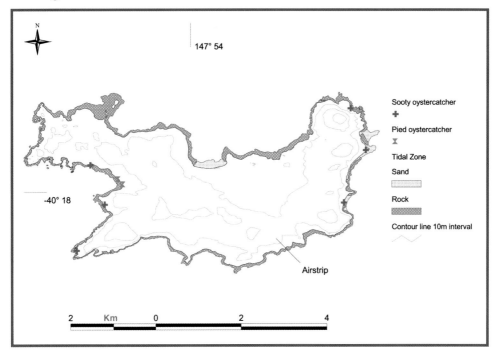

Location: 40°18'S, 147°54'E
Survey date: 30/11/86
Area: 1242.63 hectares
Status: Private property

Badger Island is an elongate, low-lying island, with little burrowing habitat for seabirds, as most of the island has limited soil overlying limestone, apart from at the northern end and southern tip where there are granite exposures. At the north, these are present as large picturesque boulders. The island is extensively grazed by cattle, sheep, Red-necked (Bennett's) Wallaby and pademelons. Buildings include a homestead, shed, windmill and extensive jetty. There is also an airstrip.

BREEDING SEABIRD SPECIES

Sooty Oystercatcher

8 pairs, 2 with 2 eggs, were located scattered around the island. Flocks of 5 and 10 birds were also seen.

Pied Oystercatcher

2 pairs were defending a section of the main beach on the western side of the island. Their nest site was found behind a massive seagrass pile on sand amongst scattered pebbles.

VEGETATION

Poa sp. and stipa dominate the far west end of the island. Elsewhere there are variations in vegetation related to the underlying rock type with *Melaleuca* and *Casuarina* scrub dominating different areas.

OTHER SEABIRD SPECIES
Black-faced Cormorant
Australian Pelican – 3 seen ashore
Silver Gull
Crested Tern – 30 on rocks at north end

MAMMALS
Native:
Red-necked (Bennett's) Wallaby – many
Tasmanian Pademelon
Introduced:
Cattle
Sheep
House Mouse
Cat
Tasmanian Devil – released in 1998/9

BIRDS
Native:
Brown Quail
Cape Barren Goose – a major breeding island
Grey Teal
White-bellied Sea-Eagle – nesting in *Macrocarpa* with a freshly laid egg
Brown Falcon
Ruddy Turnstone
Masked Lapwing
Brown Thornbill
White-fronted Chat – 3 eggs in stipa
Grey Fantail
Flame Robin
Silvereye
Introduced:
Skylark
European Greenfinch
European Goldfinch
Common Blackbird
Common Starling

REPTILES
Mountain Dragon
Metallic Skink
Ocellated skink
White's Skink
Tiger Snake
Blue-tongue Lizard
White-lipped Whip Snake

COMMENTS

Introduced flora and fauna, grazing and burning have all contributed to the destruction of natural habitat of the island and continue to do so. In 1998 or 1999 Tasmanian Devils were introduced onto the island and have become established. Unless removed, they will destroy many of the island's indigenous species. The island has geoconservation significance due to its limestone pavement and granite intrusions, which are considered representative and outstanding for Tasmania (Dixon 1996).

Little Badger Island

(Badger Island Group, page 236)

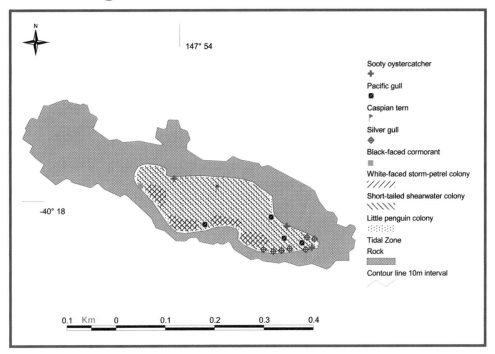

Location: 40°18'S, 147°55'E
Survey date: 28/11/86
Area: 2.51 hectares
Status: Nature Reserve

It is an elongate, granite islet orientated north-west to south-east. There is a gentle rise from its coastline to a central ridge.

BREEDING SEABIRD SPECIES

Little Penguin

100 pairs, with eggs and chicks, are scattered in burrows over most of the island. Colonies are interspersed with those of Short-tailed Shearwaters and White-faced Storm-Petrels in habitat dominated by *Poa poiformis* and *Rhagodia candolleana*.

Short-tailed Shearwater

An estimated 16 401 pairs are in burrows across the island, interspersed with Little Penguin and White-faced Storm-Petrel colonies in *Poa* sp. and *Rhagodia*.

White-faced Storm-Petrel

An estimated 36 pairs were located in four patches across the island in habitat dominated by *R. candolleana*.

Pacific Gull

4 pairs with chicks were located nesting on the southern half of the islet shoreline. Each nest was positioned next to a tussock of *P. poiformis*.

Silver Gull
7 pairs were defending the colony site on the south-east end of the island. The colony was located amongst rocks and *R. candolleana*.

Sooty Oystercatcher
3 pairs with eggs were located in *Carpobrotus rossii* patches around the shoreline.

Black-faced Cormorant
97 pairs were located in a colony on the north-west end of the islet on granite rocks and slabs with gutters bisecting the colony. There is a large accumulation of guano suggesting that the colony is in regular use. 50 birds were also roosting nearby.

Caspian Tern
1 pair at a nest scrape was located on the eastern side on a granite slab amongst *Bulbine* sp.

VEGETATION
The vegetation is dominated by *P. poiformis* with *R. candolleana* and *Tetragonia* sp..

No mammals, reptiles or other seabirds were recorded.

BIRDS
Cape Barren Goose – 2 pairs

COMMENTS
There may be a low incidence of harvesting of Short-tailed Shearwaters. This disturbance would affect the cormorants and could result in damage to the storm-petrel colony. The islet has appropriate status for its protection.

Mount Chappell Island

(Badger Island Group, page 236)

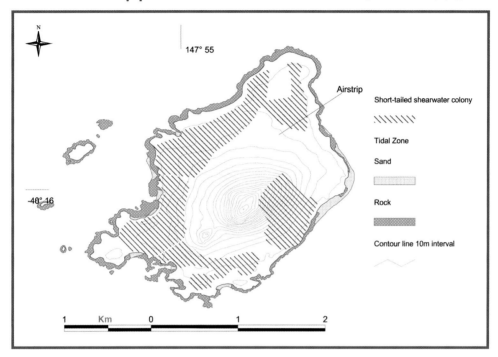

Location: 40°16'S, 147°56'E
Survey date: 29/11/86
Area: 323.26 hectares
Status: Private property

The island is ovate, dominated by the cone-shaped hill in the centre, which slopes gently to the sea. There are huts, an airstrip, fences and sheep on the island and the lesees of the island use a tractor and motorbike for transportation around the island. There are beaches on the south-west side of the island and the rocks are predominantly dolerite, with acidic intrusions, limestone bands and sandstones. There is a beach terrace at the northern end and another at the south-east side.

BREEDING SEABIRD SPECIES
Short-tailed Shearwater

An estimated 971 330 pairs of shearwaters were located nesting along the western side, the northern tip and in a patch on the eastern side of the island. The majority of the birds (estimated 80%) nest underneath the *Rhagodia* and, in many cases, are surface-nesting in shallow trenches.

VEGETATION

The vegetation of the island is dominated by introduced plants such as horehound and African boxthorn and pasture grasses. The original vegetation such as *Rhagodia* is restricted in distribution. *Atriplex* is restricted to the northern section of the island and *Acacia sophorae* is scattered around the island.

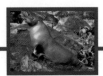

OTHER SEABIRD SPECIES
Pacific Gull
Sooty Oystercatcher
Pied Oystercatcher

MAMMALS
Introduced:
Cat
House Mouse
Rat
Sheep

BIRDS
Native:
Cape Barren Goose – 100+ pairs

Swamp Harrier
Peregrine Falcon – breeding
Hooded Plover
Masked Lapwing
Welcome Swallow – nesting
Introduced:
Common Starling – nests were located in the boxthorn

REPTILES
Bougainvilles Skink
Ocellated Skink
Metallic Skink
Three-lined Skink
Tiger Snake – very abundant.

COMMENTS
There may be a low incidence of harvesting of Short-tailed Shearwaters, but commercial harvesting ceased in 1975. The island has been slashed, ploughed, grazed and burnt. There are introduced cats, rats and mice. The island is actively farmed both as a commercial venture and as part of the management program of Cape Barren geese. The presence of cats and sheep and the expansion of the introduced plant, horehound, are all affecting the numbers of Cape Barren Geese on the island. The cats are probably also responsible for the absence of gulls and oystercatchers from the island. Little Penguins were once recorded breeding in large numbers on the island but do not at present. This island is a classic example of degradation as the result of human commercial and conservation pursuits. The management of the island should be reviewed and in particular the need to maintain exotic pastures to support over 100 Cape Barren Geese. The size and locality of the island suggest that Little Penguins, Sooty Oystercatchers, Pacific Gulls and Silver Gulls should be breeding on the island. The high reptile diversity is also a significant feature of the island.

Western side with Mt Strezlecki in background.

North West Mount Chappell Islet (Badger Island Group, page 236)

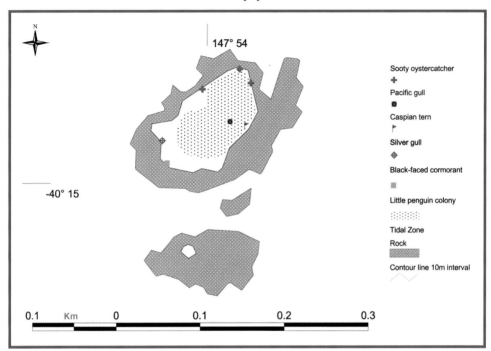

Location: 40°17'S, 147°56'E
Survey date: 29/11/86
Area: 0.71 hectares
Status: Nature Reserve

It is a flat, oval islet composed of granite and lying alongside the larger Mount Chappell island.

BREEDING SEABIRD SPECIES
Little Penguin
An estimated 50 pairs were located nesting beneath *Rhagodia* and stipa and matted succulents.

Pacific Gull
35 pairs, 11 with no eggs, 19 with 2 eggs, 4 with 3 eggs and one with 2 chicks and one egg, were located nesting by stipa on the north side of the islet.

Silver Gull
14 pairs which were defending and 78 used nest sites were located on the western side of the islet in a patch of isolated vegetation dominated by stipa and *Senecio* sp.

Sooty Oystercatcher
3 pairs, two with 2 eggs, were located on the western side of the islet.

Black-faced Cormorant
81 nests were located in a colony on the south-west side of the islet on large shoreline granite boulders. There were also 50+ adults roosting nearby the colony.

Caspian Tern

1 pair, with 2 eggs, was nesting in coarse granite grit within 5 m of the Pacific Gulls.

No mammals, reptiles or other seabirds were recorded.

BIRDS

Native:

Grey Teal – 8 chicks

Introduced:

Common Starling – common in *Rhagodia candolleana* thickets.

VEGETATION

The vegetation is dominated by *Tetragonia* sp. and stipa with *Senecio* sp. scattered around exposed granite

COMMENTS

The islet has a high species diversity but low numbers.

Inner Little Goose Island

(Badger Island Group, page 236)

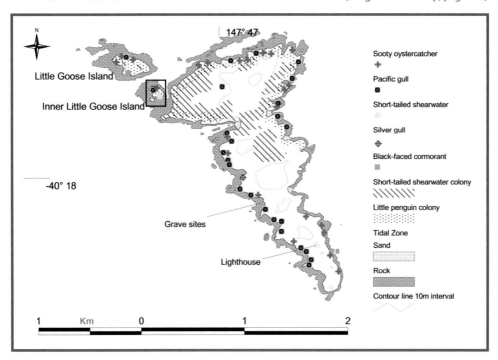

Location: 40°18'S, 147°48'E
Survey date: 28/11/86
Area: 4.5 hectares
Status: Conservation Area

It is a round granite island lying between Little Goose and Goose Islands, separated from them by a 50 metre tidal channel. There is sufficient soil in patches for burrowing birds.

BREEDING SEABIRD SPECIES

Little Penguin

An estimated 50 to 100 pairs, with eggs and chicks, were located scattered in burrows where there is sufficient soil, concentrated along the edges of the tussock patch.

Short-tailed Shearwater

An estimated 10 pairs were found in burrows where soil is sufficiently deep.

Pacific Gull

1 pair was nesting on the shoreline at the eastern side. Another 78+ were sitting on the shoreline of Goose Island, 150 metres away.

Sooty Oystercatcher

2 pairs, one with one egg, were on the shoreline on the eastern side.

VEGETATION

The vegetation is dominated by dense *Poa poiformis* across the central region of the islet with a strip of stipa, *Rhagodia candolleana* and *Tetragonia implexicoma* around the edge.

No other seabirds or mammals were recorded.

BIRDS
White-fronted Chat – nest with 3 eggs in stipa and *Rhagodia*.

REPTILES
Three-lined Skink

COMMENTS
Although the habitat is suitable for storm-petrels, none were found.

Furneaux cottage

Little Goose Island

(Badger Island Group, page 236)

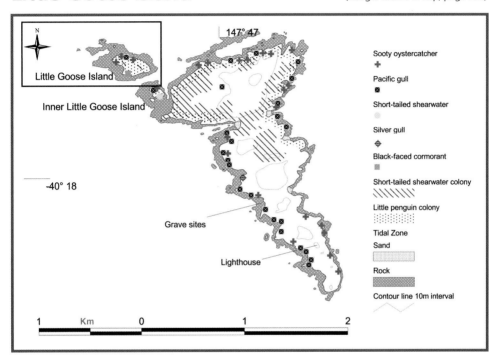

Location: 40°18'S, 147°47'E
Survey date: 28/11/86
Area: 3.64 hectares
Status: Nature Reserve

It is a round, very flat, granite islet with a depression in the centre. There are protruding rounded granite rocks all over the island, but mostly to the north end. There are only two or three very small patches of soil which support vegetation.

BREEDING SEABIRD SPECIES
Little Penguin
100 to 150 pairs were sparsely scattered in burrows over most of the island under rocks covered with *Tetragonia* sp. and in burrows under driftwood.

Short-tailed Shearwater
22 pairs were located in two patches of burrows restricted to the only areas with soil which are vegetated with *Poa poiformis* and *Senecio* sp.

Pacific Gull
42 pairs were identified and a total of 178 adults were counted. 21 of the nesting birds had 2 eggs, 8 had 1 egg, 6 had no eggs, 1 had 3 eggs, 5 had 2 chicks and 1 had a chick and an egg. Some of the nests were in the open on top of patches of vegetation, but most were next to rocks, tussocks or tiny clumps of vegetation.

Sooty Oystercatcher

6 pairs were found distributed all around the island. One nest was located in the middle of a large area of *Carpobrotus rossii*.

Black-faced Cormorant

A colony was located at the south-east end and a roost site on the northern end.

OTHER SEABIRD SPECIES

Australian Pelican –1 seen

Crested Tern – 50+ on rocks at the northern end

No mammals or other birds were recorded.

REPTILES

Metallic Skink

VEGETATION

The vegetation is dominated by *Tetragonia* sp., *Senecio* sp., *P. poiformis*, *Rhagodia candolleana* and patches of *Carpobrotus rossii*.

COMMENTS

The presence of the Black-faced Cormorant colony is significant, however the status is appropriate for its protection. The central depression is of geoconservation significance.

Goose Island
(Badger Island Group, page 236)

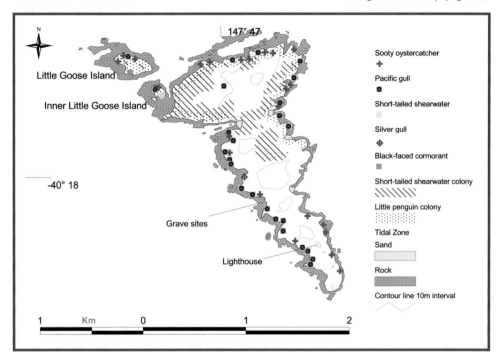

Location: 40°18'S, 147°47'E
Survey date: 28 & 29/11/86
Area: 108.58 hectares
Status: Conservation Area

It is an elongate granite island, orientated north to south with the northern end being the widest, tapering towards the southern end. The northern end has spectacular large boulders whilst the shoreline is indented with deep gullies, occasional beaches and interspersed large, jumbled boulders. The south western coastline has spectacular bays and headlands. A lighthouse, rock walls, foundations and graves of deceased lighthouse keepers are found at the southern end.

BREEDING SEABIRD SPECIES
Little Penguin
An estimated 7036 pairs were located in densities of $0.02/m^2$ in a 5 metre strip around the island. Most were scattered in burrows amongst Short-tailed Shearwaters and under granite slabs surrounded by succulents. Burrows are concentrated in the northern section on the east, north-west and south-west corners.

Short-tailed Shearwater
486 598 pairs nest in crowded colonies dominated by *Poa* sp. predominantly in the southern half of the island. Burrows are mostly very short, less than 40 cm long.

Timber track carriage way, light house at the islands south end.

Silver Gull
16 pairs, with eggs and small chicks, were located in a colony amongst a tangled pile of driftwood and boulders with patches of *Senecio* sp. on the western side of the island midway along the shoreline.

Pacific Gull
14 pairs in one colony and 32 solitary pairs with eggs, small chicks and runners were counted. The colony was located on the northern side opposite Inner Goose Island towards the centre of the island in open country with *Bulbine* and grasses. The nests were located next to individual clumps of *Poa poiformis*. The solitary pairs were located along the western and north-east shorelines. They were conspicuously absent from the eastern shore of the southern two-thirds of the island. There were a total of 445 birds counted at roost sites. These roosts had a similar distribution to the breeding sites.

Sooty Oystercatcher
26 pairs were found concentrated along the western shore, southern and south-east point and on either side of the north-east promontory.

VEGETATION
The vegetation is dominated by mats of Aizoaceae succulents, dense patches of *P. poiformis* and stipa and at the northern end by impenetrable thickets of the African boxthorn, *Lycium ferocissimum*. Less prevalent are woody plants such as *Acacia mucronata var. longifolia* and *Leptospermum parviflorus*.

No mammals or other seabirds were recorded.

BIRDS
Native:
Brown Quail
Black Swan –1 seen
Cape Barren Goose
Masked Lapwing
White-fronted Chat – nesting
Silvereye
Introduced:
House Sparrow
Common Blackbird

REPTILES
Metallic Skink
Ocellated skink

COMMENTS
The African boxthorn, which is restricted to two large patches, may spread further and completely destroy the breeding habitat for the burrowing seabirds. The past use of the island by light keepers probably still affects the diversity of seabirds using the island today. Gravestones are in a poor condition with their lettering disappearing.

Efforts should be made to eradicate the African boxthorn and to identify other factors from past human occupation that may be reducing the species diversity on the island. The historic value of the island should be assessed and maintenance of the gravestones and other historic remnants undertaken.

Beagle Island

(Badger Island Group, page 236)

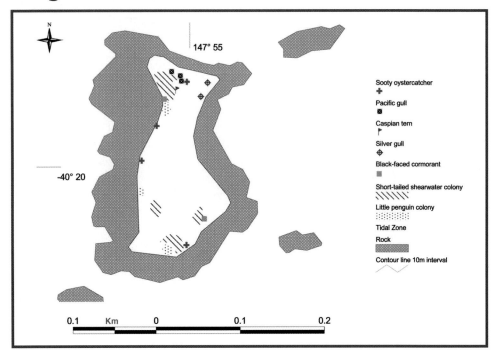

Location: 40°20'S, 147°55'E
Survey date: 28/11/86
Area: 1.24 hectares
Status: Nature Reserve

Beagle Island is a low, flat, rectangular granite island with a curved northern tip that encompasses a bay. There is a high point of bare rock at the southern end and a beach on the western side.

BREEDING SEABIRD SPECIES
Little Penguin

18 burrows were located in small patches, two of which are in association with Short-tailed Shearwaters, with the other being separate.

Short-tailed Shearwater

142 pairs were located in patches of *Rhagodia candolleana* and *Atriplex cinerea*, where there is sufficient soil for burrowing.

Pacific Gull

24 pairs, at nests with eggs, and runners, were located on the shoreline all around the island close to stipa

Silver Gull

30 pairs were located, distributed between two colonies at the northern end of the island with evidence of breeding in the form of dead chicks, old nests and dead adults.

Sooty Oystercatcher

4 pairs were located on the western side of the island on the shoreline and one pair on the eastern side.

Black-faced Cormorant

158 pairs in newly-constructed nests were located in a colony on the south-east side of the island on bare granite rocks 1.5 metres above the sea. This extensive colony stretches for approximately 40 m by 10 m along the shore. There is a large accumulation of guano suggesting it has been in use for a long time. There is also a colony of old nests on the north-west shoreline.

Caspian Tern

One nest was located at the northern end of the island in a depression in *Tetragonia* sp.

VEGETATION

The vegetation is dominated by *Rhagodia* and *A. cinerea* with patches of *Carpobrotus rossii* and *Disphyma crassifolium*.

OTHER SEABIRD SPECIES

Common Diving-Petrel – 1 dead

White-faced Storm-Petrel – 1 dead

Australian Pelican – 1

No mammals or reptiles were recorded on the island.

BIRDS

Native:
Cape Barren Goose – 1 pair

White-fronted Chat – nest with chicks

Little Grassbird

Introduced:
Common Starling – 1 nest in a crevice

COMMENTS

The status of the island is in accord with its importance as a seabird island because of the species diversity and presence of a large and old Black-faced Cormorant colony. Further work should be carried out to monitor the status of storm-petrels and diving-petrels on the island.

Tin Kettle Island Group

(Region 4, page 195)

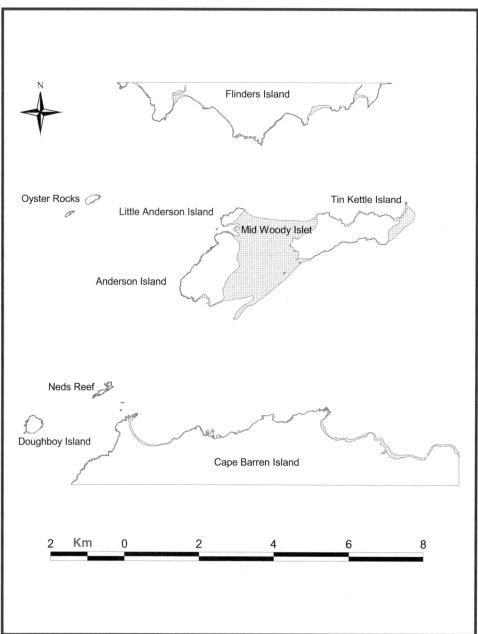

Anderson Island

(Tin Kettle Island Group, page 254)

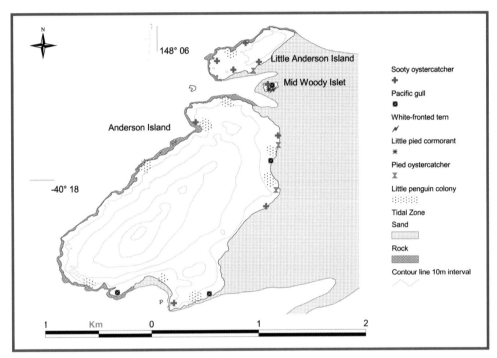

Location: 40°18'S, 148°06'E
Survey date: 2/12/86
Area: 166.49 hectares
Status: Non-allocated Crown Land, Lease

Commonly known as Woody Island, Anderson Island is elongate island, rising to a high point on the western side where there are large granite boulders. There is very little or no soil on the island and therefore virtually no burrowing bird habitat. There are eight or more freshwater soaks on the west coast. Buildings on the island include huts, shearing sheds, stockyards and fences. Sheep and cattle graze the island. The extensive mudflats at low tide form vast feeding areas for oystercatchers and other waders. A sandblow exists in the south-east.

BREEDING SEABIRD SPECIES

Little Penguin

An estimated 100 to 150 breeding pairs were located, scattered in very low numbers all over the island with four dense colonies and other isolated scattered pairs. They were nesting mostly under boulders, rarely in burrows and occasionally under stipa Such burrowing habitat is particularly sensitive to livestock grazing and trampling.

Pacific Gull

3 pairs were nesting on the eastern shoreline. All nests had eggs.

Sooty Oystercatcher

4 breeding pairs were located scattered around the shoreline. One pair had 2 eggs. Many were feeding on the vast tidal mudflats.

Pied Oystercatcher

Several nests were located on seagrass on the sandy beaches on the eastern coastline. Many were sighted feeding on the mudflats.

VEGETATION

There are a few remnant patches of *Melaleuca ericifolia* on the eastern side which become mud flats at low tide. There are also patches of stipa around the coast. The majority of the island is being used as a large grazing paddock.

COMMENTS

The island carries very few seabirds and is heavily grazed. There is stock damage to the freshwater soaks and the north-west dune area. Some areas should be fenced off to protect the soaks and dunes. The intertidal zone supports large numbers of migratory waders, a factor which should be taken into account in any management planning.

Historical records describe a large shearwater colony on the west side of the island. In the early part of last century pigs were introduced to dig up the burrows and to eat the birds and eggs.

No other seabirds were recorded.

MAMMALS

Introduced:

Sheep

Cattle

BIRDS

Native:

Brown Quail

Cape Barren Goose – dead

Grey Teal

Brown Falcon

Red-capped Plover

Masked Lapwing

White-fronted Chat

Scarlet Robin

Grey Fantail

Forest Raven – nesting in tea tree

Welcome Swallow – nesting in sheds

Little Grassbird

Silvereye

Introduced:

House sparrow

Skylark

Common Blackbird

Common Starling

REPTILES

Metallic Skink

Little Anderson Island

(Tin Kettle Island Group, page 254)

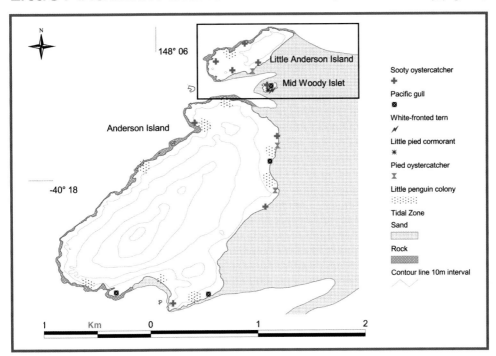

Location: 40°18'S, 148°06'E
Survey date: 2/12/86
Area: 13.4 hectares
Status: Non-allocated Crown Land

The island is joined to Anderson island at low tide. It has been grazed in the past and there are remains of fences on the eastern side. A patch on the northern side had been burnt prior to the survey and much of the island has been burnt over time. There are a few patches of scrub and some freshwater soaks. There is a flat depression in the middle of the island.

BREEDING SEABIRD SPECIES

Little Penguin
An estimated 30 pairs were located, lightly scattered around the island under the few remaining dense patches of vegetation and under rocks in the centre of the island.

Sooty Oystercatcher
4 breeding pairs, with eggs, were located scattered around the island.

Pied Oystercatcher
1 pair, with eggs, was located on a beach at the eastern side.

No mammals or other seabird species were recorded on the island.

BIRDS
Native:
Brown Quail
Cape Barren Goose
Masked Lapwing
Little Grassbird

REPTILES
Metallic Skink
Ocellated Skink
Bougainvilles Skink

VEGETATION
The vegetation is dominated by grasses mainly *Poa poiformis*. There are also several patches of scrub.

COMMENTS
The island has been devastated by fire and grazed by stock. This may cease to allow for natural recovery. The diversity of reptiles is possibly significant.

Mid Woody Islet

(Tin Kettle Island Group, page 254)

Mid Woody Islet with Flinders Island in the background.

Location: 40°18'S, 148°06'E
Survey date: 2/12/86 and 15/12/86
Area: 0.66 hectares
Status: Non-allocated Crown Land ⇑

Mid Woody Islet is a very small, round island with a central high point.

BREEDING SEABIRD SPECIES
Little Penguin

A colony with 30 burrows is concentrated on the south side of the islet in soft sand under *Atriplex cinerea* and *Rhagodia candolleana*. Other burrows are scattered around the island under rocks and scrub. Most burrows surveyed had eggs.

Pacific Gull

9 pairs with eggs were found on and around the central summit.

Sooty Oystercatcher

7 pairs with eggs were located scattered around the island.

White-fronted Tern

A nest with 1 egg and 2 nest scrapes were located on 15/12/86.

VEGETATION

The vegetation is dominated by *A. cinerea*, *R. candolleana* with *Correa alba* and *Poa poiformis* on the high ground.

OTHER SEABIRD SPECIES

Little Pied Cormorant

There were no mammals, other birds or reptiles recorded.

COMMENTS

It is a relatively unspoilt islet in a group of devastated islands and, as such, should be conserved. The status should be upgraded to conservation area.

White-fronted Tern

Tin Kettle Island

(Tin Kettle Island Group, page 254)

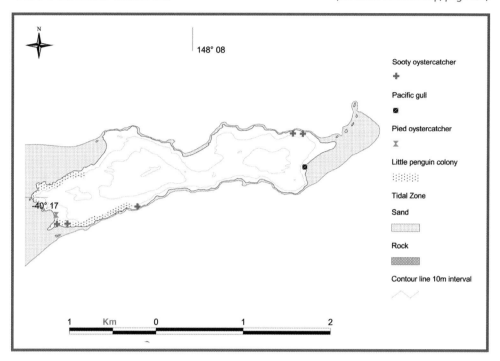

Location: 40°17'S, 148°08'E
Survey date: 2/12/86
Area: 176 hectares
Status: Non-allocated Crown Land, Lease

Tin Kettle Island is an elongate island with an irregular coastline and beaches. There are old vegetated dunes, possibly Pleistocene in age, which run east to west and dominate the central section of the island. The island is sandy with scrub, well-developed pastures, fencing and freshwater soaks. The island is farmed. Buildings include a house and shed. There are extensive mudflats at the eastern end.

BREEDING SEABIRD SPECIES

Little Penguin

An estimated 100 pairs were located, scattered in burrows over the western corner of the island and in a denser colony in the south-west section. Burrows were in sandy soil in the vegetated old dunes. There were few birds around the rest of the island. Several pairs were found near the house.

Pacific Gull

1 pair, with eggs, was nesting on the shoreline at the eastern end.

Sooty Oystercatcher

5 pairs were nesting at the north-east and south-west sides of the island.

Pied Oystercatcher
A nest was located at the south-west Little Penguin colony.

VEGETATION
Introduced grasses dominate the vegetation. There are also areas of scrub dominated by Myrtaceae and *Myoporum* species and patches of *Poa poiformis*.

COMMENTS
The island is farmed. The presence of stock may be restricting the numbers of shore breeders and burrowing penguins.

There are extensive tidal flats and rocks exposed around the island and it is therefore surprising that there are not more oystercatchers. The island is well-maintained with extensive use of electric fencing to contain stock. The lessee has also fenced freshwater soaks, which supply troughs located on the beach. This stops habitat

North side looking west.

degradation around the soaks. The old dunes could probably support more burrowing seabirds such as Little Penguins and Short-tailed Shearwaters. These ridges should be fenced off and the grazing restricted to the low-lying areas between the ridges in an effort to reduce habitat destruction. The island is also used as a commercial venture with quail shooters.

OTHER SEABIRD SPECIES
Caspian Tern
White-fronted Tern

MAMMALS
Horse
Cattle
Rat

BIRDS
Native:
Brown Quail
Cape Barren Goose
Brown Falcon
Hooded Plover
Masked Lapwing
White-fronted Chat
Scarlet Robin
Olive Whistler
Grey Fantail
Forest Raven
Little Grassbird
Silvereye
Introduced:
Skylark
European Goldfinch
Indian Peafowl (Peacock)

REPTILES
Metallic Skink

Oyster Rocks

(Tin Kettle Island Group, page 254)

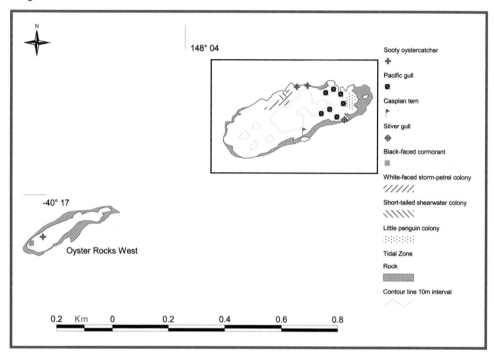

Location: 40°17'S, 148°39'E
Survey date: 2/12/86
Area: 5.31 hectares
Status: Conservation Area

It is an elongate island, orientated with the long axis east to west. There are large, round, granite outcrops with areas of bare granite.

BREEDING SEABIRD SPECIES

Little Penguin
40 pairs are concentrated along the east coast and at the eastern end, nesting under the vegetation and rocks.

Short-tailed Shearwater
10 pairs inhabit the central northern end of the island.

White-faced Storm-Petrel
An estimated 1806 pairs were located nesting under very dense matted *Poa poiformis* and also amongst stipa in several places, including the mid north coast.

Pacific Gull
9 pairs at nests with eggs, small chicks and runners were located. The birds were nesting along the eastern shore and all nests were solitary.

Silver Gull
1 pair was defending a site on the eastern side of the island.

Sooty Oystercatcher
2 pairs, one with 2 eggs, were nesting at the northern end of the island.

Caspian Tern
1 pair, with eggs, was located nesting in the centre of the island at the south side amongst *Carpobrotus rossii*.

VEGETATION
The vegetation is dominated by stipa and dense mats of *P. poiformis*.

No mammals were recorded.

OTHER SEABIRD SPECIES
White-fronted Tern – 7 were seen at sea to the north.

BIRDS
Native:
Cape Barren Goose – breeding
Masked Lapwing
White-fronted Chat
Forest Raven

REPTILES
Metallic Skink

COMMENTS
There is a high species diversity on the island and the presence of White-faced Storm-Petrels makes this an important seabird island. The lack of burning has probably allowed the extensive matted vegetation cover on which the birds rely for breeding. This island is therefore a good example of seabird breeding habitat in the absence of frequent fire and constant grazing. The status of the island is sufficient.

Burrowing habitat of central area looking west.

Oyster Rocks West

(Tin Kettle Island Group, page 254)

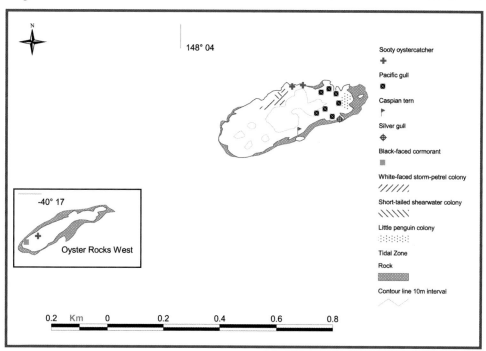

Location: 40°18'S, 148°39'E
Survey date: 2/12/86
Area: < 1 hectares
Status: Conservation Area

It is a bare reef without any vegetation.

BREEDING SEABIRD SPECIES
Sooty Oystercatcher
One pair was defending.

Black-faced Cormorant
More than 40 old nests were located on the south-west side of the reef in rock gutters. The colony was probably larger but many nests had been washed away. A large roost site was also located at the east end currently occupied by 20 birds.

OTHER SEABIRD SPECIES
Crested Tern – 35 were roosting at the west end.

Pacific Gull – 1 pair

No mammals, reptiles or other birds were recorded.

COMMENTS
The island is an important breeding and roosting site for Black-faced Cormorants.

The status of the island as a conservation area is sufficient for the protection of the seabirds.

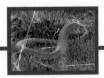

Neds Reef

(Tin Kettle Island Group, page 254)

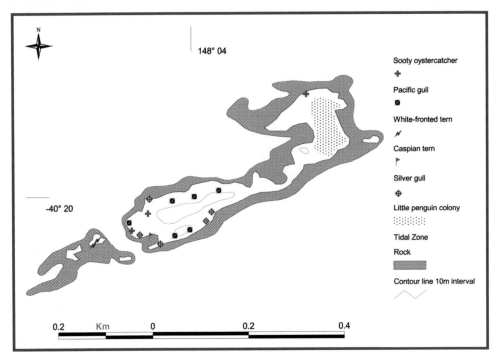

Location: 40°20'S, 148°04'E
Survey date: 1/12/86
Area: 2.78 hectares
Status: Non-allocated Crown Land

The reef comprises three granite islets. They are joined at low tide by extensive mudflats, which effectively increase the islets area 3 to 4 times. The western end of the west islet is dominated by bare granite with little soil. Beaches occur around the islets' coastlines.

BREEDING SEABIRD SPECIES
Little Penguin
30 pairs were located under rocks and thick patches of *Rhagodia candolleana*, *Tetragonia* sp. and *Senecio* sp. on the west islet. On the east islet 20 pairs were under low dense shrubs and in shallow burrows beneath dense vegetation.

Pacific Gull
6 pairs, three with chicks, were nesting on the shoreline all around the western islet.

Silver Gull
5 pairs were counted at nests on the western islet.

Sooty Oystercatcher
3 pairs, 2 with 2 eggs were nesting on the western islet and one was on the small rock to the north, which is isolated at high tide.

Caspian Tern
1 pair was nesting at the west end of the western islet on bare soil alongside a *Carpobrotus rossii* patch.

White-fronted Tern
2 pairs were defending at the west end of the western reef. Their nests were in granite cracks on bare gravel beside stipa and *Senecio* sp.

No mammals, reptiles or other seabirds were recorded.

BIRDS
Native:
Cape Barren Goose – 1 pair
Introduced:
Common Starling

VEGETATION
The western end is dominated by bare granite with stipa and *Senecio* sp. clumps with little soil. The eastern section and the eastern islet are dominated by stipa and scrub.

COMMENTS
The islet had been burnt. Fires are a major threat to the seabird breeding habitat as any removal of vegetation exposes what little soil is available for burrowing birds, increasing its vulnerability to erosion.

Doughboy Island

(Tin Kettle Island Group, page 254)

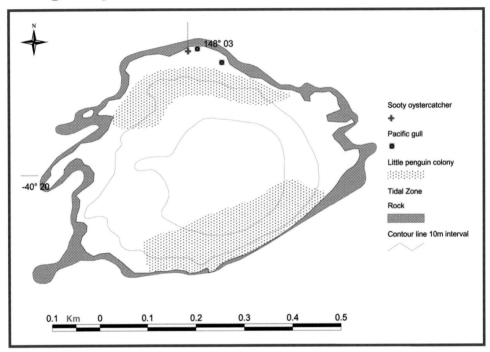

Location: 40°11'S, 148°03'E
Survey date: 1/12/86
Area: 30 hectares
Status: Non-allocated Crown Land, Lease

It is a gently undulating, dome-shaped island with bays on adjacent sides, dolerite dykes, a limestone outcrop on the north-east coast and basalt near the summit. The island has been devastated by grazing and burning but remnants of native vegetation survive in areas such as granite slabs where fires could not reach. Soils are very sparse with much gravel.

BREEDING SEABIRD SPECIES

Little Penguin

90 pairs in two major colonies were located on the northern and southern sides and 20 burrows scattered elsewhere. They nest in crevices or take advantage of the thick stipa patches that survive on the granite slabs.

Pacific Gull

2 pairs, with eggs, were located on granite slabs next to stipa

Sooty Oystercatcher

1 pair was nesting alongside the Pacific Gulls.

OTHER SEABIRD SPECIES
Pied Oystercatcher – on the beach at east end, not breeding

Silver Gull

No mammals were recorded on the island.

BIRDS
Native:
Cape Barren Goose – breeding

Introduced:
Skylark

Common Starling –1 nest in a crevice 10 cm off the ground.

REPTILES
Metallic Skink

VEGETATION
Poa sp. and stipa are dominant, but have been regularly burnt. There are small patches of remnant vegetation.

COMMENTS
This island has been devastated by irresponsible farming practices and fire.

Great Dog Island Group

(Region 4, page 195)

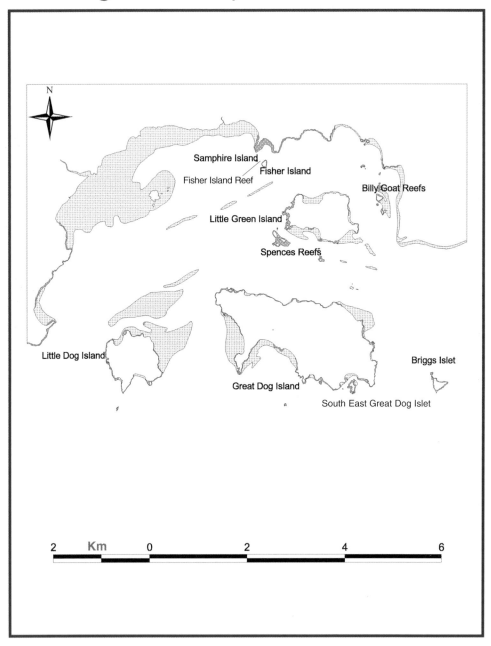

Little Dog Island

(Great Dog Island Group, page 270)

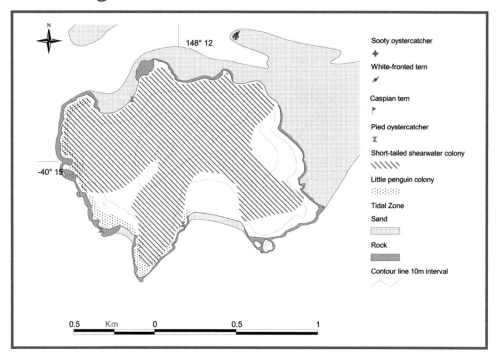

Location: 40°15'S, 148°12'E
Survey date: 3/12/86
Area: 83.01 hectares
Status: Game Reserve, Private property (annual lease)

It is a square, relatively flat, granite island orientated north-west to south-east with rocky shores apart from two sandy bays, located on the north-west and central southern coasts. It has a central ridge with bare rock or shallow ground on the edge up to the high point. There are two huts, one in the south to the north-west of the bay, and one on the north-east.

BREEDING SEABIRD SPECIES
Little Penguin

Up to 100 pairs breed here, scattered throughout the shearwater colony all around the outer edge of the island. It is difficult to gauge exact numbers due to the extent and density of the shearwaters. Islands in this area are probably unfavourable for penguins due to enclosed waters and strong tidal flow, currents making foraging inefficient. Also exposed tidal flats restrict shore access to burrows.

Short-tailed Shearwater

An estimated 563 643 pairs breed in densities of 0.479/m^2 over 70% of the island, apart from the scrub patch on the east coast, the rock and grass patches on the south eastern coast and the rocky area in the western section. They are generally in vegetation dominated by *Poa poiformis* with everlasting daisy co-dominant.

Pied Oystercatcher

Two pairs were located, one on the southern beach and one, with 2 eggs, on the north-western beach. An additional non-breeding individual was located on the west coast.

OTHER

Green and gold bell frog (K. Ziegler pers. comm., September 2000)

VEGETATION

The dominant native vegetation is *P. poiformis*. There is a large patch of *Kunzea ambigua* scrub on the east coast with very few burrows and few extensive patches of *Disphyma crassifolium* and *Tetragonia* to the west. In the low-lying ground at the back of the southern hut is an extensive area of very thick, high grasses with no burrows.

OTHER SEABIRD SPECIES

Silver Gull

Sooty Oystercatcher – 3

Australian Pelican – 10 (K. Ziegler pers. comm., September 2000)

MAMMALS

Rat – around the huts

BIRDS

Native:

Cape Barren Goose

Brown Falcon

Swamp Harrier – nest at the southern end in thick bracken and grass with three chicks. Also in the nest were one freshly-laid shearwater egg and the remains of a juvenile. Common Starling.

Lewin's Rail

White-fronted Chat

Grey Fantail

Forest Raven

Silvereye

Introduced:

Skylark

Common Blackbird

Common Starling

REPTILES

Metallic Skink

Tiger Snake – many

COMMENTS

The island's vegetation and seabird habitat is regenerating as the island has not been grazed for five to six years. There is an abundance of succulent herbs and shrubs in contrast to Little Green Island which has succulents surviving only in inaccessible rock crevices.

Little Dog Island (rock to north) (Great Dog Island Group, page 270)

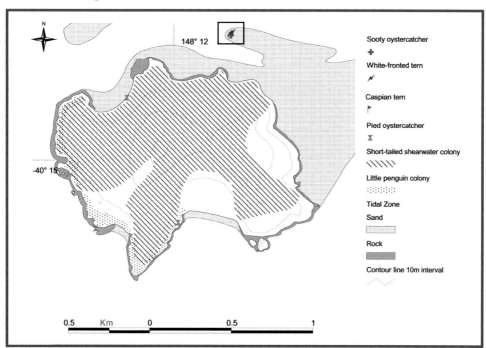

Location: 40°15'S, 148°12'E
Survey date: 9/12/86
Area: Approx. 20m²
Status: Non-allocated Crown Land

It is a tiny, half-bare granite rock, elongate in shape and oriented north to south. It is not joined to Little Dog Island at low tide.

BREEDING SEABIRD SPECIES
Sooty Oystercatcher
2 empty nest scrapes were located at the northern end.

Caspian Tern
A pair with one egg was located by the *Poa* sp. on *Carpobrotus rossii* 6 metres to the south.

White-fronted Tern
3 nest scrapes were located in the shell midden at the northern end in *C. rossii* and 6 birds were defending.

VEGETATION
The island's vegetation is dominated by *C. rossii* and *Tetragonia*, with a tiny clump of *Poa* sp. and *Coprosma*.

OTHER SEABIRD SPECIES

Black-faced Cormorant – 38 roosting

No mammals, reptiles or other birds were recorded.

COMMENTS

Although White-fronted Terns may not always breed here, they are very sensitive to disturbance when breeding. Considering this, any visits to the island should be strictly controlled. Mutton-birding on nearby islands potentially exposes this islet to disturbance and should be monitored so as to protect the White-fronted Terns' breeding habitat.

Black-faced Cormorant

Great Dog Island

(Great Dog Island Group, page 270)

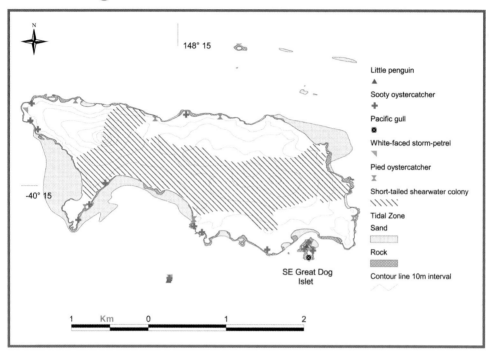

Location: 40°15'S, 148°15'E
Survey date: 8/12/86
Area: 353.84 hectares
Status: Private property

Commonly known as Big Dog Island, it is oval in shape with an irregular coastline and many granite beaches. It is relatively flat, rising to a high point known as Great Dog Hill. There are birders' shacks on the island.

BREEDING SEABIRD SPECIES
Short-tailed Shearwater

An estimated 290 000 pairs occur over much of the island but are not found in the forest habitat, on the hard ground on the north-west side or in a grass paddock on the southern side.

Dense *Poa* sp. of a shearwater colony.

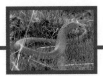

North-east beach and adjacent patch of remnant scrub.

White-faced Storm-Petrel
One burrow and the carcasses of 10 birds were found at the north-west tip of the island. The birds had been killed by cats.

Sooty Oystercatcher
6 pairs were located on the south coast, 3 on the west coast and one on the north coast.

Pied Oystercatcher
6 pairs were found scattered around the island.

VEGETATION
The vegetation is dominated by *Poa poiformis* which proliferates as a result of Short-tailed Shearwater activity in conjunction with burning. There is a mixed forest community on the north-eastern side of the island with smaller patches on the south-west and south-east points. The forest comprises *Allocasuarina verticillata*, *Allocasuarina rhomboidea*, *Eucalyptus viminalis*, *Acacia verticillata*, *Melaleuca ericifolia* and various species of *Leptospermum*. It is one of the few forests surviving on an island in the Furneaux group.

COMMENTS
The island has been severely affected by fire, grazing, Short-tailed Shearwater harvesting and introduced mammals and has still retained a large species diversity. Efforts should be made to preserve the forest on the island, reduce the use of fire and eradicate the introduced mammals. If this occurs, it is possible that the White-faced Storm-Petrels will continue to successfully recolonise the island. Little Penguins may also colonise the eastern shore, where there is suitable habitat. The island has an impressive list of reptiles.

OTHER SEABIRD SPECIES
Black-faced Cormorant

MAMMALS
Native:
Water Rat

INTRODUCED:
Cat – a control program in the early 1990s eradicated most of them, however there may still be a small population.
Mouse
Rat

BIRDS
Native:
Cape Barren Goose
Brush Bronzewing
White-bellied Sea-Eagle – pair
White-fronted Chat
Grey Fantail
Masked Lapwing
Forest Raven
Red-capped Plover
Swamp Harrier

Introduced:
European Goldfinch
Common Blackbird
Common Starling – nesting

REPTILES
Metallic Skink
Ocellated skink
Three-lined Skink
Blue-tongue Lizard
Copperhead Snake
Tiger Snake

'With the decline of sealing the mutton bird became the most important item of subsistence for resident sealers and their families. In the earliest days, the exploitation was on a wholesale scale – adult birds, fledglings and eggs were taken indiscriminately. A very large trade was done in feathers and down for upholstery. There was a particularly large trade in mutton bird fat for the Tasmanian saw-milling industry, where the liberal application of fat to the wooden skids greatly facilitated movement. A rusted boiler used for rendering the birds into fat is still evident on Great Dog Island.' D. Serventy from Bass Strait Australia's Last Frontier, 1969.

South East Great Dog Islet (Great Dog Island Group, page 270)

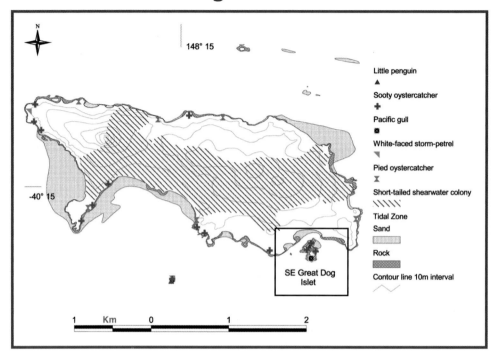

Location: 40°15'S, 148°16'E
Survey date: 7/12/86
Area: 0.6 hectares
Status: Non-allocated Crown Land ⇑

It is a small granite island, cuboid in shape with a deep cove ending in a beach on the southern side and a narrow point on the north-west side. The island is low-lying and slopes gently to the east.

BREEDING SEABIRD SPECIES
Little Penguin
An estimated 30 pairs, some with empty nests and some with eggs and chicks, are scattered in burrows over most of the island amongst storm-petrel burrows with the densest concentration in the centre of the gentle slope on the eastern side. A few were also located in rock crevices.

White-faced Storm-Petrel
An estimated 3662 burrows were located all over the islet with the highest densities on the gentle slopes of the sides of the island. The burrows are in habitat dominated by *Poa poiformis* but the highest density transect was located in deeper soil with *Tetragonia* sp. dominant.

Pacific Gull
One pair was nesting alongside the southern beach.

Sooty Oystercatcher
3 pairs, at nests with eggs, were located scattered around the island.

VEGETATION

The vegetation is dominated by *Disphyma crassifolium*, *Tetragonia* sp., *Senecio* sp. and *P. poiformis* with *Coprosma* sp. bushes scattered around.

No mammals, reptiles or other birds were recorded.

COMMENTS

This is an important seabird island because White-faced Storm-Petrels are relatively abundant here. Nature reserve status is warranted

Briggs Islet

(Great Dog Island Group, page 270)

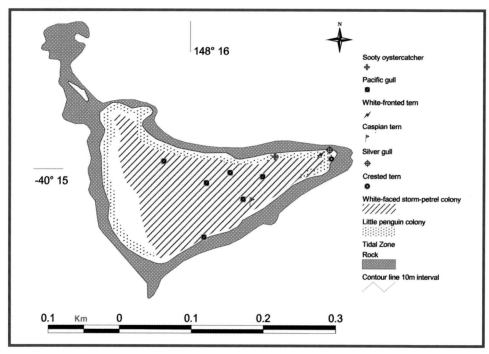

Location: 40°15'S, 148°17'E
Survey date: 5/12/86
Area: 3.41 hectares
Status: Conservation Area

A small triangular shaped islet, oriented north-east to south-west, it is low and composed of granite with a rocky shoreline.

BREEDING SEABIRD SPECIES
Little Penguin

397 pairs were located, largely confined to the 5 metre strip of stipa around the shoreline of the islet, especially on the south side. The birds nest in burrows and under rocks.

White-faced Storm-Petrel

An estimated 1500 pairs in average densities of 0.43/m² were found nesting all over the island with 90% of the birds in matted *Poa poiformis*. Where the tussocks are too thick, providing an impediment to burrowing, the density of birds declines to approximately 0.2/m².

Pacific Gull

6 pairs, 3 with no eggs, 1 with 1 egg, 1 with 2 eggs and a half-grown runner were in a colony amongst *Poa* and stipa in the centre of the island. 28 adults were counted.

Silver Gull

A colony of 208 pairs, mostly with eggs and chicks, was located on the north-east shore on the edge of the tussock habitat.

Sooty Oystercatcher

1 pair was located at the north-east point of the islet.

Caspian Tern

1 pair was nesting alongside the Pacific Gull colony.

Crested Tern

8 pairs, with one egg each, were nesting in a tight group in a bare patch of soil alongside the Silver Gulls and White-fronted Terns.

White-fronted Tern

20 pairs at nests with eggs and chicks were distributed between 2 colonies at either end of the Silver Gull colony along the edge of the *Poa* habitat.

VEGETATION

The vegetation is dominated by *P. poiformis* and stipa in a 5 m strip around the coastline. There is also some *Tetragonia*.

OTHER SEABIRD SPECIES

Black-faced Cormorant – 139 were roosting at a regular roost site on the north-east side of the island, where there is a build up of guano extending almost to the centre of the island.

Fairy Tern – 2 flew over

No mammals were recorded on the island.

BIRDS

Cape Barren Goose

REPTILES

Lizards were seen but not identified.

COMMENTS

The islet is seldom visited. There should be restricted access to the island during the breeding season as disturbance of the terns, particularly the rare White-fronted Tern, will make them more vulnerable to the Silver and Pacific Gulls. The potential for the Silver Gull colony to increase in response to the food source provided by the local fish processor at Lady Barron should be monitored, as an increase may adversely affect the island's tern colonies. The White-faced Storm-Petrel colony is significant.

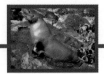

Billy Goat Reefs

(Great Dog Island Group, page 270)

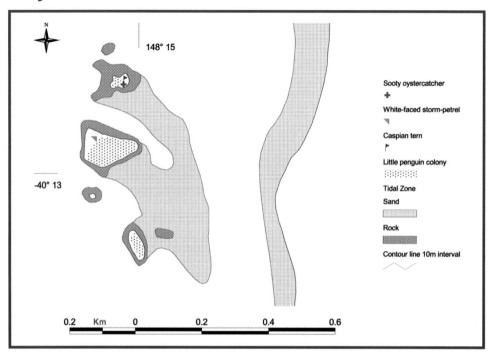

Location: 40°13'S, 148°16'E
Survey date: 7/12/86
Area: 4 islets: total = 1.31 hectares
Status: Non-allocated Crown Land ⇑

The reefs are four small islets joined at low tide. The islets lie on a north-south line and will be referred to as islets 1 to 4 starting with the northernmost islet. Islet 1 is a low flat granite islet about 0.14 hectares in area rising to 3 m above sea level. Islet 2 is the largest of the islets approximately 0.93 hectares in area and reaching a height of 15 m. Islet 3 of 0.02 hectares. is the smallest islet composed of mainly bare rock rising to 5 m above sea level. Islet 4 is the second largest islet and is 0.22 hectares in area with the highest point 10 m above sea level. It has limited soil.

BREEDING SEABIRD SPECIES
Little Penguin

149 pairs were counted, Islet 1 with 15, Islet 2 with 98, Islet 3 with 1 and Islet 4 with 35. On Islet 1, burrows are under clumps of stipa in grit and gravel. There are burrows all over Islet 2 (96 short burrows and 2 nests under rocks) with densest colonies in areas of *Poa poiformis*, *Pelargonium australe*, *Tetragonia implexicoma* and ever-lasting daisy. Islet 3 harbours only one pair in a shallow short burrow protected by a clump of *S. stipoides*. Islet 4 has two burrows with 33 pairs nesting in rock crevices or in cleared areas under 1 to 1.5 m high *Coprosma* sp.

White-faced Storm-Petrel
3 pairs were located at the northern end of Islet 2 in burrows amongst *Poa poiformis*.

Caspian Tern
One pair, with 2 eggs, was located on the high point on the west side of Islet 1 on a flat granite slab with *Disphyma crassifolium*.

Sooty Oystercatcher
One pair, with two eggs, was found under stipa on the little sand beach at the northern end.

VEGETATION
Islet 1 The vegetation is dominated by *Atriplex cinerea* and *Disphyma crassifolium* with *T. implexicoma* and scattered tussocks of stipa and *P. poiformis*.

Islet 2. The vegetation varies from areas dominated by bracken to areas of large *Coprosma repens* bushes to a mixture of *P. poiformis*, *Pelargonium*, everlasting daisy and *Tetragonia*.

Islet 3. Stipa is restricted to cracks in the rock and a few isolated *Coprosma* bushes are present.

Islet 4. Very large *Coprosma* bushes dominate the vegetation with *Pelargonium* sp and *P. poiformis* elsewhere. *Disphyma crassifolium* and *Tetragonia implexicoma* grow on the south-west side.

OTHER SEABIRD SPECIES
White-fronted Tern – breeding pairs have been recorded on Islet 1 in three separate seasons but were not there during this survey.

No mammals or other birds were recorded.

REPTILES
Metallic Skink

White's Skink

White-lipped Whip Snake

COMMENTS
Given that this small group of islets has a low number of a relatively wide variety of species and a high number of reptiles, its status should be raised to Nature Reserve. The introduced species *Coprosma repens* is spreading over the islets and destroying native vegetation. Weed control is necessary to ensure that they remain attractive breeding sites for White-fronted Terns.

Little Green Island

(Great Dog Island Group, page 270)

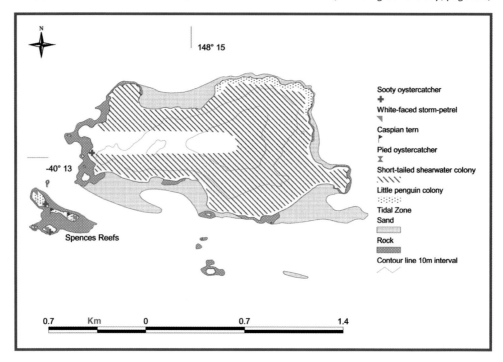

Location: 40°13'S, 148°15'E
Survey date: 7/12/86
Area: 86.62 hectares
Status: Multiple Use: Conservation Area, Private property (annual lease)

The island has an irregular coastline, is predominantly flat, composed of granite and surrounded by mudflats and reefs. About 23 hectares, which are private freehold, have been repeatedly burnt and grazed. Buildings include a shearing shed, hut and stockyards. There are navigation markers for port access to Lady Barron.

BREEDING SEABIRD SPECIES
Little Penguin
50 pairs were breeding in burrows amongst the shearwaters at the north-east corner of the island and towards the hut.

Short-tailed Shearwater
An estimated 624 158 pairs occur in high densities all over the island except for the central area stretching from east to west.

Cast iron pot, typical of those used for oil extraction from whales and seabirds.

Looking across to Big Dog Island.

Sooty Oystercatcher
One pair was located on the west side.

VEGETATION
The vegetation is dominated by tussock communities which include *Poa poiformis, Kunzea ambigua* with *Pteridium esculentum* and weed species such as *Coprosma repens* present.

COMMENTS
Short-tailed Shearwaters were commercially harvested up until 1957, with around 30,000 chicks taken annually. The Short-tailed Shearwaters are now harvested by recreational mutton birders. The island has been repeatedly burnt and grazed. The degraded state of the island can be directly attributed to the birding and farming practices. Feral cats eat both shearwaters and Little Penguins. The island's proximity to Flinders Island and to those who consider it their traditional right to harvest the birds make it unlikely that the current utilisation of the island will stop or improve, whether it is deemed legal or illegal to take Short-tailed Shearwaters or deliberately burn seabird colonies. Grazing by sheep and regular burning are incompatible with shearwater conservation.

Recently burnt shearwater nesting habitat.

OTHER SEABIRD SPECIES
Australian Pelican
Pied Oystercatcher
Pacific Gull

MAMMALS
Cat – now exterminated
House Mouse
Sheep

BIRDS
Native:
Cape Barren Goose
Red-capped Plover
Masked Lapwing
Welcome Swallow

Introduced:
Skylark
Common Blackbird

REPTILES
Metallic Skink
Tiger Snake

Spences Reefs

(Great Dog Island Group, page 270)

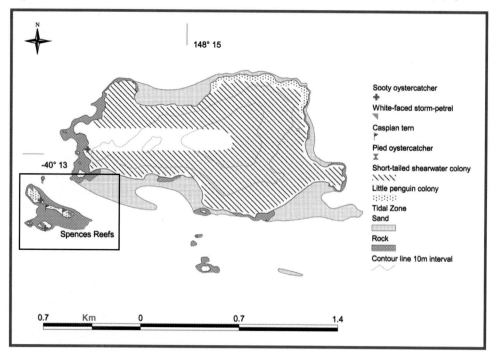

Location: 40°14′S, 148°14′E
Survey date: 7/12/86
Area: 0.65 hectares
Status: Non-allocated Crown Land ⇧

The reef comprises two islets, referred to as the west islet and the main islet. The main islet is the largest and consists of two sections, just separated at low tide. There are little sandy beaches at the eastern end and mudflats around the northern end at low tide only and a rock shore at high tide. A deep, strong-flowing channel is off the south shore. The western islet is very small and kidney-shaped.

BREEDING SEABIRD SPECIES
Little Penguin

An estimated 100 pairs breed on the reefs – 20 pairs on the west islet, 60 pairs on the eastern section of the main islet and 20 pairs on the western section. On both islets, the birds were found nesting in burrows, under the vegetation and in rock crevices all over.

White-faced Storm-Petrel

There are an estimated 2000 breeding pairs on both the islets, their burrows predominantly under the vegetation. On the west islet the birds burrow predominantly in *Senecio lautus* combined with scattered *Poa poiformis* and geranium species. There are some also under large *Coprosma* bushes. On the main islet burrows are located mainly in the areas dominated by *P. poiformis*, *Senecio* sp. and *Tetragonia implexicoma*. They also nest under *Coprosma* but are not found in any great numbers where *Rhagodia* sp. dominates.

Spences Reef.

Sooty Oystercatcher
2 pairs each with 2 eggs, were located on the main islet and one on the western islet. There were also 8 birds roosting.

Pied Oystercatcher
One pair, with 2 eggs, was located on the beach of the main islet.

Caspian Tern
One pair, with two very small chicks, was located on the main islet at the western end of the east section

VEGETATION
The vegetation on the main island is dominated by *Senecio* with *Poa poiformis* and *Tetragonia*. *Rhagodia candolleana* and *Coprosma* are extensive in the western section. On the western islet, *P. poiformis* and *Pelargonium* sp. are scattered throughout the dominant *Senecio* sp. vegetation. There are also large *Coprosma* bushes on the islet.

There were no birds, reptiles or other seabird species recorded.

MAMMALS
Introduced:
Rat

COMMENTS
The presence of dead storm-petrels, often under *Coprosma* bushes, and eggs that had been removed from their burrows indicate that rats have been active amongst the colony. On the western islet a pile of 14 storm-petrel carcasses was found under *Coprosma*. The rats should be eradicated from this important breeding site for storm-petrels and nearby islands to decrease the chances of reintroduction and predation on seabirds. The status of the reefs should be upgraded to nature reserve.

Samphire Island

(Great Dog Island Group, page 270)

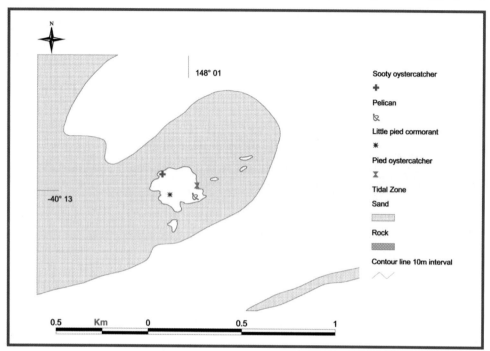

Location: 40°13'S, 148°13'E
Survey date: 9/12/86
Area: 3.29 hectares
Status: Conservation Area

It is a tiny island composed of shell grit surrounded by mudflats at low tide. The majority of the islet is flooded at high tide.

BREEDING SEABIRD SPECIES
Sooty Oystercatcher
4 pairs, with eggs, were nesting amongst *Sclerostegia arbuscula* on shell grit.

Pied Oystercatcher
4 pairs, at nests with eggs, were located in similar areas to the Sooty Oystercatcher nest sites.

Sclerostegia arbuscula dominates this island, which is almost awash at high tide.

VEGETATION

Sclerostegia arbuscula, which is inundated at high tide, is dominant around the coast. There is a patch of stipa on the north-west side.

OTHER SEABIRD SPECIES
Little Pied Cormorant

Australian Pelican

No mammals or reptiles were recorded on the island.

BIRDS
Latham's Snipe

Eastern Curlew

Sandpiper species

Ruddy Turnstone

Great knot

COMMENTS

The mudflats, which provide extensive feeding beds for various wader species, are the important habitat in this locality. Because it is composed of shell grit, surrounded by mudflats, Samphire Island is considered a unique and rare island type.

Fisher Island

(Great Dog Island Group, page 270)

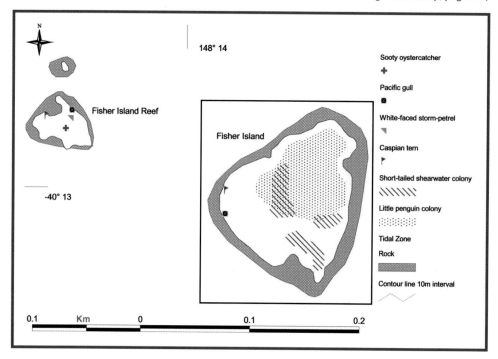

Location: 40°13'S, 148°14'E
Survey date: 9/3/76 and 9/12/86
Area: 0.9 hectares
Status: Conservation Area

It is a low granite islet rising to 5.8 metres south-west of the hut, that is situated on the mid western coast. Much of the island is bare rock interspersed with thin patches of friable soil up to a depth of 0.6 m. Shearwater burrows are found only in soil over 30 cm deep.

BREEDING SEABIRD SPECIES
Little Penguin

1 – 5 pairs were recorded in 1986 and 11 burrows in November 1999.

Nests are in *Disphyma crassifolium* amidst shearwater burrows near Potts Point.

Short-tailed Shearwater

Since 1974, the island's population has been monitored. 73 pairs were recorded in 1976, 100 – 140 pairs in 1986 and 88 pairs in November 1999. There are 3 distinct colonies, one to the north-east of the hut, and 2 in the south eastern area.

White-faced Storm-Petrel

Burrows were discovered west of north-east Boat Harbour for the first time in the 1974 – 1975 season. In the following season over 80 burrows were found. They may have been an overflow population from the colony on Spences Reef, 1.2 km to the south. None were found in the 1986 or subsequent surveys.

Pacific Gull
In 1986, one pair was located on the west coast of the island.

Caspian Tern
One pair was located on the west coast of the island.

VEGETATION
The dominant vegetation is *Poa poiformis*, the preferred habitat for the Short-tailed Shearwaters, with smaller expanses of *D. crassifolium* and *Tetragonia implexicoma*.

COMMENTS
The island's Short-tailed Shearwater population is regularly monitored by the Nature Conservation Branch of the Department of Primary Industry, Water and Environment. 52 Short-tailed Shearwater chicks were banded in March 2000.

OTHER SEABIRD SPECIES
Little Pied Cormorant

Black-faced Cormorant

Pied Oystercatcher

Sooty Oystercatcher

Silver Gull – nests on the island periodically

Crested Tern

Fairy Tern

MAMMALS
There are none resident, however occasionally water rats swim over from the adjoining mainland.

Rat – lived on the island between 1971 and 1974, when they were exterminated.

BIRDS
Cape Barren Goose

REPTILES
Southern Grass Skink – plentiful in the *Poa* sp. tussocks

White-lipped Whip Snake – vagrant, seen on very rare occasions

Copperhead Snake – vagrant, seen on very rare occasions

From Fisher Island looking north to Flinders Island.

Fisher Island Reef

(Great Dog Island Group, page 270)

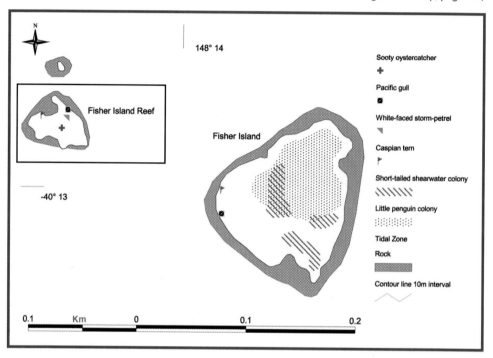

Location: 40°13'S, 148°14'E
Survey date: 9/12/86
Area: 0.10 hectares
Status: Conservation Area

A tiny granite island which is 95% covered by *Disphyma crassifolium*.

BREEDING SEABIRD SPECIES
White-faced Storm-Petrel

3 pairs were located nesting in burrows under the pigface.

Pacific Gull

1 pair was located nesting on the north-east side of the reef.

Caspian Tern

An empty nest scrape was found in *D. crassifolium* in the middle of the island. An egg had been sighted there a few days earlier.

Sooty Oystercatcher

2 pairs, one with one new chick and one with 2 eggs, were found nesting amongst *D. crassifolium*.

No mammals, reptiles or other birds were recorded.

COMMENTS

Its proximity to Lady Barron, a relatively busy port, makes the reef's surface-nesting species vulnerable to disturbance.

Vansittart Island Group

(Region 4, page 195)

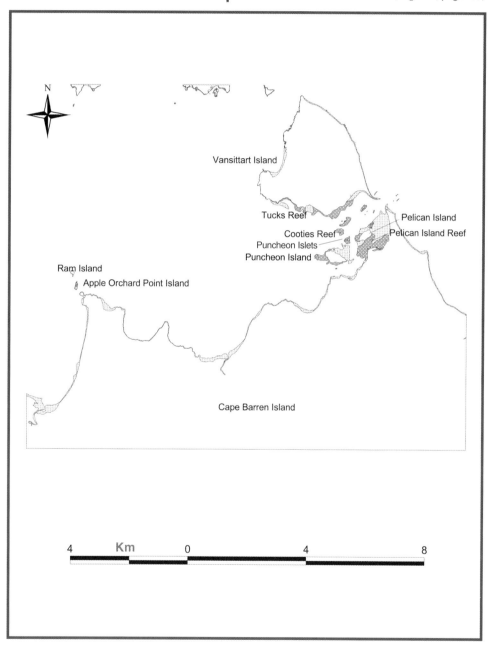

Ram island

(Vansittart Island Group, page 293)

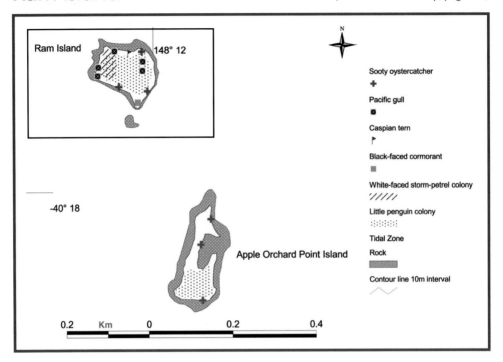

Location: 40°19'S, 148°12'E
Survey date: 2/12/86
Area: 1.01 hectares
Status: Private property

Ram Island is a small, gently rounded, oval island comprised of mudstone and dolerite. It measures approximately 60 metres from east to west and 90 metres from north to south.

BREEDING SEABIRD SPECIES
Little Penguin
30 pairs nest mostly under matted vegetation dominated by stipa, *Tetragonia* sp. and *Poa poiformis*. There are also a few burrows scattered around the island.

Pacific Gull
10 pairs, at nests with eggs, small chicks and runners, were located all over the island.

Sooty Oystercatcher
3 pairs, at nests with eggs, were located scattered around the shoreline.

White-faced Storm-Petrel
An estimated 560 pairs in average densities of $0.23/m^2$ breed on the island, predominantly on the western side.

Black-faced Cormorant
An old colony with guano, which may also be a roost site, was found on the southern coast.

View over Ram Island looking south to Cape Barren Island.

Caspian Tern
1 pair, with eggs, was located at the northern end of the island.

VEGETATION
The western side of the island is dominated by stipa with scattered *P. poiformis*, *Senecio* sp. and *Tetragonia*. On the eastern side there is a patch of *P. poiformis* in the *Senecio* sp. and *Tetragonia implexicoma*. The southern half of the eastern side is nearly all *Poa* sp. fading into *Senecio* at the northern end. There is a small patch of *Atriplex cinerea* in the south-east. *Disphyma crassifolium* grows around the fringe of the island.

BIRDS
Cape Barren Goose – 1 pair with 3 chicks.

No mammals, reptiles or other bird species were recorded.

COMMENTS
This is an important seabird island because of the diversity of species present and the presence of White-faced Storm-Petrels. The private owners should be encouraged to continue to protect the seabirds on the island.

Apple Orchard Point Island

(Vansittart Island Group, page 293)

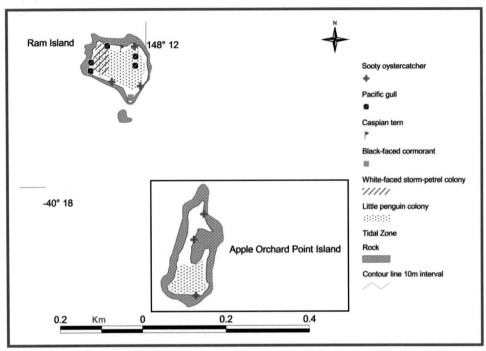

Location: 40°19'S, 148°12'E
Survey date: 3/12/86
Area: 0.85 hectares
Status: Non-allocated Crown Land

Apple Orchard Point Islet, which is just off Apple Orchard Point on the north coast of Cape Barren Island, is a very low rocky islet with a jagged shoreline and no soil. It is elongate in shape and oriented north to south. The islet is surrounded by mudflats with a small northern point, which is separated at high tide.

BREEDING SEABIRD SPECIES
Little Penguin

35 pairs were located nesting under dense vegetation in two patches with 3 pairs in the northern sub-islet and 30 pairs on the main islet.

Sooty Oystercatcher

3 pairs, on nests with eggs, were located in seagrass on the shoreline.

VEGETATION

The vegetation is dominated by the shoreline plants, *Sarcocornia* sp. and *Sclersostegia arbuscula* with *Tetragonia* sp, *Rhagodia candolleana*, stipa and *Poa poiformis* also present.

COMMENTS

This island's name is yet to be recognised by the Nomenclature Board.

OTHER SEABIRD SPECIES

Black-faced Cormorant – 40 birds were roosting.

No mammals were recorded.

BIRDS
Native:
White-fronted Chat

Little Grassbird

REPTILES
White-lipped Whip Snake

Metallic Skink

Puncheon Island

(Vansittart Island Group, page 293)

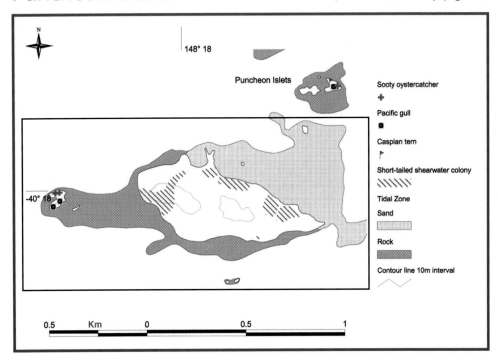

Location: 40°18', 148° 18'
Survey date: 6/12/86
Area: 17.56 hectares
Status: Private property

The main island is elongate with a broad western shore that tapers via a straight northern shoreline and indented southern shore to the pointed eastern end. It is surrounded by mudflats. There is a small islet off the western side which is connected at low tide by mud and a rocky spit. The island has been extensively burnt and grazed. Buildings include a house and numerous stone walls. The island is farmed.

BREEDING SEABIRD SPECIES

Short-tailed Shearwater

There are 7 colonies on the island with an estimated population of 49 613 pairs, all located in the valley-like depressions, where *Poa poiformis* dominates. The 4 northern colonies had been recently burnt. Two beach-washed birds were also found.

Pacific Gull

Two pairs, one with one egg and one with 2 eggs, were nesting on the shoreline of the western islet.

Puncheon Island.

Sooty Oystercatcher
Two nests were found and 4 pairs were defending on the pebbly shore of the western islet amongst dry seagrass.

VEGETATION
The main island is surrounded by mudflats at low tide and dominated by introduced grasses with *Poa* and stipa which provide the shearwater habitat. The western islet is covered in densely-matted stipa and *Tetragonia* sp. with *Sclerostegia arbuscula* dominant along the shoreline and *Sarcocornia* and *Disphyma crassifolium* also present.

Stone wall.

OTHER SEABIRD SPECIES
Southern Fulmar – beach-washed

Fairy Prion – 2 were beach-washed

Black-faced Cormorant – 30 roosting on the western islet.

No mammals were recorded.

BIRDS
Native:
Cape Barren Goose
Masked Lapwing
White-fronted Chat
Little Grassbird
Introduced:
Skylark

COMMENTS
The island is burnt, grazed and the vegetation is dominated by introduced pasture species. Unless the farming practices are altered, this island will remain unsuitable for seabird habitation.

Puncheon Islets

(Vansittart Island Group, page 293)

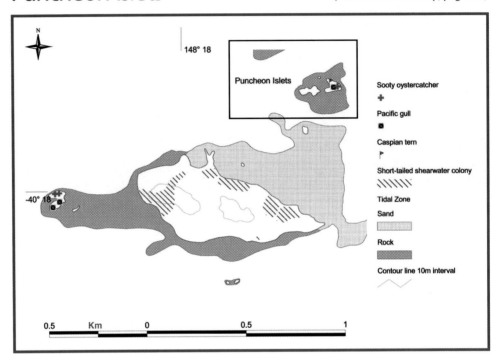

Location: 40°18'S, 148°18'E
Survey date: 6/12/86
Area: < 1 hectare
Status: Non-allocated Crown Land

This islet is to the east of the main island and is made up of two rocks joined at low tide. The islet is low, rounded and surrounded by mudflats with no burrowing habitat.

BREEDING SEABIRD SPECIES
Pacific Gull
1 pair was located on the western section.

Caspian Tern
6 pairs were located at the top of the islet on bare soil amongst *Carpobrotus rossii*.

Sooty Oystercatcher
3 pairs, with nest scrapes and no eggs, were nesting in *C. rossii* on the eastern side of the north-eastern rock and 3 pairs were on the southern rock

No other seabirds or mammals were recorded.

REPTILES
Metallic Skink
Blue-tongue Lizard

VEGETATION
The western rock is dominated by *Poa poiformis* and stipa. The eastern rock is dominated by myrtaceous scrub with *P. poiformis*.

COMMENTS
The status of these islets should be investigated as they are not connected to the freehold lease of Puncheon Island and do not appear on the Nomenclature list. They are not important seabird islets, but the vegetation and reptiles present are interesting with respect to the size and isolated nature of the islet.

Pelican Island

(Vansittart Island Group, page 293)

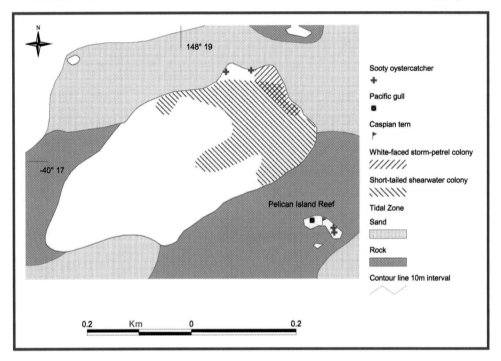

Location:	40°18'S, 148°19'E
Survey date:	6/12/86
Area:	6.67 hectares
Status:	Non-allocated Crown Land, Game Reserve

It is a rectangular shaped island with a low profile, orientated east to west. The island has sufficient soil in the eastern sector for burrows, but elsewhere the soil is shallow and rocky along a central east to west running ridge. The island was leased until 1984 for grazing.

BREEDING SEABIRD SPECIES
Short-tailed Shearwater
An estimated 32 833 pairs have colonised the eastern half of the island in very high densities in tall 100% *Poa* sp. The burrows are confined to a distinct habitat bordered by the central ridge.

White-faced Storm-Petrel
20 burrows were located alongside the Short-tailed Shearwaters at the north-east corner of the island.

Sooty Oystercatcher
Two pairs were defending on the eastern shore. 8 adults were seen.

VEGETATION
The vegetation is dominated by *Poa poiformis* and surrounded by a 5 – 10 m wide border of *Sclerostegia* sp. bushes scattered amongst *Sarcocornia* sp. and *Disphyma crassifolium*.

OTHER SEABIRD SPECIES

Pied Oystercatcher – 5 were feeding on the mudflats

MAMMALS

Red-necked (Bennett's) Wallaby

Possibly cats, as a midden of Short-tailed Shearwaters was found deep inside a clump of *Poa* sp. tussocks.

BIRDS

Black Swan – 6 pairs, one with 5 cygnets.

Cape Barren Goose

REPTILES

Skink – unidentified

COMMENTS

There may be a low incidence of harvesting of Short-tailed Shearwaters. There are introduced fauna, possibly including feral cats on the island. The island was heavily grazed in the past and this together with fire, is likely to have restricted the size of the bird colonies. The likely presence of cats on the island should be investigated and, if they are there, they should be eradicated as soon as possible.

Pelican Island Reef

(Vansittart Island Group, page 293)

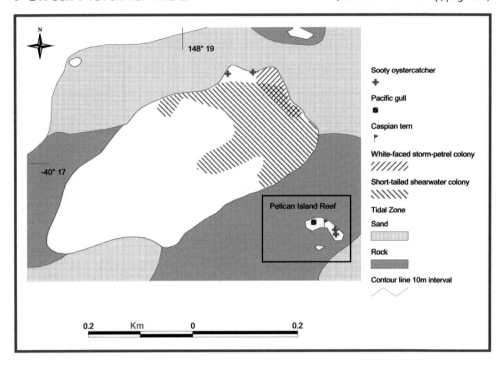

Location: 40°18'S, 148°19'E
Survey date: 6/12/86
Area: 0.13 hectares
Status: Non-allocated Crown Land

The reef consists of two tiny rocky islets offshore, which are joined to Pelican Island at low tide. These islets are vegetated with stipa and *Tetragonia* sp.

BREEDING SEABIRD SPECIES
Pacific Gull
One pair, with one egg and chick, was located on the eastern rock and one pair with one egg was located on the western rock.

Sooty Oystercatcher
Two pairs, with runners, were located on the western rock.

Caspian Tern
One pair, with a chick, was located on the western rock on a shell-grit patch with *Tetragonia* sp., 7 metres away from the Pacific Gull.

There were no mammals or other birds recorded on the reef.

REPTILES
Lizards – not identified

COMMENTS
This islet should be included in the Nomenclature Board List.

Tucks Reef

(Vansittart Island Group, page 293)

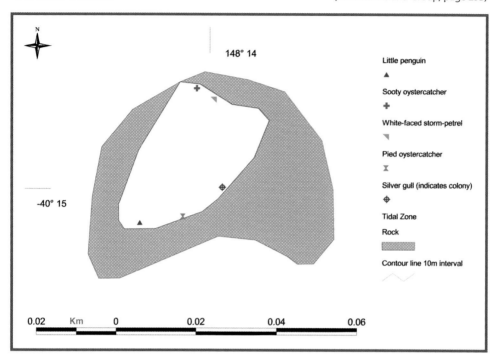

Location: 40°16'S, 148°15'E
Survey date: 8/12/86
Area: 1.14 hectares
Status: Non-allocated Crown Land

Tucks Reef is a small rocky islet dominated by *Poa* sp. and *Rhagodia* which provide good nesting habitat for seabirds.

BREEDING SEABIRD SPECIES

Little Penguin
20 pairs are scattered all over the reef in burrows, mostly under *Poa* sp.

White-faced Storm-Petrel
A small number of burrows, about 20, were found scattered throughout the island. Gulls have destroyed a lot of the nesting habitat, particularly the south side.

Silver Gull
100 pairs, some with eggs and chicks but most with runners and fledglings, were located all around the island. Many dead runners were recorded, possibly killed by heavy rain on 7/12/86.

Sooty Oystercatcher
1 pair with 2 eggs was located on the north side of the island in *Disphyma crassifolium*.

Pied Oystercatcher
1 pair, with 2 eggs, was located under a *Corposma* bush, very low down in the south-east corner.

VEGETATION
The main vegetation is *Poa poiformis* and *Rhagodia* with small patches of *Coprosma* and stipa at the southern end.

There were no mammals, reptiles or other birds recorded on the island.

COMMENTS

The Silver Gull population should be monitored to ensure that it doesn't burgeon as a result of nearby fish processing facilities providing a more abundant food source. The ecological balance of the island needs to be maintained to protect the White-faced Storm-Petrel colony.

Vansittart Island

(Vansittart Island Group, page 293)

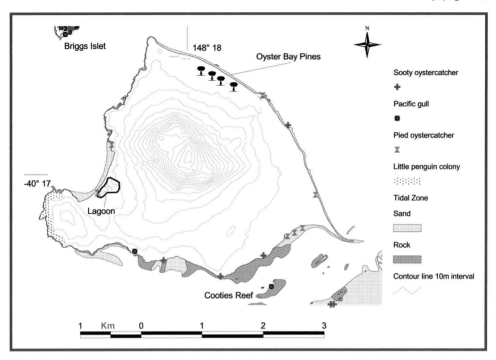

Location: 40°16'S, 148°17'E
Survey dates: 6 & 7/12/86
Area: 807.31 hectares
Status: Multiple Land Tenure: Private property, Lease

Commonly known as Gun Carriage Island, Vansittart is a triangular island dominated by a centrally-located hill with steep flanks and a pointed summit. The south-east and western sides of the island are dominated by beaches. The north western end has small coves and beaches. The eastern side is a long exposed beach with granite headlands ending at the south-east point which lies close to Cape Barren Island. There are lagoons on the island and mudflats abound on the southern side. The vegetation has been devastated by fires and clearing, which was carried out with bulldozers and chains. This clearing destroyed stands of Oyster Bay Pine which are succeeded in places by tea tree. The island is grazed by cattle. There are farm buildings and stockyards present on the island.

Head stones near the west side settlement.

BREEDING SEABIRD SPECIES

Little Penguin
An estimated 150 pairs were nesting amongst granite boulders and under rocks on the south-west side. 10 pairs were found at the end of the west beach near the cattle yards and homestead.

Pacific Gull
One old nest located on the south side of the island.

Sooty Oystercatcher
4 pairs at nests with eggs were found scattered on the beaches on the south and eastern sides.

Pied Oystercatcher
7 pairs with empty nests, and chicks and eggs, were located on sandy beaches scattered around the island.

OTHER SEABIRD SPECIES
Australian Pelican

MAMMALS
Native:
There are possibly echidnas, as indicated by the presence of diggings.

Red-necked (Bennett's) Wallaby – now extinct from the island

Tasmanian Pademelon – now extinct from the island

Introduced:
Cattle

BIRDS
Native:
Black Swan – nesting
Cape Barren Goose
Australian Shelduck
Grey Teal
Chestnut Teal
White-faced Heron
Great Egret
Red-capped Plover
Yellow-tailed Black-Cockatoo
Brown Thornbill
New Holland Honeyeater
White-fronted Chat
Scarlet Robin
Olive Whistler
Grey Fantail
Forest Raven
Beautiful Firetail
Silvereye
Dusky robin

Introduced:
House Sparrow
Skylark
European Goldfinch
Common Blackbird
Common Starling

REPTILES
Tiger Snake
Southern Grass Skink
Metallic Skink
Bougainvilles Skink

OTHERS
Frog – species unidentified

VEGETATION

Much of the original vegetation has been destroyed, including a lot of Oyster Bay pines. *Leptopermum* sp is regrowing on the eastern side, which has been frequently burnt in the past.

COMMENTS

The island is grazed by cattle and there has been regular land clearing, involving fire. Oystercatchers are in low numbers even though there are extensive feeding grounds close by. This is probably as a result of the cattle travelling and feeding along the beaches. The eastern side has been fenced for five years and regeneration of native vegetation is excellent. The area is now excluded from managment fires which were used to stimulate a green pick for grazing. The coastal vegetation is now regenerating well with high species diversity including Oyster Bay pines. (Karen Ziegler pers. comm.) The island is a refuge for Cape Barren geese but otherwise is unimportant as a seabird island. The presence of high numbers of other species of birds and vegetation such as Oyster Bay pines suggests that an attempt to save the last areas of natural scrub would be worth pursuing. The large tidal delta (11 x 8 km), formed due to the transport of sediment eastward through Franklin Sound by fast tidal currents, which encounter swells from Tasman Sea, are considered of outstanding significance for Tasmania (Dixon, 1996).

Cooties Reef

(Vansittart Island Group, page 293)

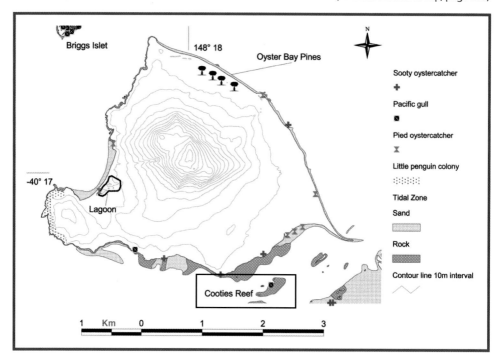

Location: 40°18'S, 148°19'E
Survey date: 6/12/86
Area: 0.15 hectares
Status: Crown Land

Cooties Reef is a long narrow islet in the channel between Vansittart and Cape Barren Islands, mostly surrounded by sand.

BREEDING SEABIRD SPECIES
Pacific Gull
15 pairs, 10 empty nests, 2 with 1 egg, 1 with 2 eggs, 2 with half-grown runners are largely confined to the *Rhagodia* sp. in the south.

VEGETATION
The dominant vegetation is *Tetragonia*, with *Rhagodia* and scattered stipa in the west. *Sarcocornia* dominates the shoreline. There is some *Bulbine* sp. on the south shore.

OTHER SEABIRD SPECIES
Black-faced Cormorant – 20+ were located at a regular roost site in vegetation on the south side.

Australian Pelican – 10

No mammals were recorded on the reef.

BIRDS
Ruddy Turnstone – 6

REPTILES
Metallic Skink

Babel Island Group

(Region 4, page 195)

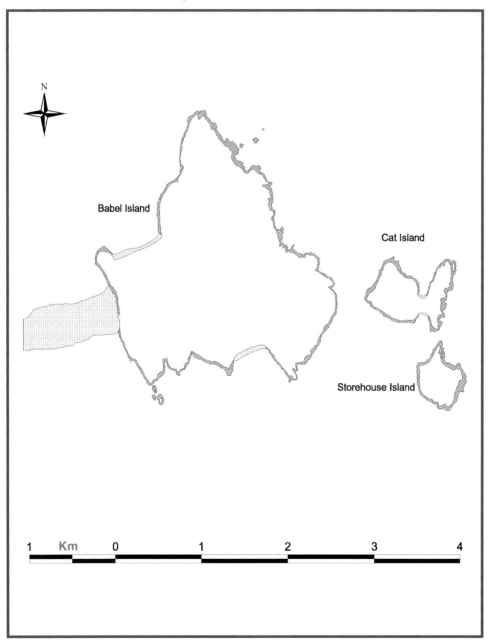

Babel Island

(Babel Island Group, page 311)

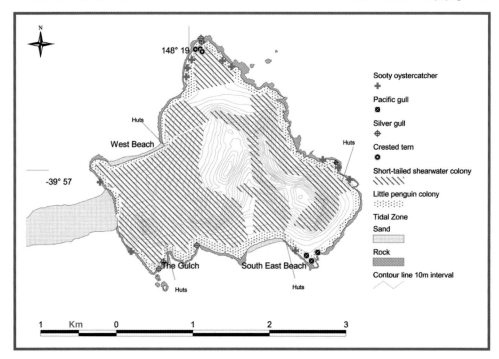

Location: 39°56'S, 148°19'E
Survey date: 4 – 5/12/86
Area: 440.98 hectares
Status: Private property

Babel Island is a relatively large granite island with sandy soil and a central hill which slopes gently to the shoreline. The island is pointed towards the north and broad at the south. There are two large beaches, West Beach and South East Beach, with a boulder-strewn granite rock shoreline elsewhere except for a sandy spit at the south-west side. There are several buildings along the east coast and at West Beach, the Gulch and South East Beach. An old monument stands at the northern tip.

BREEDING SEABIRD SPECIES

Little Penguin

19 733 pairs occur on the island, burrowing and nesting under boulders in *Rhagodia* and *Tetragonia*

With several million birds returning to burrows ashore, it is not unusual to encounter vast flocks nearby Babel Island.

North-east shoreline.

all around the island in a 10-20 metre wide strip except along the south-western sandy shoreline. One of the more concentrated areas occurs to the north of West Beach where densities reach $0.1/m^2$ in a 10-20 metre wide boulder strip with *Tetragonia* cover. The other is on the east-south-east corner of the island, which has an estimated 3000 pairs in a 10 metre coastal strip of *Rhagodia* with burrow densities estimated to be $0.06/m^2$. There are obvious penguin landing spots at regular intervals all along the coast.

Short-tailed Shearwater

With an estimated 2 825 287 pairs, Babel Island has the largest Short-tailed Shearwater breeding colony in Australia, harbouring 12.3% of the national population. The colony covers almost the entire island including the summit and under the myrtaceous scrub. The combination of sandy granitic soil and vegetation dominated by *Poa poiformis* provides an ideal nesting environment for the shearwaters, with the greatest densities of $0.88/m^2$ occurring in the north. Less extensive colonies also occur in *Rhagodia*, *Tetragonia* and scrub throughout the island.

Silver Gull

407 freshly-built nests occur amongst the granite boulders and stipa on the island's coast and a further 473 individual adult birds were counted. There are 4 major colonies, 2 located at the south-east point, one at the northern point and the other, where only 4 pairs had built nests on the eastern side of the island.

Of the nests, 206 had no eggs, 86 had one egg, 98 had 2 eggs and 17 had 3 eggs.

Pacific Gull
One nest with 3 eggs was located on the south-east point of the island on a rocky outcrop with *Rhagodia*.

Sooty Oystercatcher
12 pairs were located scattered around the shoreline.

Crested Tern
574 nests were counted, 12 of which had 2 eggs with the remainder having single eggs.

The colony occurs on the northern point of the island on bare granite with the nests in narrow soil leads with sparse *Carpobrotus rossii* on the edge of the island's vegetation.

VEGETATION
The vegetation is dominated by *Poa poiformis* with *Rhagodia candolleana*, *Tetragonia implexicoma* and scrub dominated by myrtaceous species elsewhere. Introduced boxthorn, *Lycium ferocissimum* is common, although attempts have been made to eradicate it.

COMMENTS
There is an extensive list of introduced mammals, birds and flora on the island, which could adversely affect the island's natural values, if not

OTHER SEABIRD SPECIES
Black-faced Cormorant – 60 were counted at a regular roost site on the reef at the south western tip.

Pied Oystercatcher – on the southern coast.

Crested Tern – many were roosting next to the north-east gull colony.

MAMMALS
Native:
Red-necked (Bennett's) Wallaby
Tasmanian Pademelon
Introduced:
House Mouse
Cat

BIRDS
Native:
Brown Quail
White-faced Heron
Brown Falcon
White-bellied Sea-Eagle
Peregrine Falcon – breeding on the east cliffs
Nankeen Kestrel
Masked Lapwing
Grey Fantail
Black Currawong
Forest Raven
Silvereye
Introduced:
House Sparrow
Common Blackbird
Common Starling
Red Junglefowl (Chicken)

REPTILES
Metallic Skink
Three-lined Skink
White's Skink
Blue-tongue Lizard
Tiger Snake

monitored and controlled. Harvesting of Short-tailed Shearwaters continues and the island still gets burnt.

Babel Island is a significant seabird breeding island containing the largest Short-tailed Shearwater colony in the world, a major Little Penguin colony, a large Crested Tern colony and a large population of Silver Gulls. In addition there is an extensive list of reptiles and birds native to the island. Efforts to rehabilitate the island should be continued.

Sheltered cove at the south end.

'All of us feel edgey about snakes. For as long as anyone can remember people have been getting bitten during birding. There isn't much you can do about it, but I worry about snakes because I've been bitten so many times. The first time I was bitten Mum lit three fires and they knew to come out to the island quick-smart because that means "snake-bite". The second time there was a plane neaby so everyone pulled the sheets off their beds and wrote "snake-bite" on them and took them to the rocks. The last time I got bitten ...everyone came down from the rookery while I was rushed to hospital. As soon as I was alright, I went back to the island. The others felt better when they saw me, but I had to force myself to go back to the rookery.' Vernon Graham, from Return to the Islands, 1983

Cat Island

(Babel Island Group, page 311)

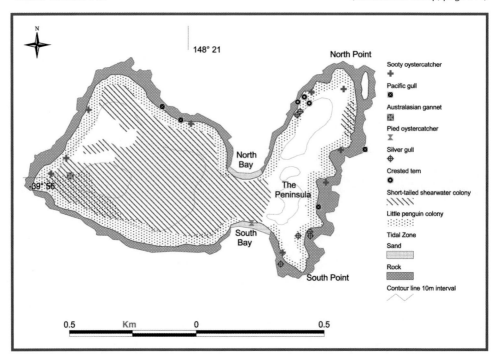

Location: 39°57'S, 148°21'E
Survey date: 4/12/86
Area: 38.82 hectares
Status: Conservation Area ⇧

Cat Island is a granite island, fish-like in shape with the eastern and western areas joined by a flat, narrow isthmus with sandy bays on the southern and northern sides. The coastline is predominantly granite slabs and boulders and the central area is flat with a ridge on the north-west side. Most of the coastline is rocky but not precipitous although there are steep cliffs on the eastern side. There are relatively large patches of bare hard ground on the north-western side, which are unsuitable for burrowing birds.

BREEDING SEABIRD SPECIES
Little Penguin

An estimated 2800 pairs occur all around the coastline under boulders and in burrows amongst *Poa poiformis*. They are also scattered over the island in burrows. The highest densities of $0.1/m^2$ occur on the south-west shore, where they nest amongst boulders and *Poa* to the top of the ridge. The largest colony is on the eastern side where the birds land via a smooth, sloping granite slab and nest mainly in or under rocks and boulders along the coastline in a 20 m strip.

Short-tailed Shearwater

An estimated 286 436 pairs occur in relatively dense colonies. Burrows are generally restricted to

In this region of Australia, crested terns seldom form larger colonies than have those on Cat Island.

areas of *P. poiformis* with an average density of 0.63/m² except in the south eastern section where the shearwaters burrow in lower densities amongst *Rhagodia candolleana* and *Atriplex cinerea*. The burrowing habitat is restricted by extensive areas of bare rock in the north western section, in part a legacy of the 1984 fire.

Pacific Gull

38 breeding pairs were counted, one on an empty nest, the others on nests with eggs and chicks. The main colony of 33 nests is at the south end of the island in association with *Tetragonia implexicoma*, *Senecio lautus* and scattered stipa. 2 and 3 pairs respectively were counted in two subsidiary colonies on the north side of the peninsula.

Silver Gull

One colony of 18 nests occurs on the south eastern point. 40 adults were counted and nests with eggs, chicks and runners were present. Old nests also occur midway along the peninsula. In the second colony, 60 adults were counted. Eggs and chicks were present but no nest count was made because of the potential disturbance. This colony is located in stipa amongst granite boulders next to the Crested Tern colony at the northern end of the peninsula.

Sooty Oystercatcher

10 pairs, some with empty nests and some with eggs, were counted, scattered around the island on patches of soil with sparse vegetation on granite slabs.

Pied Oystercatcher

One pair was located on a nest with eggs on a sandy beach on the flat area joining the two sections of the island.

Australasian Gannet

In the 1986 survey one bird on an egg was located amongst decoys deployed at the old gannet colony. This was perhaps the last of the large colony of between 5000 and 10 000 pairs that once existed here. More recent reports suggest that there are no gannets now breeding here, possibly as a result of White-bellied Sea-Eagle predation.

Crested Tern

A colony of 1011 pairs occurs on the north-western coast of the peninsula. An estimated 600 pairs were on eggs. A further 150 were estimated to be roosting in the same site.

VEGETATION

The vegetation is dominated by *P. poiformis* which covers most of the centre of the island, *Tetragonia implexicoma*, *Atriplex cinerea* and *Rhagodia candolleana*. On the northern end of the peninsula where the soil is thin and subject to salt spray, there are mats of *Disphyma crassifolium*.

Adult Gannets attending young chicks. Exploitation has caused the extinction of this colony.

OTHER SEABIRD SPECIES

Black-faced Cormorant – there is a large roost site on a granite slab at the north-eastern side of the peninsula..

MAMMALS

The largely nocturnal Water Rat has been recorded on the island but was not seen during this visit.

One dead Australian Fur Seal was found on a northern beach.

BIRDS

Native:

Brown Quail

Cape Barren Goose – 5 adults were seen

White-bellied Sea-Eagles that nest on Babel Island have been responsible in the past for preying on gannets.

Swamp Harrier

Brown Falcon

Nankeen Kestrel

Ruddy Turnstone

Hooded Plover – 1 nest

Black Currawong

Forest Raven

Silvereye

Richard's Pipit

Little Grassbird

Introduced:

Common Starling

House Sparrow

REPTILES

White's Skink

Tiger Snake

COMMENTS

Because of the seabird diversity and potential threats, Cat Island should be upgraded to nature reserve status. The island is accessible by boat with easy landings and is frequently visited because of its reputation as a safe anchorage. Short-tailed Shearwaters are occasionally poached. The Australian Maritime Safety Authority regularly visits the island via helicopter to maintain the navigation light. In 1984 an escaped fire caused extensive damage to the Short-tailed Shearwater and Australasian Gannet breeding areas. Nesting habitat has still not fully recovered. The depletion of the Cat Island gannet colony by humans, originally for use as rock lobster bait and subsequently by fire and disturbance over the last 70 years combined with probable sea-eagle and Tiger Snake predation has led to this colony being completely destroyed. There were an estimated 5000 to 10 000 birds when first described in 1908. It was probably the largest colony of this species of Australasian Gannet and one of only four colonies in Tasmanian waters and nine in Australian waters. Of these, only three contain significant numbers of breeding pairs and are increasing in size. The attempt from 1988-1998 by the then Department of Parks Wildlife and Heritage to encourage Australasian gannets to continue to breed on the island by deploying decoys was not successful because of disturbance by sea-eagles. Further efforts should be made to maintain the decoy system and deter the eagles. The vegetation in the vicinity of the gannet colony should be appropriately managed to retain suitable habitat as an incentive for the gannets to resume nesting, as birds still visit the locality.

Storehouse Island

(Babel Island Group, page 311)

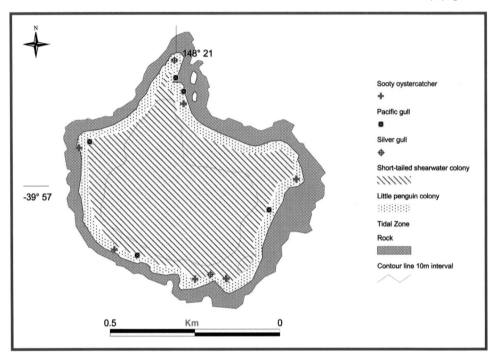

Location: 39°57'S, 148°22'E
Survey date: 4/12/86
Area: 19.8 hectares
Status: Conservation Area

Storehouse Island is a circular granite island with a pointed northern end which is mainly bare rock. There is a boulder beach on the northern shore, a small sandy beach on the north-west shore and granite slabs at the east side.

BREEDING SEABIRD SPECIES
Little Penguin

An estimated 11 210 pairs nest in densities of 0.43/m^2 under rocks along the eastern shore in a 5 metre strip and are scattered in lower densities around the island also in a 5 metre strip.

Short-tailed Shearwater

An estimated 317,071 pairs occur in densities averaging 1.04/m^2 over the entire island except for the rocky northern point.

Pacific Gull

11 pairs were found at nests with eggs and chicks. Solitary nests were located on the shoreline all around the island on bare granite in association with *Carpobrotus rossii*.

Silver Gull

3 pairs were located at freshly-built nests on both the northern and southern coasts.

Lookiing west past Babel Island towards Flinders Island.

Sooty Oystercatcher
7 pairs were located on nests and eggs scattered around the shoreline on bare granite in association with *C. rossii*.

VEGETATION
The island has a gently undulating landscape with extensive expanses of *Poa poiformis*. The largest patch of *Rhagodia*-dominated scrub occurs at the northern end. There is a strip of *Disphyma crassifolium*, *Tetragonia* sp. and *Rhagodia* sp. around the perimeter of the island.

COMMENTS
There may be a low incidence of harvesting of Short-tailed Shearwaters. The presence of extensive Short-tailed Shearwater and Little Penguin colonies justify the conservation area status.

OTHER SEABIRD SPECIES
Crested Tern – 23 roosting on rocks.

No mammals were found, but it is thought that water rats frequent the island.

BIRDS
Brown Quail

REPTILES
Metallic Skink

Tiger Snake

Sisters Island Group

(Region 4, page 195)

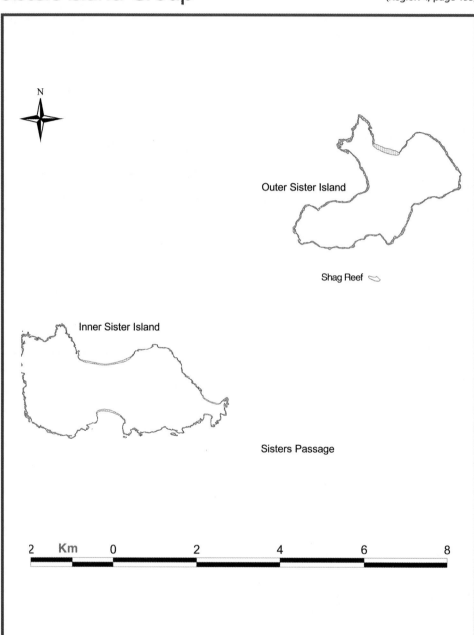

Inner Sister Island

(Sisters Island Group, page 322)

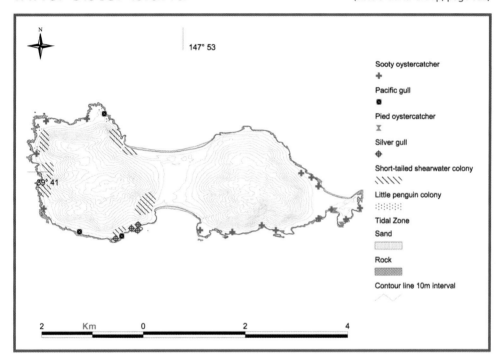

Location: 39°42'S, 147°54'E
Survey date: 16, 17, 19/12/86
Area: 747.74 hectares
Status: Conservation Area

It is an elongate island with an irregular coastline forming large bays on the north and south sides. The island is grazed. A house and associated farm infrastructure still exist on the island.

BREEDING SEABIRD SPECIES
Little Penguin
50 pairs were distributed between five small colonies on the north-west end of the island.

Short-tailed Shearwater
An estimated 44 367 pairs inhabit seven colonies, the largest being 2.9 hectares. They are scattered around the western side of the island in association with pasture, thistles and some native vegetation including *Carpobrotus rossii* and *Poa poiformis*.

Pacific Gull
Three solitary nests were located on the western side of the island.

Silver Gull
4 small colonies, containing a total of 32 pairs with eggs and chicks, were located close together on the south shore overlooking the southern bay.

Sooty Oystercatcher
22 pairs, with eggs and chicks, were located scattered around the shoreline, but concentrated on the southern shore.

Pied Oystercatcher
One pair was defending the south-east point.

VEGETATION
The vegetation is dominated by introduced grasses and weeds. Thistles cover much of the island. There are small patches of scrub remaining and *Myoporum insulare* is now regenerating in the tussock grassland. *Poa labillardieri* is the most common *Poa* species. Boxthorn is a problem around the house and shed.

COMMENTS
The Short-tailed Shearwaters are harvested. The island is grazed by sheep. There are extensive areas of introduced pasture and thistles. The high species diversity of seabirds, but low numbers for an island of this size suggests that the management procedures on the island are contributing little towards seabird conservation. Management practices such as fencing sheep away from the western side of the island where the burrowing penguins, Short-tailed Shearwaters, Silver Gulls and Pacific Gulls survive, should be considered. The faunal diversity on the island is extensive, with 2 species of marsupial (presumably native to the island), 4 species of reptile, 7 species of seabird (6 breeding) and at least 20 species of terrestrial bird recorded. It is the only known offshore island with bandicoots. The management procedures for the island must be reviewed and farming practices must be brought into line with conservation objectives. The two are not incompatible.

OTHER SEABIRD SPECIES
Black-faced Cormorant

MAMMALS
Native:
Eastern Barred Bandicoot
Tasmanian Pademelon
Introduced:
Sheep
Hare

BIRDS
Native:
Brown Quail
Cape Barren Goose
White-faced Heron
White-bellied Sea-Eagle
Wedge-tailed Eagle
Brown Falcon
Masked Lapwing
Fan-tailed cuckoo
Tasmanian Scrubwren
Brown Thornbill
Crescent Honeyeater
White-fronted Chat
Flame Robin
Grey Fantail
Forest Raven
Silvereye
Olive Whistler
Introduced:
House Sparrow
Common Starling

REPTILES
Three-lined Skink
White's Skink
Tiger Snake
White-lipped Whip Snake

Outer Sister Island

(Sisters Island Group, page 322)

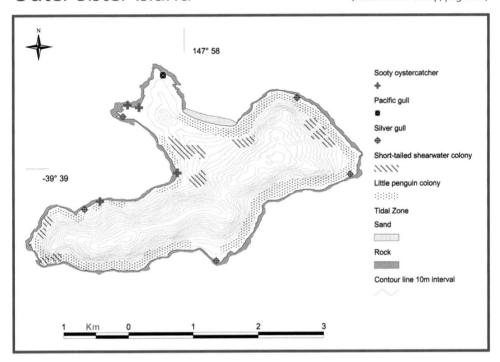

Location: 39°39'S, 147°59'E
Survey date: 16/12/86
Area: 545.07 hectares.
Status: Muttonbird Reserve, Lease

It is a granite island with an irregular coastline forming large bays on the north-west and north-east sides. There is a dolerite outcrop on the southern coast.

BREEDING SEABIRD SPECIES
Little Penguin

An estimated 2350 pairs breed all around the shoreline, preferring areas where there is adequate breeding habitat in conjunction with slabs of granite which provide them with easy access. The birds nest under dense vegetation and boulders covered by stipa They also use burrows, which are located in the same areas as those of the Short-tailed Shearwaters. The only shoreline without penguins is that on the northern peninsula.

Short-tailed Shearwater

There are 8 colonies in the northern half of the island with an estimated 20 566 pairs. Burrows were generally close to the shoreline. Densities were limited by the dry sandy soils which are prone to collapsing if there are too many burrows. Hence, the colonies are densest in the wetter gullies.

Pacific Gull

1 pair was located on the northern peninsula.

Silver Gull

25 pairs, with eggs and chicks, were distributed around the island.

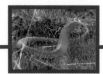

Rugged peaks and dense vegetation at the eastern end.

Sooty Oystercatcher
4 breeding pairs were located, scattered around the shoreline. There were also 4 individual adults together on the north-western beach.

VEGETATION
The vegetation is dominated by introduced grasses and weeds. Stipa dominates the coast.

COMMENTS
The Short-tailed Shearwaters are harvested. The island is grazed by sheep and has feral cats and mice on it. There are extensive areas of introduced pasture and thistles. Although cats and snakes, which prey on Little Penguins, inhabit this island, there are more penguins than on Inner Sister where there are fewer predators.

The faunal diversity on the island is extensive with one species of marsupial (presumably native to the island), 5 species of reptile (including the rare combination of all the snake species found in Tasmania), 6 species of seabird (5 breeding) and at least 12 species of terrestrial bird recorded. The recommendations for Inner sister Island apply here and, as the islands are similar with respect to faunal assemblage and locality and are farmed by the same lessee, the future of both should be considered jointly.

OTHER SEABIRD SPECIES
Black-faced Cormorant

MAMMALS
Native:
Tasmanian Pademelon
Introduced:
House Mouse
Cat
Sheep

BIRDS
Native:
Brown Quail
Cape Barren Goose
White-bellied Sea-Eagle – breeding.
Brown Falcon
Lewin's Rail
White-fronted Chat
Grey Fantail
Forest Raven
Silvereye
Masked Lapwing
Introduced:
House Sparrow
Common Starling

REPTILES
White's Skink
Metallic Skink
Tiger Snake
White-lipped Whip Snake
Copperhead Snake

Shag Reef

(Sisters Island Group, page 322)

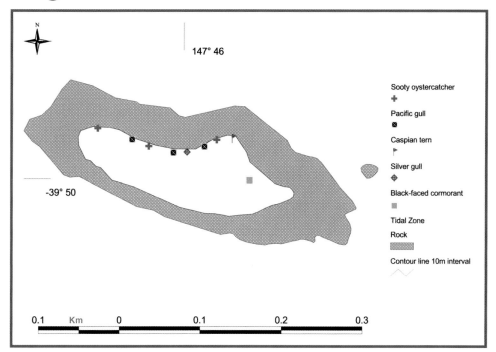

Location: 39°50'S, 147°46'E
Survey date: 16/12/86
Area: 1.24 hectares
Status: Non-allocated Crown Land ⇑

It is a low, boulder-strewn, granite, elongate reef, oriented east to west with very little vegetation. It is mostly wave-washed. An old wreck can be seen off the west side.

BREEDING SEABIRD SPECIES
Pacific Gull
3 pairs were sighted on the north coast.

Silver Gull
One pair was recorded and 15 were seen feeding.

Sooty Oystercatcher
3 pairs, with one small chick each, were located on the central east and south-west coasts.

Black-faced Cormorant
500–600 creched to fully-fledged chicks were counted. About 30 nests still contained small chicks, several just hatched. A very old colony, as indicated by the thick guano, also exists on the reef.

Caspian Tern
One pair, with an egg, were nesting in the shell grit on the north coast.

VEGETATION
Some succulent species occur in the rock crevices.

Black-faced Cormorant nesting.

OTHER SEABIRD SPECIES

Australian Pelican – 1

No mammals, reptiles or other birds were recorded.

COMMENTS

This is an island of moderate seabird significance due to its large Black-faced Cormorant colony. Its status should be upgraded to nature reserve to ensure its protection.

Sentinel Island Group

(Region 4, page 195)

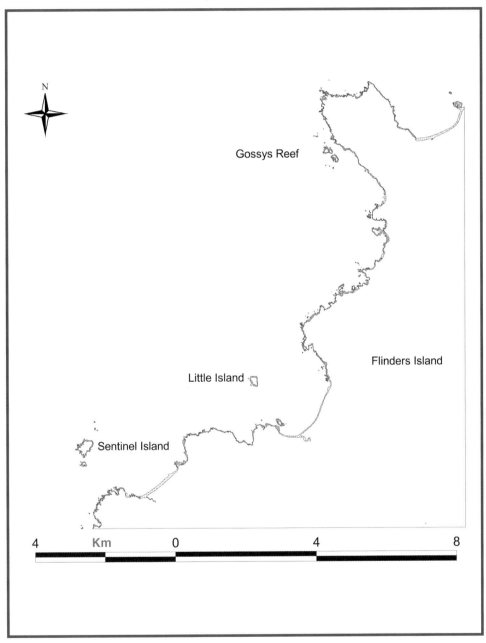

Sentinel Island

(Sentinel Island Group, page 329)

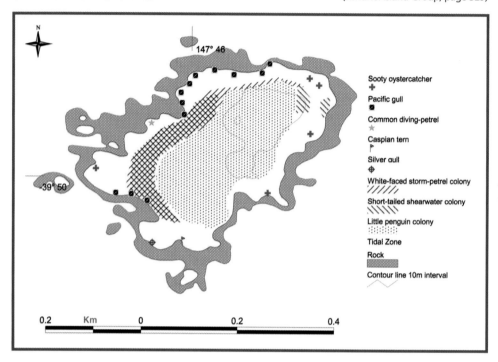

Location: 39°50'S, 144°46'E
Survey date: 15/12/86
Area: 10 hectares
Status: Non-allocated Crown Land ⇧

It is a granite island with a rounded south-west side, pointed southern side and straight eastern and northern sides resulting in a rounded cuboid appearance. The centre is dominated by a huge rock mass or sentinel. To the west there is a flat area and a gently sloping area to the east. The island was grazed up until 1985. There is a hut on the island.

BREEDING SEABIRD SPECIES
Little Penguin

An estimated 900 pairs are lightly scattered all over the island, in densities of between 0.01/m^2 to 0.2/m^2 mostly in burrows. The higher densities occur in vegetation dominated by *Poa* and *Rhagodia*.

Short-tailed Shearwater

1322 pairs inhabit three colonies located adjacent to the sentinel on the eastern and western sides. In addition, burrows are lightly scattered all over the island. Densities range from 0.04/m^2 on the eastern side in *Bulbine*, *Tetragonia* and *Rhagodia* to 0.2/m^2 in the main colony to the west of the sentinel in *Rhagodia*, *Poa* and *Tetragonia*.

Common Diving-Petrel

One adult was located in a burrow on the west side of the island and there were remains in Pacific Gull pellets. It is probable that there are more birds scattered around the island.

White-faced Storm-Petrel
An estimated 1261 pairs were located scattered all over the island, but concentrated on the western side. They are scarce in the main Short-tailed Shearwater colony.

Pacific Gull
12 pairs were located generally at empty nests but 2 were with eggs and 2 with chicks. They nest in solitary pairs on the north-west, south-west sides of the island.

Silver Gull
10 pairs were located in a colony on the southern shore.

Sooty Oystercatcher
5 pairs, only one with one egg, were concentrated on the eastern side of the island.

Caspian Tern
1 pair was located at a nest alongside the Silver Gull colony.

VEGETATION
The western side of the island is dominated by stipa and *Rhagodia* sp. and *Tetragonia, Bulbine* and *Pelargonium*. In sections of the west side there is a thick mat of *Atriplex cinerea*, stipa and *Rhagodia* sp.

MAMMALS
Dead sheep

BIRDS
Native:
Cape Barren Goose
Masked Lapwing
White-fronted Chat
Little Grassbird

REPTILES
Metallic Skink

COMMENTS
The proximity to Flinders Island, the traditional use of the island for sheep grazing and presence of Short-tailed Shearwaters all make the island vulnerable to poaching and vandalism. Harvesting of the Short-tailed Shearwaters and pre-harvest burning, a common practice, will affect the numbers and distribution of all the burrowing birds.

The diversity of seabirds breeding on the island justifies the status being upgraded to nature reserve.

Little Island

(Sentinel Island Group, page 329)

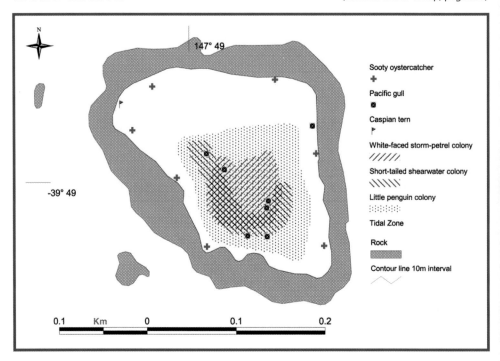

Location: 39°49'S, 147°49'E
Survey date: 15/12/86
Area: 2.94 hectares
Status: Non-allocated Crown Land ⇑

It is a very low, flat long islet of granite in Killiecrankie Bay.

BREEDING SEABIRD SPECIES

Little Penguin

An estimated 360 pairs in very low densities were located scattered around the island, interspersed with shearwaters and storm-petrels.

Short-tailed Shearwater

An estimated 1144 pairs were located scattered lightly around the island, predominantly in the north to south running gutters in *Poa*, *Tetragonia* and *Rhagodia*.

White-faced Storm-Petrel

An estimated 1530 pairs were located scattered around the island, interspersed with shearwaters and penguins.

Pacific Gull

7 pairs were counted around the shearwater colony.

Sooty Oystercatcher

7 pairs, one with a chick and one with 2 eggs, were counted around the coastline.

Caspian Tern

1 pair, with a quarter-grown chick, was found on the western side.

OTHER SEABIRD SPECIES

Black-faced Cormorant – 9 were sitting on the south eastern rocky shore.

No mammals or reptiles were recorded.

BIRDS
Native:
Cape Barren Goose

Introduced:
Skylark

VEGETATION
The vegetation is dominated by a combination of *Poa*, *Tetragonia* and *Rhagodia* with bare rock outcrops.

COMMENTS
This small island supports a high diversity of seabirds, which justifies the increasing of its status.

Gossys Reef

(Sentinel Island Group, page 329)

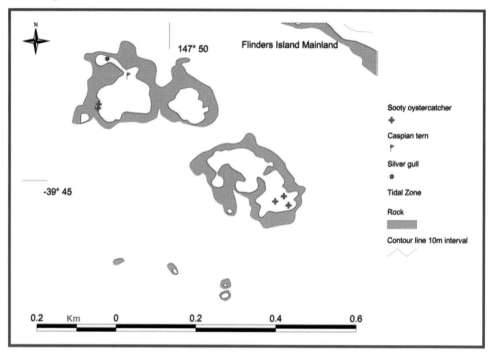

Location: 39°46'S, 147°51'E
Survey date: 15/12/86
Area: 1.52 hectares
Status: Non-allocated Crown Land

Gossys Reef is a very small granite islet composed of two main boulders referred to as north and south rock. There is no soil on the islet.

BREEDING SEABIRD SPECIES
Sooty Oystercatcher
5 pairs, with eggs and chicks, were located. 2 were on the north rock and 3 on the south.

Silver Gull
1 pair was defending its territory on the north rock.

Caspian Tern
1 pair was defending its territory on the north rock.

VEGETATION
There are small patches of *Disphyma crassifolium*, *Tetragonia implexicoma* and stipa in cracks on the north rock.

OTHER SEABIRD SPECIES
Black-faced Cormorant
Crested Tern

No mammals or other birds were recorded.

REPTILES
Metallic Skink

COMMENTS
With such a small diversity of species, the status of Gossys Reef is appropriate.

Pasco Island Group

(Region 4, page 195)

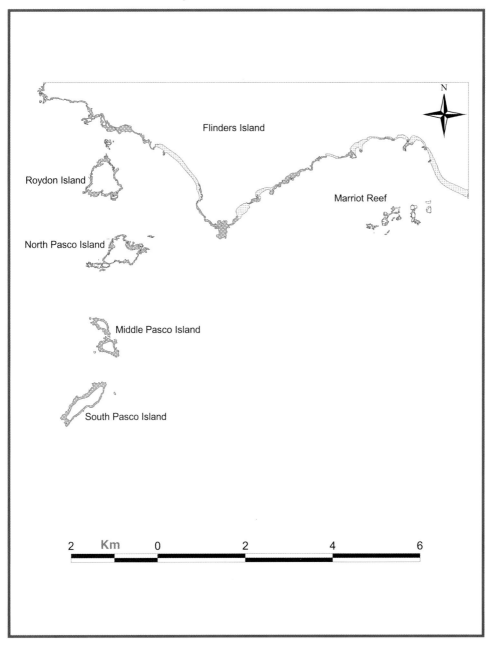

Roydon Island

(Pasco Island Group, page 335)

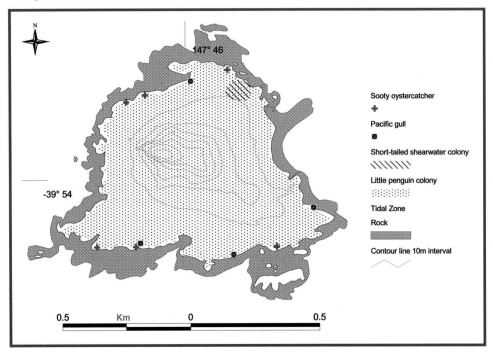

Location: 39°54'S, 147°46'E
Survey date: 14/12/86
Area: 36.68 hectares
Status: Non-allocated Crown Land, Lease

It is an elongate island with an irregularly shaped shoreline and orientated north to south. The island has a hill in the middle which has a steep cliff-like north-western side and a flat top. The base of the hill slopes gently towards the sea and the soil is sandy in these regions.

BREEDING SEABIRD SPECIES
Little Penguin
An estimated 2062 pairs breed on the island. There were signs of penguins coming ashore all around the island, especially on the north-west side, where there were well-formed runways through the dense vegetation to the slopes of the hill. The birds nest in crevices, under scrub and a few in burrows up to the top of the island. There were surprisingly few signs of burrows in relation to the well-formed pathways and other signs of activity around the island.

Short-tailed Shearwater
There are probably fewer than 10 pairs. Only a few dead birds were found.

Pacific Gull
4 pairs, at nests with chicks, were located scattered all around the island.

Summit of the island.

Sooty Oystercatcher
6 pairs, most with eggs, were located scattered around the island.

VEGETATION
The eastern side is dominated by boxthorn, stipa and *Atriplex cinerea*. The western side is dominated by boxthorn. There is scrub up the hillsides to the summit.

COMMENTS
The island is leased for conservation reasons.

OTHER SEABIRD SPECIES
Black-faced Cormorant
Crested Tern – 12 birds

No mammals were recorded.

BIRDS
Native:
Brown Quail
Cape Barren Goose
Brown Falcon
Little Grassbird
Silvereye

Introduced:
Skylark
House Sparrow
Common Blackbird
Common Starling

REPTILES
White's Skink
Metallic Skink

Middle Pasco Islands – north & south (Pasco Island Group, page 335)

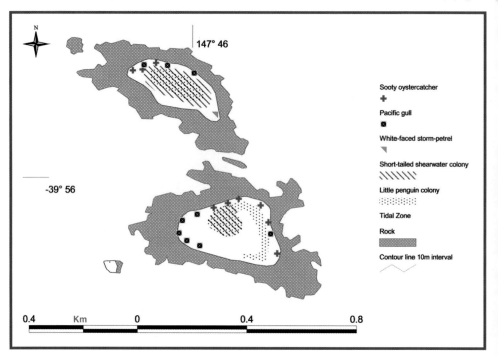

Location: 39°57'S, 147°46'E
Survey date: 14/12/86
Area: 8.37 hectares
Status: Non-allocated Crown Land, Lease

Middle Pasco comprises two islands of similar size. The southern island is a very low, round island with a depression in the centre containing several swamps. It is granite with bare outcrops and scattered boulders and a rocky shoreline, There are small beaches on the north-east and south-east ends. The northern island is undulating, low and rounded with all-granite rock shores and a lot of bare rock outcrops and boulders throughout. The southern end of the island has no suitable soil for burrowing.

BREEDING SEABIRD SPECIES
Little Penguin

An estimated 882 pairs were scattered all over the south island in burrows and under *Tetragonia* sp.-covered rocks in average densities of 0.02/m². Very few burrows were located on the north to north-east shore because of unsuitable habitat. There were also very few burrows in the saltbush habitat which is concentrated at the western end. An estimated 437 pairs were located scattered all over the northern island in the same densities as the southern island but with very few burrows at the northern and southern extremites of vegetation. Most were in burrows but some were also under *Tetragonia*-covered rocks.

Short-tailed Shearwater

An estimated 9702 pairs were located on the southern island in average densities of 0.2/m². They have a similar distribution to the Little Penguins with most birds in burrows, and a few nesting on the surface under dense patches of saltbush. They prefer *Rhagodia, Senecio* and *Tetragonia*-dominated habitat. On the northern island 1620 burrows were sparsely scattered in densities of 0.1/m² in vegetation dominated by *Atriplex, Senecio* and *Poa*. Burrow density was the lowest in the *Atriplex*.

White-faced Storm-Petrel

Only one burrow was located at the edge of the saltbush on the northern island just at the southern edge of the burrowing habitat.

Pacific Gull

6 pairs, each with empty nests, were located all around the southern island and 3 pairs, 2 with runners, were located on the north-eastern coast of the northern island.

Sooty Oystercatcher

6 pairs, one with one egg, were located on the northern end of the southern island and 3 pairs, 2 with eggs were located around the coast of the northern island.

VEGETATION

The dominant vegetation on the northern island is *Atriplex cinerea*, which is co-dominant with *Senecio* sp. in places. Elsewhere there are patches of *Senecio* sp. co-dominant with *Poa poiformis*. The southern island's vegetation is dominated by patches of *A. cinerea*, mallow bush and a combination of *Senecio* and *Poa* with scattered stipa

There were no mammals or other seabird species recorded.

BIRDS

Native:
Cape Barren Goose – on both islands
Forest Raven – on the northern island
Little Grassbird – on the northern island

REPTILES

None were recorded but the islands provide excellent habitat.

COMMENTS

The fate of the single storm-petrel burrow on the northern island should be monitored to see if a new colony is forming.

North Pasco Island

(Pasco Island Group, page 335)

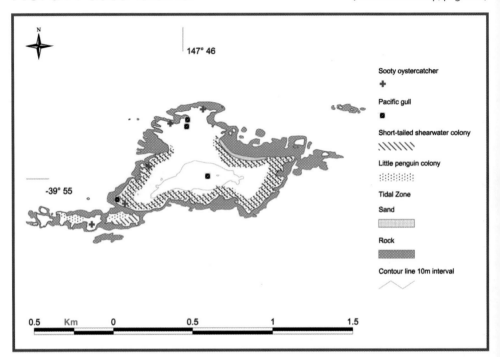

Location: 39°55′S, 147°47′E
Survey date: 14/12/86
Area: 27.8 hectares
Status: Non-allocated Crown Land, Lease ⇑

It is a granite island, roughly cuboid in shape with an irregularly shaped shoreline. The island is low and flat with a slightly raised central area and is mainly composed of hard rock – granite and limestone under vegetation. There is sandy soft soil behind the beach area on the north-west shore. There are two rocky islets off the south-west point. Grazing has ceased on the island and the scrub appears to be recovering. However, the severely compacted soil caused by grazing, is still evident. The northern patch of stipa had been burnt during the previous year.

BREEDING SEABIRD SPECIES
Little Penguin

An estimated 2514 pairs were found in a 30 m wide strip around the island nesting under vegetation and in burrows. This strip of birds stretched from the north-east area around the southern shore to the north-west area. The latter had the highest densities of penguins, where they were breeding in soft soil behind the dune area along the main beach. 22 pairs were located under rocks on the two rocky islets to the south-west.

Short-tailed Shearwater

An estimated 675 pairs were located scattered around the island in the same areas as the Little Penguins but outnumbered by them. There was

also a patch of birds in dense *Poa poiformis* at the south-west end of the island. 50 pairs were located under dense vegetation on the larger rocky islet.

Pacific Gull
18 pairs, at nests with egg and chicks, were found scattered all around the island with a colony of 13 pairs in the centre of the island on the high ground. 3 pairs were on the rocky islets

Sooty Oystercatcher
4 pairs, with eggs, were located scattered around the island. One was located on the southern end of the main south-west rocky islet.

VEGETATION
The vegetation is dominated by scattered *P. poiformis* and *Correa* sp. There is a patch of thick *P. poiformis* on the south-west side and stipa at the northern end. The larger rocky islet off the south-west side of the island is dominated by *Rhagodia candolleana* with scattered stipa

COMMENTS
There is severe compaction of soil by sheep and erosion, which limits the distribution of burrowing birds. If the status of the part of the island, that is currently non-allocated crown land was upgraded to conservation area, thereby giving it an opportunity to recover, the numbers of Little Penguins and Short-tailed Shearwaters would most likely increase. White-faced Storm-Petrels may also colonise the island. There are large numbers of terrestrial birds and two reptiles surviving on the island, which would also benefit from increased status.

OTHER SEABIRD SPECIES
Black-faced Cormorant

Australian Pelican

Crested Tern – 39 birds were roosting on the south-east point

No mammals were recorded.

BIRDS
Native:
Brown Quail

Cape Barren Goose

Brown Falcon

Swamp Harrier

Lewin's Rail

Whimbrel

Hooded Plover

Masked Lapwing

White-fronted Chat

Little Grassbird

Silvereye

Introduced:
Skylark

Common Starling

REPTILES
White's Skink

White-lipped Whip Snake

South Pasco Island

(Pasco Island Group, page 335)

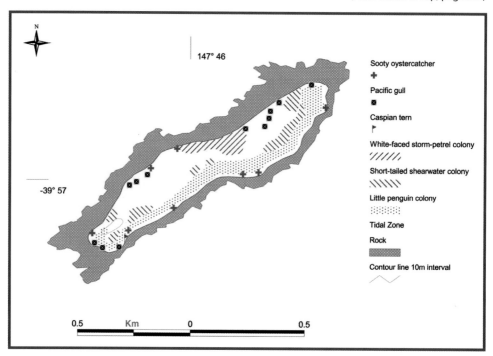

Location: 39°57'S, 147°46'E
Survey date: 14/12/86
Area: 20.66 hectares
Status: Non-allocated Crown Land, Lease ⇑

It is a flat, elongate island with a pointed southern end. 60% is devoid of or has little soil.

BREEDING SEABIRD SPECIES
Little Penguin

An estimated 5249 pairs were found concentrated in a 10 m wide strip around the island nesting under driftwood and boulders which were covered by *Tetragonia* sp. Burrows are also present in association with Short-tailed Shearwaters all over the island where the soil depth is adequate. Densities range from $0.05/m^2$ in the coast strip to $0.02/m^2$ in north-eastern shearwater colony.

Short-tailed Shearwater

There were six major colonies with densities ranging from $0.03/m^2$ to $0.2/m^2$, the largest and densest colony being in the north-east. An estimated 4842 pairs inhabit the island. The colonies at the west end of the island are in *Rhagodia* sp. and *Senecio* sp. At the east end they are dominated by *Atriplex cinerea* and *Poa poiformis*. The central area is devoid of birds on the southern side and there are only a few scattered birds on the northern side.

White-faced Storm-Petrel
Up to 1000 pairs inhabit the island, concentrated at the west end in low *P. poiformis* and at the east end in scattered *P. poiformis* with other grasses present. Both these areas have shallow soils and as a result the birds are in low densities.

Pacific Gull
17 pairs, with eggs and chicks, were located scattered all around the island, but mainly concentrated on the northern side.

Sooty Oystercatcher
8 pairs, 3 with eggs, were located scattered around the island.

Caspian Tern
1 pair, with a large chick, was found at its nest on the south-eastern side.

VEGETATION
The vegetation is dominated by *P. poiformis* and other grasses with patches of *A. cinerea*. On the east end there is a small patch of remnant scrub.

COMMENTS
The compaction of the soil by sheep limits the distribution of burrowing birds. If the status of the part of the island, that is currently non-allocated crown land was raised and stock grazing prohibited, the numbers of breeding seabirds, particularly White-faced Storm-Petrels, would most likely increase.

OTHER SEABIRD SPECIES
Black-faced Cormorant – roost site on the eastern end of the island, used irregularly

MAMMALS
Sheep

BIRDS
Native:
Brown Quail
Brown Falcon
Little Grassbird
Introduced:
Common Starling
Skylark

REPTILES
White's Skink

Marriot Reef

(Pasco Island Group, page 335)

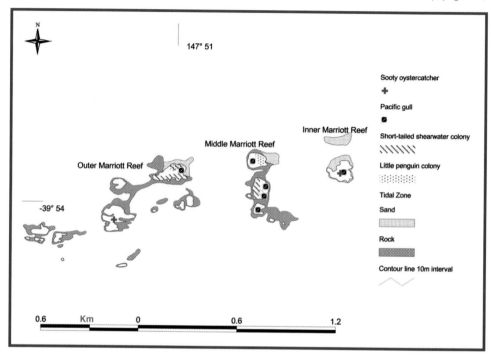

Location: 39°57'S, 147°46'E
Survey date: 14/12/86
Area: 3.4 hectares
Status: Non-allocated Crown Land

Marriot Reef is a small islet group with three main granite islets, which are vegetated and support breeding seabirds. There are five other rocks in the group, which have scant vegetation and only one pair of Sooty Oystercatchers on one of them.

BREEDING SEABIRD SPECIES
Little Penguin
15 pairs were located nesting under stipa on the north-west shore of the outer reef.

Short-tailed Shearwater
An estimated 8174 pairs were concentrated in a patch of *Poa poiformis* on the east side of the outer and middle islets.

Pacific Gull
13 pairs were located scattered all over the outer island and one pair with chicks on the middle islet.

Sooty Oystercatcher
2 pairs were located on the coast of the outer islet.

VEGETATION
The vegetation of the outer islet is dominated by *P. poiformis* and *Tetragonia* sp. The dominant vegetation of the inner reef is *P. poiformis* with halophytic shrubs on the northern side.

Marriot Reef.

COMMENTS
The group supports a low diversity and abundance of seabirds.

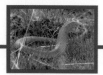

Prime Seal Island Group

(Region 4, page 195)

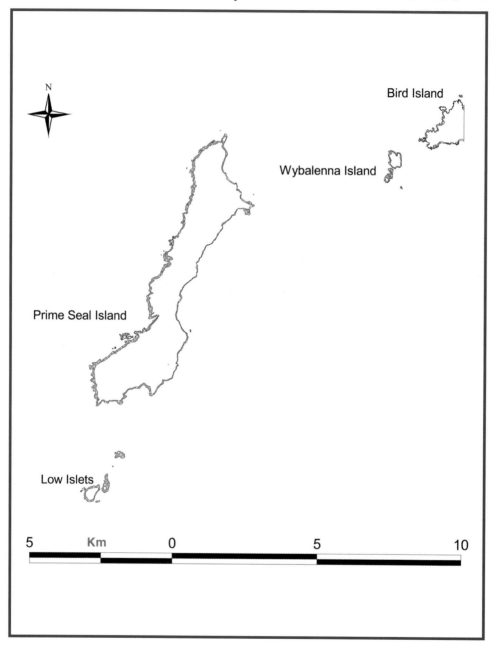

Prime Seal Island

(Prime Seal Island Group, page 346)

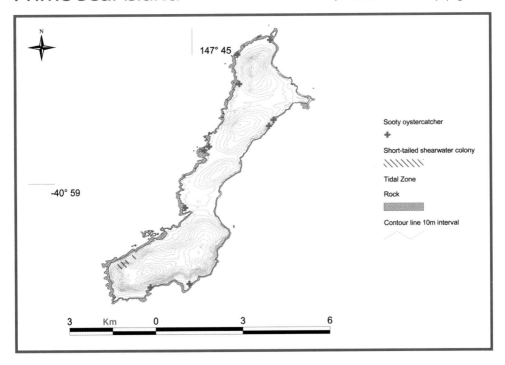

Location: 40°05'S, 147°45'E
Survey date: 11/12/86
Area: 1220.758 hectares
Status: Non-allocated Crown Land, Lease

It is a large, elongate island with a high central ridge. The island is extensively grazed by cattle, sheep and pademelons. It is dominated by limestone overlying granite with soils deep enough for extensive colony formation by burrowing seabirds. There are, however, no Little Penguins and relatively few Short-tailed Shearwaters on the island. There is a dolerite outcrop on the northern coast and a large sandblow on the southern coast. There are significant historical features including a rock overhang cave site, Mannalargenna Cave, which is believed to have been occupied 8,000 years ago.

BREEDING SEABIRD SPECIES

Short-tailed Shearwater

There are an estimated 2850 pairs in one colony at the south-west tip of the island located on a steep slope which comprises bare soil, extensive *Tetragonia* and barley grass and thistle in places. The colony has been divided by a large eroded area where there is only the occasional burrow left. The colony continues at the northern edge in a thin strip up the hill. This is also badly eroded and the density of birds is very low.

Sooty Oystercatcher
9 pairs were located around the coast.

VEGETATION
Tetragonia dominates the south-west tip, however the rest of the island is dominated by introduced grasses, predominantly barley grass, and thistle. On the south-east there is thick tea tree scrub and some remnant *Allocasuarina stricta* adjoining the pasture at the northern end of the island. There is an extensive *Myoporum insulare* shrubland on the island, one of the most extensive in Tasmania and the only known occurrence of *Eucalyptus ovata* on an outer Furneaux Island.

OTHER SEABIRD SPECIES
Little Penguin – beach-washed
Black-browed Albatross – beach-washed
Bullers Albatross – beach-washed
Pied Oystercatcher
Silver Gull
Crested Tern
Australian Pelican

MAMMALS
Native:
Tasmanian Pademelon – extremely common
Introduced:
Cat – common
House Mouse
Sheep
Cattle

BIRDS
Native:
Cape Barren Goose
White-bellied Sea-Eagle – a nest was located on the steep eastern slope amongst thick *Melaleuca* scrub.
White-faced Heron
Brown Falcon
Ruddy Turnstone
Hooded Plover
Banded lapwing
Masked Lapwing
Brush bronzewing
Fan-tailed Cuckoo
Tasmanian Scrubwren
Brown Thornbill
Eastern Spinebill
White-fronted Chat
Grey Fantail
Black Currawong
Forest Raven
Beautiful Firetail – lots of nests in tea tree
Little Grassbird
Silvereye
Introduced:
Indian Peafowl (Peacock)
Skylark
European Goldfinch
Common Blackbird – 1 male seen
Common Starling

REPTILES
Tiger Snake
Metallic Skink
Three-lined Skink

COMMENTS

Habitat destruction and erosion are prolific, as the result of overgrazing and burning.

The island's karst features, particularly the caves, are considered of outstanding significance at the State level (Dixon, 1996). The lack of Little Penguins is puzzling. The erosion affecting the Short-tailed Shearwater colony should be investigated to instigate reclamation or at least monitoring. The karst system should be protected.

Sand blow erosion on south side.

Northern end.

James Backhouse on the presence of wallabies and seals on Prime Seal island

'The wallaby abounds here. Several were killed by the natives who accompanied us. Some of these people only eat the male animals, others only the females. we were unable to learn the reason of this, but they strictly adhere to the practise. It was (Prime Seal Is.) formerly the resort of vast herds of Fur Seals, but they have nearly forsaken both it and many of the neighbouring ones, in consequence of the slaughter committed among them by the sealers.' in Fowler 1980.

Low Islets

(Prime Seal Island Group, page 346)

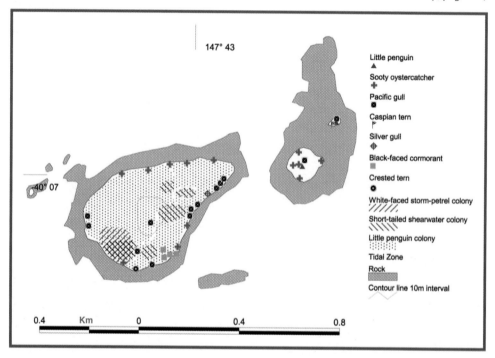

Location: 40°08'S, 147°43'E
Survey date: 11/12/86
Area: 0.86 hectares (North) 1.28 hectares (South)
Status: Non-allocated Crown Land ⇑

Low Islets are two low, flat granite islets lying south of Prime Seal Island. There are many rocky outcrops and areas devoid of soil.

BREEDING SEABIRD SPECIES
Little Penguin

Burrows were found all over the main island in low densities (0.03/m^2) and under *Tetragonia*-covered driftwood, boat wreckage and rocks on the smaller island. An estimated 63 burows were found on the larger island and 28 on the smaller island.

Short-tailed Shearwater

There are four small colonies in scattered *Poa*, *Senecio* and *Bulbine* on the larger island with an average density of 0.5/m^2. An estimated 900 pairs are spread between the colonies. Elsewhere there are few burrows due to lack of suitable soil conditions.

White-faced Storm-Petrel

There are an estimated 13 pairs located within the shearwater colony on the south-western end of the main island under low, matted *Poa*. Several burrows had been partially dug out, possibly by shearwaters, with eggs abandoned within.

View over Low Islets to Prime Seal Island.

Sooty Oystercatcher
7 pairs, two with eggs, were located nesting in shell grit around the main island's coast and 5 were on the smaller islet.

Pacific Gull
24 nests were counted scattered around the coastlines of the islets. One colony of 15 pairs was nesting in the interior of the larger island on bare granite surrounded by *Poa*.

Silver Gull
15 pairs were scattered to the east of the Crested Tern colony on the main island.

Caspian Tern
One pair, with 2 new chicks, was located on the smaller islet in the centre of a *Tetragonia* patch.

Crested Tern
25 pairs, most with eggs and some with chicks, were located on the southern end of the main island. 150 individuals were roosting by the water away from the colony.

Black-faced Cormorant
4 nests were located at a roost site with 200 individuals on the south-eastern coast of the main island.

OTHER SEABIRD SPECIES

Australian Pelican – 3 were located on east side shore near the Black-faced Cormorant roost site.

MAMMALS
Cattle
Sheep
Horse

BIRDS
Native:
Brown Quail
Cape Barren Goose
Ruddy Turnstone
White-fronted Chat
Red-capped Plover
Introduced:
Skylark
Common Starling

REPTILES
White-lipped Snake
Metallic Skink

Shipwreckage, Low Islet.

VEGETATION

The vegetation is dominated by *Poa poiformis*, *Atriplex cinerea*, *Tetragonia* sp. and *Rhagodia* sp.

COMMENTS

The presence of cattle, sheep and horses grazing on the main islet probably affects the breeding potential of the seabirds. There is confusion between this islet group and the Low Islets next to Clarke Island. The latter is a nature reserve. It appears as if the Land Use Map of Tasmania has listed both groups in the official list of reserves. Consequently it is unclear as to the status of the northern group of islets. This northern groups should be considered as one islet group and its status raised to that of a nature reserve to ensure full protection of the high diversity of seabirds.

Wybalenna Island

(Prime Seal Island Group, page 346)

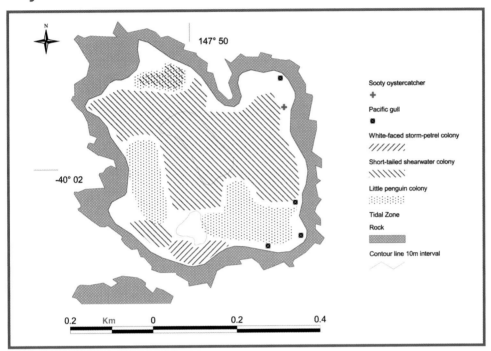

Location: 40°02'S, 147°50'E
Survey date: 12/12/86
Area: Total: 16.12 hectares
Status: Conservation Area

Wybalenna Island comprises four round granite islands: the main islet, two smaller ones of similar size and a rock to the north. The main islet is low and gently undulating with small beaches in the south, north and north-west with the rest of the shoreline being composed of granite rock.

BREEDING SEABIRD SPECIES
Little Penguin

The main island harbours 440 pairs which nest with shearwaters in *Tetragonia* and *Rhagodia*. There are thirty pairs on the middle islet and 50 on the southern islet.

Short-tailed Shearwater

An estimated 766 pairs burrow under *Tetragonia* and *Rhagodia* over the main island in densities of up to $0.25/m^2$ and 50 pairs burrow under *Poa* on the middle islet.

White-faced Storm-Petrel

A colony of up to 50 pairs was located at the south-west end of the main island amongst stipa in granite cracks.

Pacific Gull

16 pairs were on the coast of the southern islet, 4 on the main islet and one on the northern islet.

Granite rock shoreline.

Silver Gull
A colony of 36 nests was located on the granite slope near *Poa* and *Carpobrotus rossii* on the middle islet and one pair was located on the southern islet. Most nests were empty.

Sooty Oystercatcher
One pair was located on each islet.

Black-faced Cormorant
A large long-term colony of 19 nests, thick with guano build-up, was located on the eastern side of the southern islet.

VEGETATION
The middle islet is dominated by *Poa* and *Tetragonia* with a quarter of the vegetated area having enough soil for burrowing.

The southern islet is dominated by bare granite slab in the centre with vegetation elsewhere comprising chiefly *Senecio* sp., stipa, *Tetragonia* and *Rhagodia*.

The main island is dominated by *Poa* with patches of scrub and small patches dominated by stipa, *Tetragonia* and *Rhagodia*.

OTHER SEABIRD SPECIES
Caspian Tern – flew over main island
White-fronted Tern – 2 pairs were sitting on the shore of the middle islet.

BIRDS
Native:
Brown Quail – nest with 11 eggs under *Poa*
Cape Barren Goose
White-fronted Chat
Introduced:
Skylark
Common Starling

REPTILES
Metallic Skink – on main island

COMMENTS
It is interesting to note that Pacific Gulls tend to favour nesting on small islets adjacent to larger ones, a phenomenon also applicable to North Mount Chappell Islet, Little Badger and Little Goose Islands. The island's shearwaters are subjected to high rates of exploitation.

Bird Island

(Prime Seal Island Group, page 346)

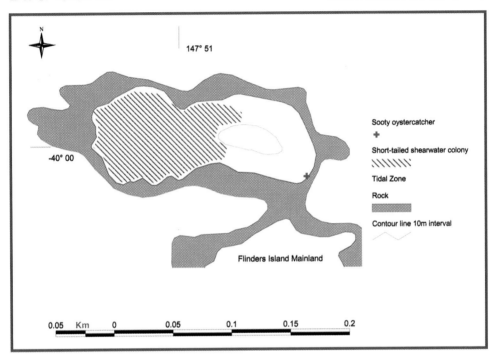

Location: 40°00'S, 147°52'E
Survey date: 12/12/86
Area: 1.51 hectares
Status: Non-allocated Crown Land ⇑

Bird Island is an elongate, granite island orientated north to south with sloping sides from a low central ridge with a spectacular karst system. It is joined to Flinders Island at low tide.

BREEDING SEABIRD SPECIES
Short-tailed Shearwater
3550 pairs burrow all over the island, excluding the eastern side, where the soil is insufficient. *Atriplex cinerea* is the dominant vegetation in these areas.

Sooty Oystercatcher
1 pair was located at the south-east corner in a crevice in the granite.

No other fauna was found on the island.

VEGETATION
There is *Atriplex cinerea* all over the island

COMMENTS
The proximity of the island to Flinders Island and the ease of access at low tide increase the Short-tailed Shearwaters' vulnerability to poaching and the island's susceptibility to vandalism. The karst system, which is of representative and outstanding significance for Tasmania (Dixon, 1996), is susceptible to trampling and should be protected.

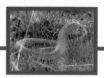

Dense cover of *Atriplex cinerea*.

Big Green Island Group

(Region 4, page 195)

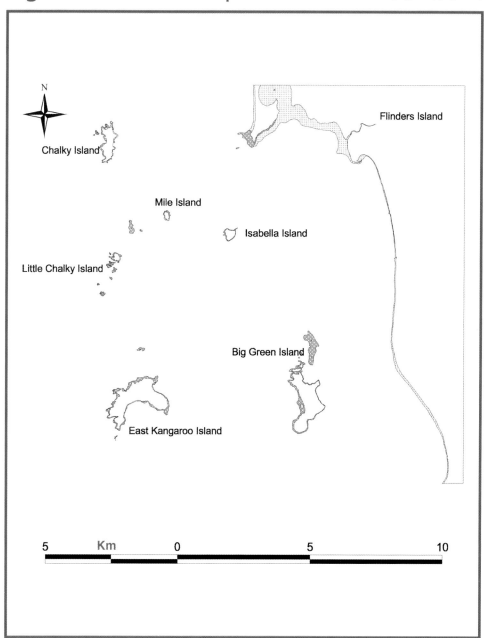

East Kangaroo Island

(Big Green Island Group, page 357)

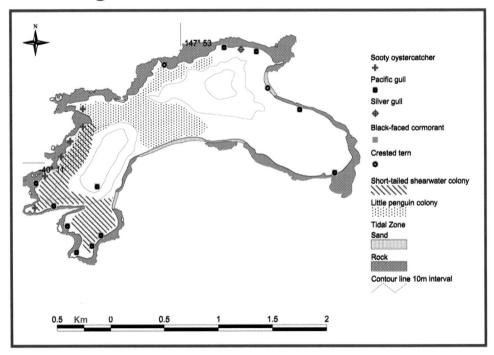

Location: 40°11'S, 147°54'E
Survey date: 9 & 10/12/86
Area: 156.90 hectares
Status: Nature Reserve

It is a crescent-shaped island with hillocks at both the northern and south-eastern ends. There is a large bay on the southern side and a deep gulch cutting into the south-east end. The island is comprised of gently undulating limestone with granite outcrops and dolerite dykes. There is a hut on the island. Sheep have been grazed here for decades but, due to poor grazing conditions, have all been recently shot. There are large erosion blowouts in the dunes, which are taking over much of the seabird breeding habitat, particularly on the western side.

BREEDING SEABIRD SPECIES
Little Penguin

Penguins are scattered all over the island, the highest densities occurring on the western shoreline in a 30 m wide strip of stipa and *Tetragonia* sp., where they are nesting in burrows. However, elsewhere on the island, they nest on the surface under dense stipa

Short-tailed Shearwater

The main colony stretches from the southern side over the neck to the western side, down to the southern point and towards the north in a 30 to 150 m wide strip. On the southern side, the birds are burrowing in pasture habitat with medium to very high densities of thistle. On the western side they are utilising similar habitat to the penguins.

Pacific Gull
13 pairs at nests, with eggs and chicks, were counted. A communal colony of 5 pairs was located in the centre of the island in pasture habitat with scattered stipa 8 pairs of solitary nests were scattered around the shoreline with most pairs at the southern tip.

Silver Gull
253 nests, most empty, others with eggs and chicks, were counted. The colony extends 100 metres along the granite outcrops on the north shore with scattered stipa

It is located alongside the Crested Tern colony with some nests actually positioned in the middle of the Crested Tern colony. Many nearly-fledged runners were present.

Sooty Oystercatcher
6 pairs were located on the western shore.

Crested Tern
3222 pairs were counted in a colony on the north shore. The colony was in two separate parts - one on the west side with 2879 pairs and one on the east with 343 pairs. 5% of the pairs had 2 eggs and there were 60 newly hatched chicks. Most of the nests were on open granite in cracks amongst sparse *Bulbine* and *Poa*.

VEGETATION
The vegetation is dominated by 70% introduced pasture which is in poor condition and has been invaded by thistles. The remaining natural vegetation is dominated by stipa and *Tetragonia* sp.

Introduced thistles.

OTHER SEABIRD SPECIES
Black-faced Cormorant
Australian pelican – on north-east shore
Pied Oystercatcher

MAMMALS
House Mouse

BIRDS
Native:
Cape Barren Goose – a significant breeding site
White-faced Heron
Ruddy Turnstone

Hooded Plover
Masked Lapwing
White-fronted Chat
Little Grassbird

Introduced:
House Sparrow
Common Blackbird
Common Starling – 1 nest in a crevice 10 cm off the ground

REPTILES
Metallic Skink
White's Skink

COMMENTS

The island has been overgrazed, which has resulted in severe erosion, with loss of vegetation and proliferation of thistles. Part of the island was reserved and has been managed to protect the Cape Barren Goose, based on the perceived benefit of grazed paddocks to the geese. A combination of poor management practices and the proliferation of geese has led to the severe degradation of the island's flora and fauna. The lease has been suspended for five years and will then be reviewed. The shearwater colony areas have been sown with rye and clover and the central north has been fenced and revegetated.

Introduced South African boxthorn with Cape Barren Goose.

Big Green Island

(Big Green Island Group, page 357)

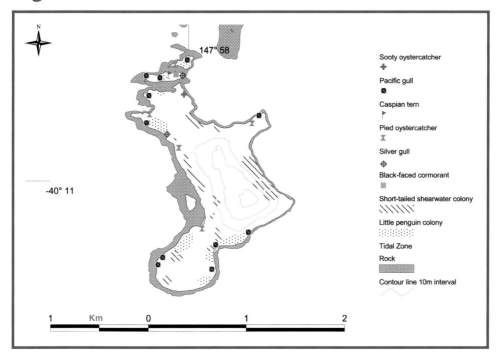

Location: 40°11'S, 147°58'E
Survey date: 7/12/86
Area: 122 hectares.
Status: Nature Reserve, Lease

An elongate island with an irregular coastline forming 5 large bays. There are 5 islets off the northern tip. The main island is mostly granite with two outcrops of dolerite, 100-200 metres wide, on the eastern side. There are also extensive limestone outcrops from the island's interior extending to the south-east. The island has been regularly burnt and grazed. There are two houses and a hut to support the farming practices. There is also an airstrip. The smaller islands are comprised of granite with shingle and grit beaches.

BREEDING SEABIRD SPECIES
Little Penguin

A 50 to 100 m wide strip of degraded penguin habitat in stipa was located on the north-western side. Birds were sheltering under the stipa Several dead penguins were found at the south-western colony. Five small colonies harbouring less than 300 pairs of birds, were located on the north-west and south-east coasts of the main island. The two larger islands to the north harbour 50 pairs each.

Short-tailed Shearwater

An estimated 22 051 pairs inhabit 15 distinct colonies. Densities range from 0.2/m^2 in areas where stipa has been regularly burnt to 0.45/m^2 in the east coast colonies, which occur on gently sloping

pasture over limestone. Burrows within the eastern colonies are deep in light soil. The northern colonies occur in stipa, shell grit and sandy soil.

Pacific Gull
11 pairs were located at nests with eggs and chicks on four of the islets. Solitary pairs were scattered around the main island.

Silver Gull
1 pair, with two eggs, was located on the islet closest to the main island.

Sooty Oystercatcher
31 nests, most with eggs, were counted on all the islets and on the shoreline of the main island.

Pied Oystercatcher
5 pairs, most with eggs, were counted on three of the islets and on the main island.

Sheep hoofprint through roof of shearwater burrow.

Black-faced Cormorant
27 nests were located on the second most northerly islet on bare rock on the north-west tip.

West shore south to Mt Chappell Island.

Extreme low tide foraging habitat for waders.

Caspian Tern

3 pairs were located – two on the islet closest to the mainland, one with 2 chicks, and one pair on the second most northerly islet.

VEGETATION

The vegetation on the main island is dominated by introduced grasses and weeds. Thistles and African boxthorn, *Lycium ferrocissimum*, have invaded most of the western side of the island. stipa has been regularly burnt, resulting in the loss of penguin habitat. The two larger islets to the north are dominated by stipa and boxthorn. The smaller islets are granite with shingle and grit beaches dominated by *Bulbine* sp. with some stipa, *Tetragonia* and *Sclerostegia* sp. on the most northerly.

COMMENTS

The island has been extensively and adversely affected by grazing, introduced vegetation and fire. The fact that there are eight seabird species breeding on the tiny northern islets compared to only five on the main island would appear to

Hut on the central east side.

indicate that disturbance and degradation on the main island are having an adverse effect on seabird breeding patterns. Twelve sites of historical significance were identified by an archaeological survey in 1991 (Sim and Stuart 1991). They should be properly documented and protected. There has been a rat control program put in place by the lessee.

OTHER SEABIRD SPECIES

Australian Pelican – 5 were sitting on the second most northerly islet.

Fairy Tern – 16 pairs were defending on the northern shingle spit of the second most northerly islet.

MAMMALS

Rat – very common

BIRDS

Native:

Cape Barren Goose – a significant breeding site

Eastern Curlew

White-fronted Chat

Forest Raven

REPTILES

Metallic Skink

Bougainvilles Skink

Little Chalky Island

(Big Green Island Group, page 357)

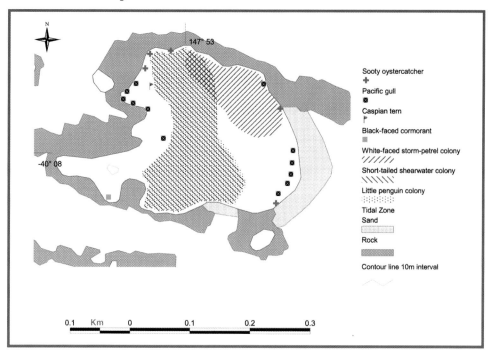

Location: 40°08'S, 147°53'E
Survey date: 11/12/86
Area: 5.15 hectares
Status: Non-allocated Crown Land ⇑

It is a small, flat, granite island with granite knolls.

BREEDING SEABIRD SPECIES

Little Penguin

An estimated 1985 penguins were nesting in burrows and under dense vegetation all over the island. They usually burrow in association with the storm-petrels and Short-tailed Shearwaters.

Short-tailed Shearwater

An estimated 2055 pairs were located burrowing mainly in the central northern part of the island where there is sufficient soil and the vegetation is dominated by *Poa poiformis*.

White-faced Storm-Petrel

An estimated 5151 pairs were found nesting all over the island with the highest densities recorded in areas of matted *Tetragonia* and *Rhagodia candolleana*.

Pacific Gull

22 pairs, some with empty nests and some with eggs and/or chicks, were concentrated on the south-east and north-west coasts. 62 adult Pacific Gulls were counted.

Sooty Oystercatcher

5 pairs, only one with eggs, were found around the island's shoreline.

Black-faced Cormorant

248 nests were located on the south-west shore on bare granite rocks and boulders. There were over 100 adults roosting and 100-200 fledged juveniles. There were two more old sites, adjacent to each other on the north-east shore, which had been used the previous season. There were 154 nests in one colony and 58 nests in the other.

Caspian Tern

One pair, with 2 eggs, was located on the high granite point in the centre of the northern section amongst sparse *Carpobrotus rossii*.

VEGETATION

The vegetation is dominated by *Poa poiformis*, *Tetragonia*, *Rhagodia candolleana* and *Atriplex cinerea*. The relative proportions of these species vary. Mallow bush is dominant in the southern third of the island.

Black-faced Cormorants amongst guano-coated boulders.

OTHER SEABIRD SPECIES

Pied Oystercatcher – single birds only, not breeding

No mammals were recorded.

BIRDS

Native:
Cape Barren Goose – breeding
Ruddy Turnstone – 8
Red-capped Plover
Little Grassbird
Silvereye

Introduced:
Common Starling

REPTILES
Metallic Skink

COMMENTS

This is an important seabird island in terms of numbers and species diversity. The status of the island should be raised to nature reserve and access limited during the breeding season.

Chalky Island

(Big Green Island Group, page 357)

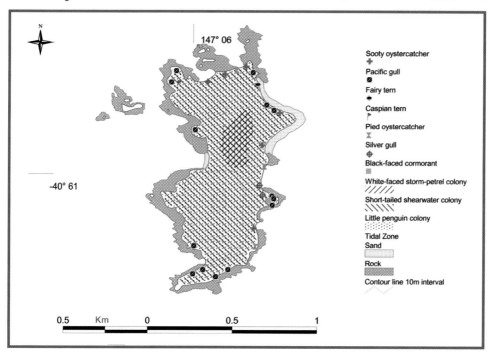

Location: 40°05'S, 147°53'E
Survey date: 11/12/86
Area: 41.31 hectares
Status: Conservation Area

Chalky Island is a granite island with dolerite dykes and limestone outcrops in places. It is an elongate, narrow island, gently undulating and sloping towards the east and west. The shoreline is composed of granite boulders with sandy beaches on both sides of the island and a sandy spit and mudflats at the northern end.

BREEDING SEABIRD SPECIES
Little Penguin

An estimated 21 218 pairs occur in burrows and under vegetation all over the island. They burrow in association with a few pairs of storm-petrels and high numbers of Short-tailed Shearwaters. The highest concentration was found in stipa along the shoreline of the island on the western and eastern sides. There were few birds in the northern area where shrubs dominate the vegetation. On the western side the birds were nesting under stipa and *Atriplex cinerea*. On the eastern side they were nesting under thick *Tetragonia implexicoma* and under rocks and boulders. There were only a few birds utilising the southern extremity of the island.

Short-tailed Shearwater

An estimated 215 524 pairs are confined mainly to the southern section of the island and in a 20 m strip up the eastern side as far as the start of the sandy beach.

White-faced Storm-Petrel
A colony of an estimated 37 767 pairs was located in the central portion of the island where *Poa poiformis* and *Rhagodia* sp. are dominant. The occasional Little Penguin and shearwater was also found nesting in this area.

Pacific Gull
12 pairs at nests with eggs and chicks, were counted. Some were concentrated on the southern tip of the island and others scattered over the rest of the island.

Silver Gull
2 pairs, defending nest sites with large chicks, were located alongside the cormorant colony.

Sooty Oystercatcher
6 pairs were located, scattered on the northern and eastern shorelines.

Pied Oystercatcher
3 pairs, with eggs and chicks, were located scattered on the northern and eastern shoreline.

Black-faced Cormorant
60 pairs, with fledged chicks, were located in a colony on the eastern shore on bare granite boulders. There may be more than 60 pairs, as it appeared as if some old nests had been washed away by high seas. As is the case for other islands, the cormorants move within and between islands utilising fresh habitat for nest building.

Caspian Tern
1 pair was sighted at the north-western end of the island.

Fairy Tern
89 pairs, with eggs and chicks, were counted on a sandy beach on the north-eastern side of the island.

VEGETATION
The dominant vegetation varies over the island. The southern third of the island is dominated by *Rhagodia candolleana* with *T. implexicoma* and stipa less common. The middle of the island is dominated by *P. poiformis* and *Rhagodia candolleana* whilst the northern area is dominated by halophytic shrubs growing on a sandy substrate, which may have been an old dune system.

No mammals or other seabird species were found on the island

BIRDS
Native:
Brown Quail
Cape Barren Goose
Brown Falcon
Little Grassbird
Silvereye

Introduced:
Common Blackbird
Common Starling

REPTILES
Metallic Skink
White-lipped Whip Snake
White's Skink
Tiger Snake

COMMENTS
Fairy Terns are extremely vulnerable to disturbance and are listed as rare under the Tasmanian *Threatened Species Protection Act 1995*, which should give them maximum protection. This is an important seabird island in terms of numbers and species diversity and should be upgraded to nature reserve status. Access to the island should be limited from August to April to ensure the undisturbed breeding and fledging of birds.

Mile Island

(Big Green Island Group, page 357)

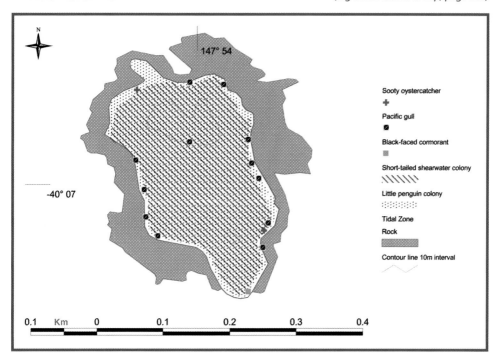

Location: 40°07'S, 147°55'E
Survey date: 10/12/86
Area: 4.05 hectares.
Status: Conservation Area ↟

It is a low, elongate island rising to a gently rounded centre.

BREEDING SEABIRD SPECIES
Little Penguin
An estimated 609 pairs were found all over the island in low numbers nesting in association with Short-tailed Shearwaters.

Short-tailed Shearwater
An estimated 21 874 pairs were located all over the island with the lowest densities where the Rhagodia is thickest.

Pacific Gull
21 pairs at nests with eggs, chicks and runners were located scattered around the perimeter of the island and in a central colony.

Sooty Oystercatcher
3 pairs were located scattered around the perimeter of the island.

Black-faced Cormorant
8 pairs nesting on bare granite were located at the southern end of the eastern shoreline.

VEGETATION
The vegetation in the southern section is dominated by Atriplex cinerea and Tetragonia implexicoma whilst the rest is dominated by Poa poiformis with areas in the north of bare granite.

Vegetation dominated by *Atriplex* looking to the north.

There were no mammals or other seabird species recorded.

BIRDS
Native:
Brown Quail
Cape Barren Goose
Hooded Plover
Silvereye
Introduced:
Common Starling

REPTILES
Metallic Skink

COMMENTS
The degree of protection afforded to the island by its conservation area status should be evaluated. Given that Mile Island has a greater diversity of seabirds than its neighbour, Isabella Island, which has nature reserve status, the former should also be conferred nature reserve status to ensure that it is fully protected.

Isabella Island

(Big Green Island Group, page 357)

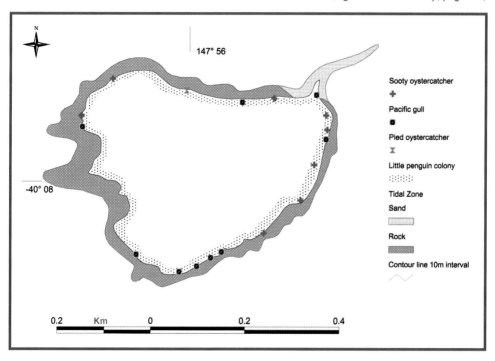

Location: 40°07'S, 147°56'E
Survey date: 10/12/86
Area: 11.4 hectares
Status: Nature Reserve

A low, flat, circular island. The shoreline is composed of jagged dolerite rocks with small patches of shell grit and piles of seagrass. Above the shoreline there is a raised cobblestone beach encircling the island. The interior is flat with depressions and vegetated with grasses, which grow on shallow litter. There is a sand spit off the north-east tip.

BREEDING SEABIRD SPECIES
Little Penguin
An estimated 1171 pairs were located on the island, with 90% of them nesting under the 10 metre wide strip of stipa around the circumference of the island.

Pacific Gull
9 pairs, at nests with eggs and chicks, were located scattered around the perimeter of the island.

Sooty Oystercatcher
8 pairs were located scattered around the perimeter of the island.

Pied Oystercatcher
One pair was located on the northern shoreline.

VEGETATION
The interior is vegetated with grasses, *Poa poiformis*, *Rhagodia* and *Acaena* sp. which grow on shallow litter. There is a 10 metre wide strip of thick stipa around the island on the cobblestone platform.

Old cobblestone beach now partially vegetated

OTHER SEABIRD SPECIES
Caspian Tern

No mammals were recorded on the island.

BIRDS
Native:
Cape Barren Goose – 1 pair with 3 fledged chicks.
Ruddy Turnstone
Masked Lapwing
Introduced:
Common Starling

REPTILES
White-lipped Whip Snake
Metallic Skink
White's Skink

COMMENTS
As a nature reserve, this island has appropriate status to ensure its full protection.

North East Islands – Region 5

(Regions, page 44)

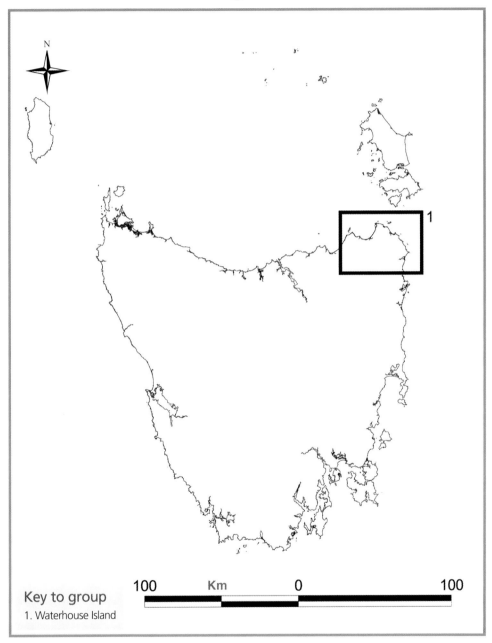

Key to group
1. Waterhouse Island

Waterhouse Island Group

(Region 5, page 373)

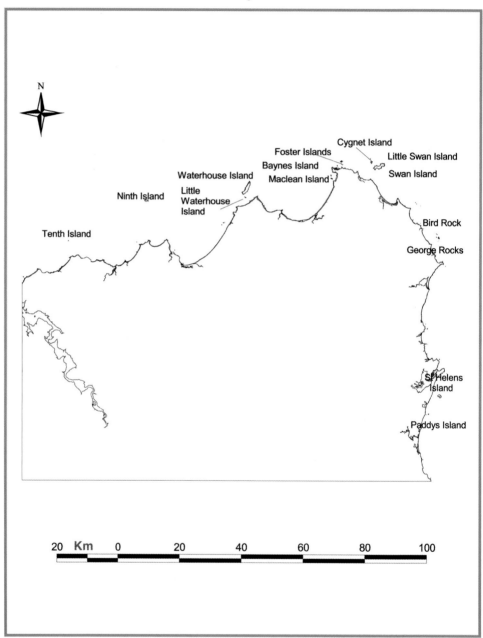

Tenth Island

(Waterhouse Island Group, page 374)

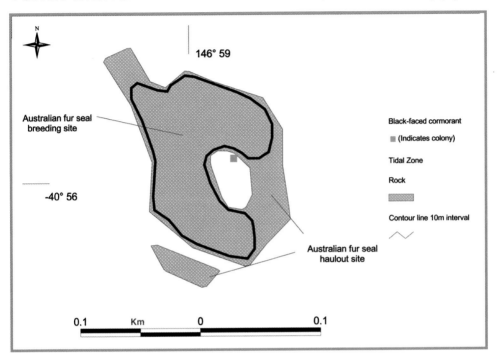

Location: 40°57'S, 146°59'E
Survey date: 3/12/85
Area: 0.09 hectares
Status: Nature Reserve

Locally known as Barrenjoey, it is an elongate island which rises to a summit at the southern end. No vegetation is present. Much of the island is wave-washed in severe storms.

BREEDING SEABIRD SPECIES
Black-faced Cormorant

15 - 20 old nests were located. The nests, which are bowls in dry guano with a lining of seaweed, were located on the north-west side of the island, just below the summit. The size of the colony and breeding timetable are probably restricted by the seals that breed all over the island and at the colony site. The seals start breeding in mid-November.

COMMENTS

The island is a significant Australian Fur Seal breeding colony, accessible by boat from nearby holiday villages and the Tamar River. Its accessibility and the presence of a seal breeding colony make it a popular site for tourists, with at least two registered charter boats in operation. If tourists approach within 100 metres of the island between November and February, it is likely that the seals will be driven into the water, which can cause pups to be killed by stampeding adults. The island is wave-washed in severe storms and this can have

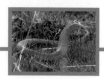

A hazardous place for young seals when high seas prevail.

OTHER SEABIRD SPECIES
Little Penguin – roost here
Pacific Gull – 8
Silver Gull – 15
Crested Tern – 1

MAMMALS
Up to 400 Australian Fur Seal pups are born each year on the island, but many drown in storms.

BIRDS
White-faced Heron

No reptiles were recorded.

a big impact on seal pup survivial, should this occur when they are very young in December and January. Under its management plan, permits are required for visits to the island and tourist operators are required to abide by the Department of Primary Industry, Water and Environment seal watching guidelines.

Ninth Island

(Waterhouse Island Group, page 374)

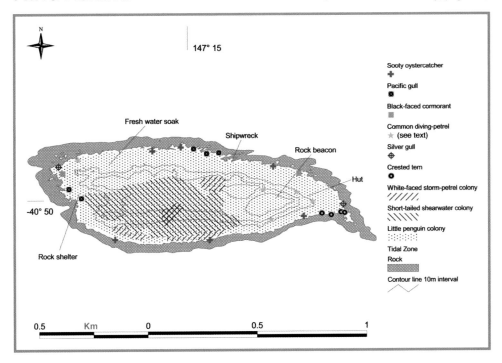

Location:	40°50'S, 147°16'E
Survey date:	3/12/85, 1-31/7/95, 10/11/99 and 26 – 28/11/99
Area:	31.74 hectares
Status:	Multiple land tenure: Private property and Non-allocated Crown Land ↑

Ninth Island is oval in shape, oriented west to east with an easily accessible shoreline which rises to a plateau. The dominant vegetation is fireweed, under which most of the burrowing birds live. Apart from much of the plateau and possibly half the coastal plain, there is adequate soil for burrowing, but it is very loose, sandy and easily damaged. There is a hut, rock shelter and freshwater soak on the island and a large wreck in the intertidal zone on the mid northern coast.

BREEDING SEABIRD SPECIES
Little Penguin

An estimated 3913 pairs (10/11/99) breed all over the island, with the highest densities occurring in the eastern sector. Their burrows were located predominantly amongst *Senecio* and *Poa*. Some also nest in rock crevices.

Short-tailed Shearwater

An estimated 6509 pairs (10/11/99) breed in low densities all over the island apart from the shore flats and plateau area, where there is insufficient soil. Their burrows are mainly in *Senecio* sp.

Common Diving-Petrel

An estimated 10 000 – 15 000 pairs were located nesting amongst steep rock faces surrounding the plateau, but particularly concentrated in the south-east and north-west areas.

White-faced Storm-Petrel

An estimated 14,312 pairs were located in four colonies. The main colony was located in the central western area amongst *Senecio* in sandy loose soil with three smaller colonies on the north-east and south perimeters of the shearwater colony.

Pacific Gull

5 pairs, 3 with runners and 2 with one egg each, were nesting in the plateau area at the west end and on the north-east coastline.

Silver Gull

There are two discrete Silver Gull colonies on the island: the western colony has 231 nests and the eastern spit colony has 362 nests.

Sooty Oystercatcher

6 pairs were located scattered around the coast.

Western shoreline ship wreckage.

Coastal slopes grading to extensive intertidal zone.

Black-faced Cormorant

422 pairs, mostly with chicks and a few with eggs, were nesting in several colonies on the island. The main colonies were located at the east end spit by the tern colony, and a smaller colony of 60 nests was located at the west end. Nests were built from seaweed and dead *Senecio* stems. 29 dead chicks of varying ages were also found. In 1999 there were several cormorant nests on the wreck.

Crested Tern

1800 pairs were nesting predominantly at the east end of the spit on the rocky shore extending into the *Senecio* sp. Nearly all were single nestings. 90% were with one egg only and 10% were with hatchlings still in the nest.

No other seabird species were recorded.

MAMMALS
Rabbit – present in low numbers

BIRDS
Native:
Brown Quail
Cape Barren Goose – 20 adults and at least 5 lots of goslings
Brown Falcon
Little Grassbird – numerous
Introduced:
Skylark
Common Starling – many were nesting in rock crevices, competing with the storm-petrels.

REPTILES
Southern Grass Skink

VEGETATION
It is dominated by *Senecio* sp.

COMMENTS
Ninth Island, which is partially privately owned, harbours a broad diversity of seabird species, which have been adversely impacted on in the past ten years by frequent fires, grazing and more recently by the *Iron Baron* oil spill in July 1995, which is estimated to have killed between 2000 and 6000 Little Penguins. A long-term monitoring program of the seabird population is carried out by the Tasmanian Museum and Art Gallery and the Nature Conservation Branch of the Department of Primary Industries, Water and Environment.

Eastern end Crested Tern and Silver Gull colony.

Waterhouse Island

(Waterhouse Island Group, page 374)

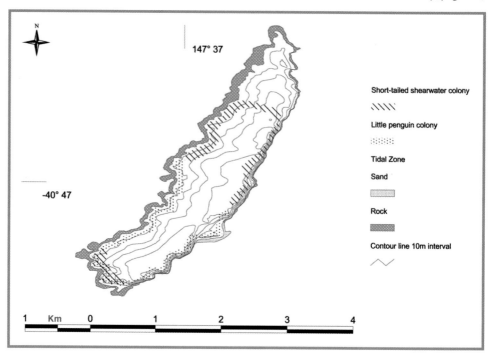

Location: 40°48'S, 147°37'E
Survey date: 5/12/86
Area: 287 hectares
Status: Lease, Non-allocated Crown Land at the north-east end.

It is a large rectangular island with steep sides, cliffs and eroded Holocene dunes. The top of the island is flat, sloping gently to the west-north-west. There is a spectacular beach and sand spit on the mid east coast, otherwise it is rocky. Buildings include a well-maintained homestead, farm sheds, fences and dry stone walls. There is an airstrip.

BREEDING SEABIRD SPECIES
Little Penguin

An estimated 200 pairs were located, scattered around the shoreline and up the steep sides into the Short-tailed Shearwater colonies. A colony extended from the western Short-tailed Shearwater colony along the slope and back to the jetty. Nests were found in the stone walls that surround the paddocks.

Short-tailed Shearwater

An estimated 15,000 in two colonies were present, one mid-way across the south end of the island (90 m by 60 m in size) and the other on the western side (200 m by 60 m in size). The southern colony is dominated by slender thistle and introduced grasses and stretches from the edge of the plateau up a gentle slope. At least half of the colony is no longer occupied because of erosion of the soft sandy soil that

Erosion at shearwater colony.

the birds burrow in. The burrows are deep. The western colony is also devastated by erosion with less than half of the former colony occupied. The south facing slope of this colony is the most badly eroded.

OTHER SEABIRD SPECIES
Australian Pelican – 3 on the eastern beach
Crested Tern – 18 on the eastern beach

MAMMALS
Introduced:
Fallow Deer – about 70
Cat
House Mouse
Sheep

BIRDS
Native:
Cape Barren Goose – 20 pairs
Hooded plover – 2 pairs seen

VEGETATION
The vegetation is dominated by introduced grasses, which are maintained as paddocks for sheep grazing. Introduced conifers, ornamental garden plants and boxthorn are also present.

Introduced:
Red Junglefowl (Chicken) – 2 hens and 1 rooster
Common Starling – breeding

REPTILES
Metallic Skink
Ocellated skink
Three-lined Skink
Bougainvilles Skink
White's Skink
Snake – not identified

COMMENTS

The status of the island is affected by the presence of introduced mammals, namely sheep, fallow deer, cats and house mice. The island is heavily grazed by the sheep and fallow deer. This grazing results in erosion and has devastated the Short-tailed Shearwater colonies. The natural habitat has been removed from the plateau by farming practices. The present owners of the island have fenced off the Short-tailed Shearwater colonies to prevent further grazing by sheep and deer and have rejuvenated many degraded areas. A full vegetation survey could be undertaken to ascertain what is left of the natural habitats of the island. The sand spit formed by refraction of the westerly swell around both ends of the island is considered to be of geomorphological significance for the State (Dixon, 1996). The island also has an exceptionally high reptile diversity.

Little Waterhouse Island

(Waterhouse Island Group, page 374)

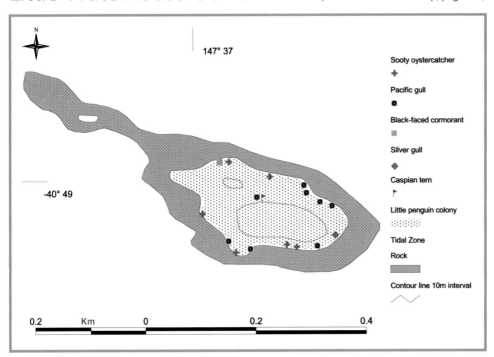

Location: 40°49'S, 147°38'E
Survey date: 5/12/85
Area: 2.5 hectares
Status: Non-allocated Crown Land

This is an oval shaped island with a low summit marked by a rock cairn. There is very little soil cover and no burrowing habitat, with most of the island consisting of bare rock.

BREEDING SEABIRD SPECIES
Little Penguin
20 pairs were located breeding in low densities under *Poa poiformis*, as there is no habitat for burrowing.

Pacific Gull
50 pairs, with eggs in nests and runners present, were located in solitary nests all around the periphery

Little Waterhouse Island

of the island and a colony of 30 or more birds was at the centre around the rock cairn. Many runners were dead or dying, some were lying in piles of 2 or 3 birds, drenched by a recent heavy rain storm.

Silver Gull
40 pairs, many incubating eggs, were located on an exposed bare rock and deep within a thick patch of *Stipa* sp. on the eastern end of the island.

Sooty Oystercatcher
6 pairs, incubating eggs, were found at solitary nests around the shoreline of the island.

Black-faced Cormorant
173 pairs, at nests with very small chicks, were located in a colony on the central north shore approximately 5 m up from the shoreline on a sloping rock. Nests were constructed of seaweed and *Poa poiformis*. Many chicks were dead, probably as a result of the previous 48 hours of intense rain. 500+ adults and mature juveniles were roosting nearby.

Caspian Tern
1 pair, incubating an egg, was located near the main Pacific Gull colony. The adults were defending the nest.

VEGETATION
The vegetation is dominated by *P. poiformis*, *Rhagodia candolleana* and *Disphyma crassifolium*.

OTHER SEABIRD SPECIES
Australian Pelican – 3 on the shore near the Black-faced Cormorant colony

No mammals or reptiles were recorded.

BIRDS
Native:
White-fronted Chat – breeding

Cape Barren Goose – 5 adults were seen and 1 old nest located.

Introduced:
Common Starling – over 20 pairs were breeding

COMMENTS
The island is accessible by boat from nearby fishing ports and the Cape Portland recreational areas. Because the island is a significant Pacific Gull and Black-faced Cormorant breeding site, its status should be upgraded to a nature reserve.

Maclean Island
(Waterhouse Island Group, page 374)

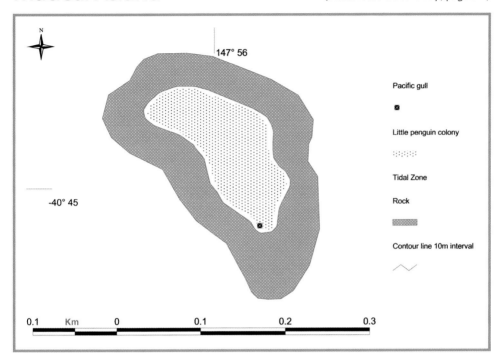

Location: 40°46'S, 147°56'E
Survey date: 17/12/85
Area: 1.11 hectares
Status: Non-allocated Crown Land

Maclean Island comprises two islets separated by a dry rock gulch. Each is round, with a jagged rocky shoreline and a flat top.

BREEDING SEABIRD SPECIES
Little Penguin
Up to 50 pairs, with eggs and chicks, were located under vegetation all over the island, although not fully utilising available vegetation cover. There are no suitable burrow conditions.

Pacific Gull
One pair with a three-quarter grown chick was located on the south end islet.

Silver Gull colony, south end.

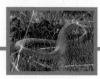

OTHER SEABIRD SPECIES
Common Diving-Petrel – 1 beach-washed

Silver Gull – all over island, but not defensive

Black-faced Cormorant – 4 roosting

No mammals or reptiles were recorded.

BIRDS
Cape Barren Goose – nesting

VEGETATION
It is dominated by stipa and *Poa poiformis*, matted with *Tetragonia*.

COMMENTS
This is a small pristine island with a low seabird diversity.

Baynes Island

(Waterhouse Island Group, page 374)

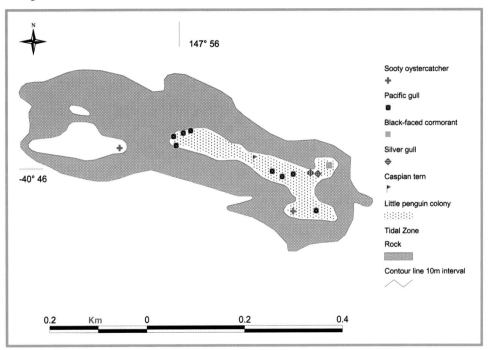

Location: 40°46'S, 147°56'E
Survey date: 17/12/85
Area: 1.62 hectares
Status: Non-allocated Crown Land ⇑

Baynes Island is a group of 3 islets all joined at low tide. The mid two are joined by a raised pebble beach neck, awash in high seas only. The northernmost section is the highest point.

BREEDING SEABIRD SPECIES
Little Penguin
Up to 50 pairs occur, scattered sparsely over the island under vegetation and in short burrows, wherever soil is adequate.

Pacific Gull
8 pairs, with eggs, were located at nests by stipa clumps throughout the island. A dead chick was also found.

Silver Gull
The main colony was found at the base of the southern slope on the easterly aspect amongst *Poa poiformis* and contained 91 nests. A dead chick was also found.

Sooty Oystercatcher
2 pairs were located, one on northern islet and one on southern coast.

Black-faced Cormorant
9 pairs were located at the south-east corner on the rocky shore. Their nests were made of seaweed, largely washed away. Over thirty were roosting at the northern end.

Caspian Tern

One pair with one egg, was located in *Bulbine* sp. on top of the central ridge 10 metres from the Pacific Gull nest. A dead adult was also found.

OTHER SEABIRD SPECIES

Australian Pelican – flew by

Crested Tern – dead

Fairy Tern – 2 flew by

No mammals or reptiles were recorded.

BIRDS

Cape Barren Goose – 1

Grey Teal – 1 nest with 7 eggs, 1 with chick

VEGETATION

Poa poiformis and stipa dominate the island's vegetation with some *Bulbine* sp. present.

COMMENTS

The presence of the Black-faced Cormorant colony warrants a status upgrade.

Cygnet Island

(Waterhouse Island Group, page 374)

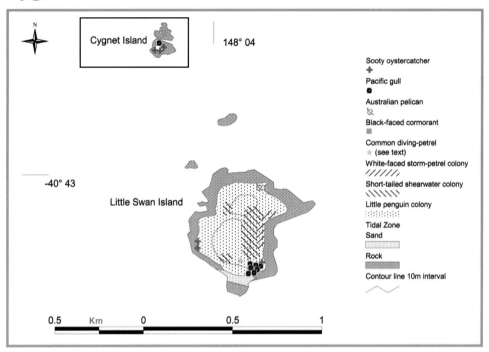

Location: 40°43'S, 148°04'E
Survey date: 17/12/85
Area: <0.5 hectares
Status: Non-allocated Crown Land ⇑

This small islet is a very low, elongate, rocky reef orientated north to south. The presence of debris over the entire reef suggests that it is completely wave-washed at times.

BREEDING SEABIRD SPECIES
Pacific Gull
One pair with one half-grown, partly downy chick was located in the centre of the reef.

Sooty Oystercatcher
3 pairs were nesting around the island. One nest, with an egg and a chick was located, the other 2 pairs were defending quite aggressively.

Black-faced Cormorant
46 nests of seaweed were located on the bare rock at the north-east end. They were newly constructed, but appeared not to have been used. Over 130 adults were also sitting on the shoreline on the south-west side.

VEGETATION
Only three species of vegetation were found on the reef, all succulents, the dominant being *Sarcocornia* sp.

OTHER SEABIRDS

Australian Pelican – 4 adults were on the reef.

Silver Gull – 4 flew about, then departed.

No other fauna was recorded.

COMMENTS

The status of the island needs to be upgraded, given the importance of the Black-faced Cormorant colony.

Australian Pelican

Foster Islands

(Waterhouse Island Group, page 374)

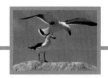

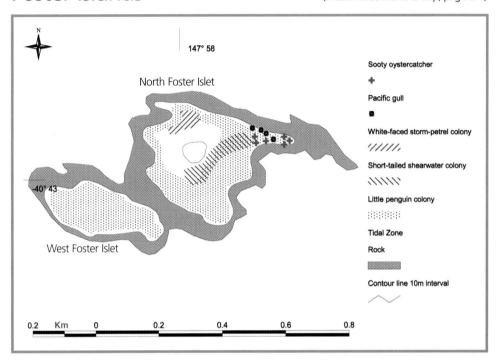

Location: 40°44'S, 147°58'E
Survey date: 17/12/85
Area: 48 hectares
Status: Nature Reserve

Foster Islands are two islets referred to as North Foster Islet and West Foster Islet, and a small offshore rock. They are connected at low tide.

North Foster Islet is square in shape. There are two pebble beaches on the east side and a spit protruding in a north-east direction. The highest point on the islet lies midway across the island towards the western side. 90% of the islet has very shallow soil lying over dolerite slabs. The soils vary from true loamy soils on the west side to sandy soils on the east side, where the Short-tailed Shearwater colony is found. West Foster Islet is round and low, rising to a central high point. The soil and vegetation are similar to North Foster Islet. It is connected to the main islet by a pebbly channel that is exposed at low tide and 1 to 2 m deep at high tide. The third islet is little more than a rock.

BREEDING SEABIRD SPECIES
Little Penguin

An estimated 300 pairs inhabit the main islet, scattered in a band around the shoreline and concentrated on the eastern side amongst the Short-tailed Shearwater colony. Burrows are short, extending to the bottom of the soil profile. Many dead chicks and flooded burrows were also located. On West Foster Islet, an estimated 100

pairs inhabit burrows where soil is adequate and also nest under *Disphyma crassifolium*.

Short-tailed Shearwater
An estimated 2880 pairs were located in a colony on the eastern side in an area of sandy soil dominated by *Poa poiformis* and *Senecio* sp.

White-faced Storm-Petrel
A very small colony of up to 20 burrows was located in the stipa area of the north-east tip of the northern islet and very sparsely scattered along the north side to the west in sandy grey soil. The birds in this colony were on eggs. The main colony of up to 100 pairs is in the north of the northern islet to the west of the of the shearwater colony. Here birds are in shallow U-shaped burrows in poor soil conditions.

Pacific Gull
4 pairs were at nests clumped on the north-east end of the north island near the white-faced storm petrel colony. Chicks could not be located and were probably affected by recent storms.

Sooty Oystercatcher
6 pairs were located defending their territories on the north-east tip of the north island. No nests were found, probably due to recent heavy rain.

Australian Pelican
They are known to breed on the islands, but were not recorded during this survey.

VEGETATION
On the northern islet, the area from the western shoreline to the low peak is dominated by thick stipa sp. and other grasses. The eastern side of the islet is dominated by shrubs and grasses. The north-east point area is comprised of sandy grey soil, dominated by stipa sp. African boxthorn is also prevalent. The western half of the west island is very rocky, with low vegetation dominated by stipa sp., whilst the eastern section is dominated by scrub. *D. crassifolium* dominates the shoreline.

Australian Pelican adults and a tight aggregation of large chicks.

No mammals or other seabirds were recorded.

BIRDS
Native:
Cape Barren Goose – 2 pairs with chicks
Chestnut Teal – 4 seen
Little Grassbird – numerous
Introduced:
Skylark
Common Starling – hundreds breeding in the boxthorn

REPTILES
Metallic Skink – on the northern islet.

COMMENTS
Destruction of nests and small chicks by heavy rain can be a problem to seabirds breeding throughout this region. The survey coincided with a period of heavy rain at a critical stage for surface-nesting seabirds, so may not be representative. The introduced boxthorn allows a large population of European starlings to breed. Whilst their presence here may have little local impact, the nesting and roosting habitat provided by the boxthorn provides an opportunity for this species to prosper and the plants should be removed. Australian Pelicans use Foster Islands as a partner breeding site to Little Swan Island. Because of its proximity to mainland Tasmania, access by summer holiday-makers could create problems for the breeding birds. Pelicans particularly, are susceptible to disturbance and may abandon their nests if approached by humans. People need to be alerted to their vulnerability and the diversity of other species breeding here. The management plan for this island group recommends access by permit only.

Little Swan Island

(Waterhouse Island Group, page 374)

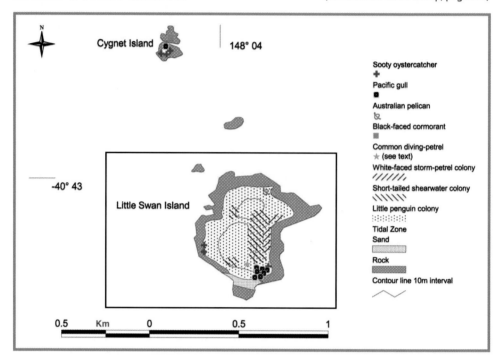

Location: 40°43'S, 148°05'E
Survey date: 17/12/85
Area: 12.64 hectares
Status: Non-allocated Crown Land ↑

A cuboid island adjacent to Swan Island, it has shingle and sandy beaches on the east side and a small sand spit on the southern end. The rest of the shoreline is composed of jagged rocks. The island rises gently to a low flat central area with a high point at the south-west end.

BREEDING SEABIRD SPECIES
Little Penguin

An estimated 300 pairs were located in burrows around the island, with the highest densities in the east within the Short-tailed Shearwater habitat. Numbers are probably restricted by the limited amount of suitable burrowing habitat.

Short-tailed Shearwater

An estimated 2400 pairs were primarily restricted to three colonies, The density of burrows appears to be restricted by shallow soil.

Common Diving-Petrel

The carcasses of 3 adults were found, indicating they may breed here. The short length of time spent ashore and the size of the island made the detection of burrows difficult, as many occurred in low densities within the white-faced storm petrel colonies.

White-faced Storm-Petrel habitat in central area of island.

White-faced Storm-Petrel

An estimated 5000 burrows were located in high densities in stipa sp. with sandy, dark soil. Lower density colonies were located in patches of *Tetragonia* along the south-east coast near the shoreline. The colonies occurred in dense patches scattered around the fringes of the Short-tailed Shearwater colonies with only a few burrows elsewhere.

Pacific Gull

7 pairs at nests were located in a colony at the south-east end of the island.

Sooty Oystercatcher

3 pairs were recorded at their nests on the rocky coastline on the western shore and on rocky points on the eastern side.

Australian Pelican

4 nests, 3 containing one egg each, were located near the eastern pebble beach at the north-east end of the island. The nests were in *Poa poiformis,* sheltered by large boxthorns. The *P. poiformis* was used to line the nests. There was evidence that the birds had been recently disturbed: one egg had been rolled out of a nest, another had claw marks on it and one nest had a wet bowl, possibly from a broken egg.

VEGETATION

The island is sparsely covered in low vegetation, mainly native tussock grassland, which includes *P. poiformis*, *Tetragonia implexicoma* and stipa.

No other seabirds, mammals or birds were recorded.

REPTILES
Metallic Skink

COMMENTS
The island is located within 1.4 km of Swan Island and, like Swan, it is accessible by boat from nearby fishing ports and holiday villages. There are many landing sites. Short-tailed Shearwaters, possibly several thousand each year, are legally and illegally taken during March and April. Colonies are in soft sand and excessive birding can result in loss of breeding habitat as burrows collapse. The pelican colony is one of only four in Tasmanian waters and the furthest south in Australia that this species breeds. The potential disturbance of the pelicans by visitors is great. In addition these birds are easily driven off their nests by activities at sea such as abalone boats and recreational boats using the anchorage. Because this island is integral to the conservation of Australian Pelicans in Tasmanian waters and has a high seabird diversity, its status should be upgraded to nature reserve.

Shoreline rocks.

Swan Island

(Waterhouse Island Group, page 374)

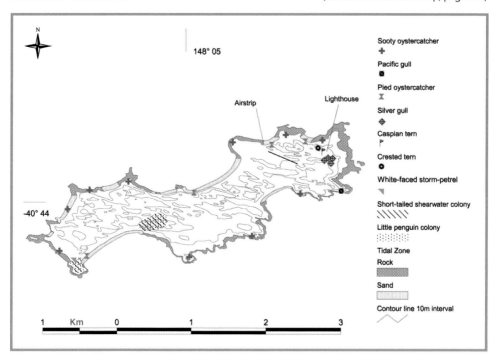

Location: 40°44'S, 148°06'E
Area: 239 hectares
Survey date: 17/12/85
Status: Multiple Land Tenure: Private property, Non-allocated Crown Land

Swan Island is a rectangular granite island with a rocky shore line, interspersed with long sandy beaches on both the northern and southern sides and dunes 10 to 20 metres high. There are parallel Holocene dunes at the south-east end. There is an automated lighthouse, several houses and an airstrip at the north-east end of the island. Most of the buildings are built from the local granite. There are vehicle tracks on the island.

BREEDING SEABIRD SPECIES

Little Penguin

50 pairs were located in burrows at the south-east point of the island in association with Short-tailed Shearwaters and White-faced Storm-Petrels. They are concentrated on the south side of the point where there is access to the sea via a beach and low rocks.

Short-tailed Shearwater

An estimated 8580 pairs are divided between two colonies on the island, the larger one of 7500 pairs on the south side of the island and a smaller colony of 1080 pairs at the south-east end. The latter is in soft sandy soil dominated by *Poa poiformis* and *Carpobrotus rossii*. The larger colony is on a south-

facing dune which is 90% *P. poiformis* growing in sandy soil. The colony extends from just above the base of the dune to the summit.

Pacific Gull
1 pair was located on the south-east end of the island.

Silver Gull
90 pairs with eggs and chicks were located in a colony alongside the Crested Tern colony at the north-east end below the lighthouse. The Silver Gulls took advantage of any disturbance of the Crested Tern colony to attempt to prey on the tern chicks and eggs.

Sooty Oystercatcher
10 pairs were located at their nests on the rocky coastline on the northern and southern shores. Generally one pair occupies each rocky headland separated by beaches.

Pied Oystercatcher
5 pairs were located at their nests on sandy beaches on the north and south coastlines.

Caspian Tern
One pair, with a big chick, was located at the north-east end in association with the Crested Tern colony.

Crested Tern
169 pairs with eggs and chicks were located at the north-east end below the lighthouse tower. Counts of eggs and chicks were prevented by Silver Gull predation, the result of observer disturbance.

VEGETATION
The island has extensive scrub belts dominated by *Myoporum* sp. and paddocks that have been grazed by sheep and cattle in the past.

No other seabird species were recorded on the island.

MAMMALS
Introduced:
Sheep – 1
Goat – several
Rabbit – numerous
House Mouse – numerous
Cat – one cat was brought onto the island in 1983 by a Department of Transport lightkeeper and never seen again.

BIRDS
Native:
Cape Barren Goose - 1 pair on a nest
Red-capped Plover – 1 pair
Hooded plover – 1 pair seen. A chick was amongst rocks on the shoreline.
Masked Lapwing – 2 pairs seen

REPTILES
Metallic Skink
White's Skink
Bougainvilles Skink
Tiger Snake – 8 seen

COMMENTS
The island is accessible by boat from nearby fishing ports and holiday villages. The landing sites are sandy, gently sloping beaches and low rocks, but the notoriously changeable local weather conditions and tides restrict access via boat to the island. However, Short-tailed Shearwaters are poached. These colonies are in soft sand and excessive birding can result in loss of breeding habitat through soil damage. The Crested Tern

colony is extremely vulnerable to disturbance by humans: Silver Gulls are quick to take advantage of any unattended eggs or chicks. Rabbits, mice and cats cause considerable damage especially during the frequent droughts in this area of Bass Strait. The presence of the owners of the island has resulted in many of these factors being addressed. Rabbits appear to have been eradicated.

The high diversity of seabirds and relatively large populations makes this an important island. The island has however had a long and continuous human presence, which has resulted in 5 species of introduced mammals. The sheep, goats and rabbits may soon be eradicated from the island. The eradication of house mice is improbable until new methods of extermination are developed. Large tracts of the island have been cleared in the past for pasture but these are now restricted to areas around the houses. The airstrip is maintained. The negative impact of past and present human presence on the island is adequately compensated for by the efforts of the current owners to maintain and improve the natural environment. Their presence has been the most effective deterrent to poaching, burning and other modes of vandalism often seen on Bass Strait islands.

'Captain William Moriarty surveyed the area and recommended that lighthouses should be built on Swan, Goose and Deal Islands in eastern Bass Strait to mark the Melbourne to Hobart shipping lane. Discussion went on for years without result until the authorities were shocked into action by two disastrous shipwrecks in 1845. The Mary, with 17 people, was lost near Flinders Island and the Cataraqui *crashed into a reef just of King Island during a gale, killing 360 people, mainly British emigrants on their way to Melbourne. A 28 metre lighthouse was subsequently built in late 1845 on Swan Island and the following year another was built on Goose Island.'* from Edgecombe J., 1986.

Bird Rock

(Waterhouse Island Group, page 374)

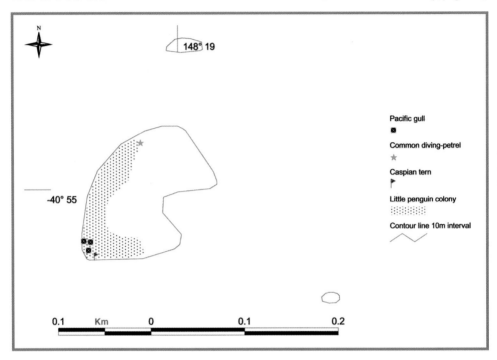

Location: 40°56'S, 148°20'E
Survey date: 9/12/85
Area: 0.9 hectares
Status: Non-allocated Crown Land ⇑

Lying to the north-west of George Rocks, Bird Rock is a rectangular-shaped, granite island with a small gulch on the eastern side. In the centre, it rises to about 10 m high.

BREEDING SEABIRD SPECIES
Little Penguin
Up to 50 pairs were estimated to inhabit burrows on the south-west side of the island, where soil depth permits.

Common Diving-Petrel
50 – 100 pairs nest mostly beneath *Tetragonia implexicoma* and rocks at the north-west end of the island and in lower densities in burrows.

Pacific Gull
3 pairs, one nest with 3 eggs and one with 2 eggs were found on the south-west end of the island in association with *Poa poiformis*.

Caspian Tern
1 pair and a full grown chick, not yet fledged lying next to a dry bowl nest, were sighted on the south-west end of the island in association with *P. poiformis* close to the Pacific Gulls' nests.

OTHER SEABIRD SPECIES

Black-faced Cormorant – 30 roosting

Great Cormorant – 6 roosting

Sooty Oystercatcher – 1 pair

No mammals or reptiles were recorded.

BIRDS

Ruddy Turnstone – 6 birds were on the water's edge.

VEGETATION

Around the raised centre of the island, the vegetation is dominated by *T. implexicoma* with *Poa poiformis* around the edges.

COMMENTS

Bird Rock should be officially incorporated into George Rocks Nature Reserve.

George Rocks

(Waterhouse Island Group, page 374)

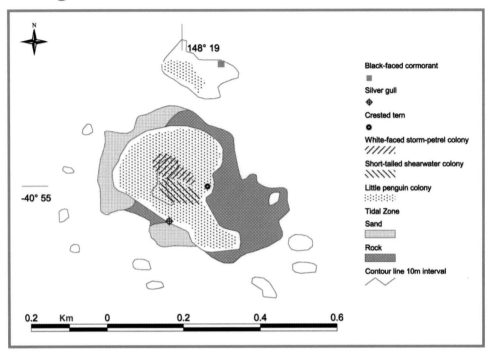

Location: 40°55'S, 148°20'E
Survey dates: 16/11/77 and 9/12/85
Area: A total of 3 islands = 7 hectares in area.
Status: Nature Reserve

George Rocks comprises three islets and many granite reefs and boulders. The islands and reefs, which form a U-shaped bay open to the south-west, provide a natural anchorage sheltered from most of the prevailing winds.

BREEDING SEABIRD SPECIES
Little Penguin

100 pairs were located on the main island in scattered burrows and rock crevices in association with *Poa poiformis*. On the smaller northern island, there is little soil suitable for burrowing and about 50 birds nest under dense *Poa* clumps.

Short-tailed Shearwater

100 pairs nest in a restricted central area of the main island in a gulch facing east. This was the only area where sufficient soil would permit burrowing.

White-faced Storm-Petrel

Napier (1978) records 200 breeding pairs of White-faced Storm-Petrels, spread sparsely over most of the main island. None were recorded in the 1985 survey.

Silver Gull

About 100 pairs nest each year on the western end of the south-western gulch. 30 birds were

From the northern islet a view over the main island.

also sighted on the shoreline possibly nesting on a rock off the south of the island.

Black-faced Cormorant
15 nests, with attending adults but no eggs, were restricted to granite ledges at the north-east end of the outer northern island. 5 were also seen roosting.

Crested Tern
This species is believed to breed irregularly on the main island (Napier, 1978).

VEGETATION
The islands are vegetated with predominantly *P. poiformis* and *Carpobrotus rossii* with *Tetragonia* scattered amongst boulders and slabs of granite.

OTHER SEABIRD SPECIES
Great Cormorant – 6 were roosting

Sooty Oystercatcher – 2 pairs, one on the main island and one on Bird Rock

MAMMALS
Introduced:

Rabbit

Rat

REPTILES
Metallic Skink. – common

COMMENTS

The islets are easily accessible by boat from the nearby camping grounds of Ansons Bay and Mt William National Park and holiday shacks, in addition to being a popular anchorage for fishing boats. Landing on the islands is easy at a number of locations. As a result Short-tailed Shearwaters are occasionally poached. The Black-faced Cormorant and Crested Tern colonies are vulnerable to disturbance by visitors for much of the breeding season.

Comparison of the records collected by Napier in 1977 (*Corella* seabird series) and Brothers show that White-faced Storm-Petrels have disappeared off both Bird and George Rocks, perhaps as a consequence of rat destruction. Black-faced Cormorants were also absent during the 1985 survey, but local movements of colonies of cormorants are often observed. In fact the birds may have moved to the outer north islets where they were nesting in 1985 but not in earlier years. Crested Terns were recorded by Napier as breeding irregularly on George Rocks and regularly on Bird Island but not recorded at all in the 1985 survey. Silver Gulls were also not breeding in 1985 but had in earlier years. Species such as these regularly move colonies. There should be efforts made to eradicate introduced mammals from the island. The tourist potential of the island also justifies an increase in the direct management of the people who visit.

Spectacular granite rock formations at the south end.

St Helens Island

(Waterhouse Island Group, page 374)

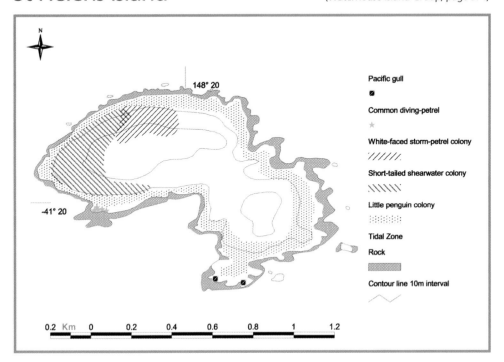

Location: 41°21′S, 148°20′E
Survey date: 21 – 23/1/88
Area: 51 hectares
Status: Conservation Area ↑

It is a kidney-shaped, granite island with a steeply sloping rocky shoreline and no beaches. Its uneven terrain is covered by sandy loam and rises to two plateau-like summits, each about 30 metres above sea level.

BREEDING SEABIRD SPECIES
Little Penguin

Over 5000 pairs, with half-grown to fully-fledged chicks, were scattered all over the island, usually under matted vegetation, in dense *Poa* or amidst boulders right by the shoreline.

Short-tailed Shearwater

Over 3000 pairs, many with chicks, were nesting in low density amongst the *Poa* and patches of bracken and *Rhagodia* that form a strip around the western side of the island. Burrows are short and steep and mainly on the gentle northern slopes.

Common Diving-Petrel

An estimated 10 pairs were recorded breeding on the island. Several pairs were found in burrows beneath dense matted *Poa* by the rock shoreline at the south-west tip.

White-faced Storm-Petrel

An estimated 10 000 pairs, many with chicks, were concentrated in discrete colonies in very dense scrub with unvegetated ground beneath, mainly in

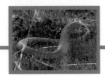

the north-western section of the island. They nest in a range of habitats, from matted *Poa* with scattered scrub patches to very dense 100% scrub with an open understorey.

Pacific Gull
2 pairs were recorded on the southern coast. 2 pairs were also sighted in the north-east area, but were not nesting.

VEGETATION
Much of the larger north-west section is dominated by *Acacia sophorae*, *Leptospermum scoparium* and *Leucopogon parviflorus* scrub above a coastal periphery of thick *Poa poiformis*. Exposure and fires have markedly reduced scrub coverage on the smaller south-east portion where *Poa* predominates. Bracken fern, *Pteridium esculentum*, is common throughout and a thick covering of *Tetragonia implexicoma* is present above the shoreline in the gulch area.

Southern shore to west, Tasmanian mainland in background.

COMMENTS
The island's seabird habitat has been burnt in the past and is also subjected to severe rabbit grazing, which occurs when conditions are favourable. Because of the diversity and abundance of seabirds, its status should be reviewed to protect the habitat and its potential as a tourist destination.

OTHER SEABIRD SPECIES
Australasian Gannet – feeding offshore

Black-faced Cormorant – single birds were occasionally sitting on rocks on the western side

Sooty Oystercatcher

Caspian Tern – feeding offshore

Silver Gull – a colony with an estimated 250 pairs was recorded on the south-east corner of the island in 1977, but none were seen during the 1988 survey.

MAMMALS
Rabbit – 7 seen

BIRDS
Native:
Brown Quail

White-bellied Sea-Eagle – flew over

Swamp Harrier

Brown Goshawk

Lewin's Rail

Silvereye

Introduced:
Common Blackbird

Common Starling

REPTILES
Metallic Skink

Paddys Island

(Waterhouse Island Group, page 374)

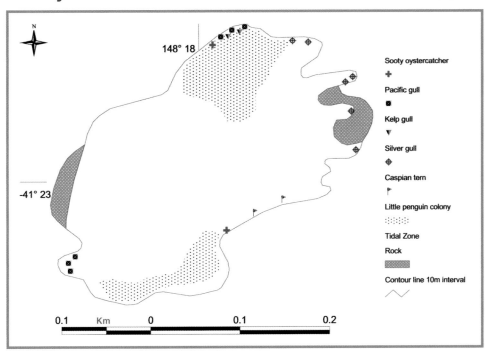

Location: 41°24'S, 148°18'E
Survey date: 8/12/91
Area: 4.6 hectares
Status: Non-allocated Crown Land ⇑

The island is half vegetated and half bare granite. The eastern side has an extensive bare rock shoreline, which extends almost to the centre of the island. It is a flat island with low elevation.

BREEDING SEABIRD SPECIES
Little Penguin

39 pairs were counted. Most were located in short burrows beneath *Poa poiformis* with the greatest concentration at the north-west end and elsewhere largely towards the south end. Some were also found nesting under granite boulders. Numbers were restricted by the lack of nesting habitat.

Pacific Gull

39 pairs, 26 with no eggs, 3 with 1 egg, 3 with 2 eggs, 1 with 3 eggs, 3 with 1 chick, 1 with 2 chicks, 1 with 1 chick and 1 egg, 1 with 1 egg and 2 chicks, were counted. The main colony was located on the north-eastern coast with a few nests in the south-west.

Silver Gull

54 pairs, 19 with no eggs, 8 with 1 egg, 13 with 2 eggs, 8 with 3 eggs, 2 with 1 chick, 1 egg, 1 with 2 chicks 1 egg, 1 with 2 chicks, 1 with 1 chick and one dead chick and 1 with 2 broken eggs, were mainly located amongst large granite boulders on the eastern and northern coasts. They tend to nest in association with *P. poiformis*.

Kelp Gull
2 pairs were recorded. Each had one egg. The two nests were located just south of the main Pacific Gull colony in the north-east.

Sooty Oystercatcher
2 pairs were sighted, 1 on the central east coast with 2 broken eggs and the other on the north-west with 2 eggs.

Caspian Tern
2 pairs, 1 with 2 three-quarter grown chicks and 1 with an egg and one three-quarter grown chick, were located on the central eastern coast.

VEGETATION
Vegetation is dominated by *Poa poiformis* with scattered *Disphyma crassifolium*, *Atriplex cinerea* and introduced grasses.

OTHER SEABIRD SPECIES
Black-faced Cormorant – 80 were roosting both at the northern and south-west ends of the island.

Crested Tern – 1 dead juvenile

No mammals, reptiles or other birds were recorded.

COMMENTS
As it is one of very few islands on which Kelp Gulls breed, an upgrading of status is justified.

East Coast Islands – Region 6

(Regions, page 44)

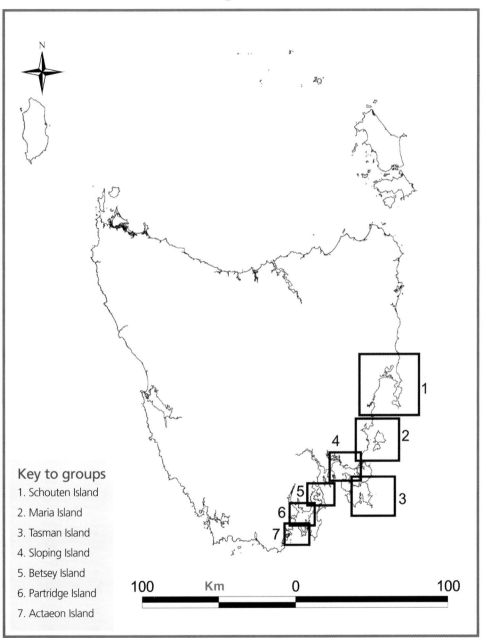

Key to groups
1. Schouten Island
2. Maria Island
3. Tasman Island
4. Sloping Island
5. Betsey Island
6. Partridge Island
7. Actaeon Island

Schouten Island Group

(Region 6, page 409)

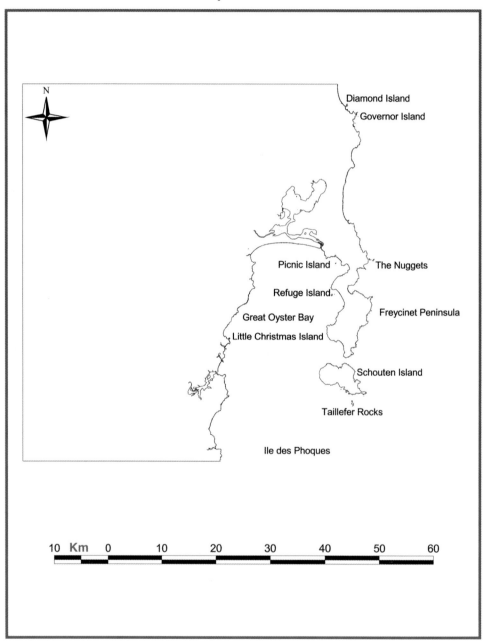

The Nuggets

(Schouten Island Group, page 410)

Location: 42°07'S, 148°22'E
Survey date: 28/10/78
Area: 6.76 hectares
Status: Non-allocated Crown Land ⇑

Lying to the east of Freycinet Peninsula, the Nuggets is a group of four granite islets, the easternmost of which is composed entirely of bare rock and supports no vegetation or breeding seabirds. The other three are referred to as the west, central and east islets. The west, the largest of the group is 300 metres long by 150 metres wide and 30 metres high. The other two are similar in size, about 15 metres long by 100 metres wide by 30 metres high. The central nugget has a flat top circled by cliffs on all sides. The others have cliffs only on their southern and south-western faces, with steep slopes on their remaining sides.

BREEDING SEABIRD SPECIES
Little Penguin

About 100 pairs breed here, 50 pairs on each of the west and east islets. On the former, breeding occurs mainly in burrows less than 1 metre long but occasionally also under *Poa* tussocks. The soil is shallow and burrows often twist to avoid rocks. Two nests were located in rock crevices. All nests contained two eggs or two small chicks. On east islet, the birds were breeding mainly in rock crevices. All nests were constructed of *Poa* and again contained two eggs or two small chicks.

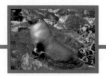

West cliffs.

Short-tailed Shearwater
An estimated total of 50 – 60 pairs breed on all islets.

Fairy Prion
10 nests were located in rock crevices on the western side of the east islet.

Common Diving-Petrel
7 burrows, only one inhabited, were found on the east islet. 12 carcasses were found on the central islet, but no live birds were found, despite intensive searching.

Pacific Gull
14 nests, constructed of *Poa*, were found amongst tussocks on the central islet.

White-faced Storm-Petrel
An estimated 250 pairs occur only on the central islet, where burrows are mainly under *Poa*, with some occurring in areas covered by *Carpobrotus rossii*. All burrows had been freshly excavated and a few contained single birds incubating their eggs.

Black-faced Cormorant
An estimated 5 – 10 pairs were found on the central islet. Their nests, constructed of seaweed and twigs, were located on a ledge 3 metres below the top of the cliff, about 0.3 metres apart.

Caspian Tern
An estimated 10 – 15 pairs breed on the central islet amongst the *C. rossii*.

OTHER SEABIRD SPECIES

Great Cormorant

Silver Gull – 25 pairs were found breeding on the central islet during a visit in 1974.

Kelp Gull

Crested Tern

No mammals were recorded.

BIRDS
Native:

White-bellied Sea-Eagle

Swamp Harrier

Peregrine Falcon

Forest Raven

VEGETATION

The dominant vegetation on the west islet is *C. rossii, Tetragonia implexicoma, Poa poiformis, Leucopogon parviflorus* and *Allocasuarina verticillata* and on the central islet, *C. rossii, Poa poiformis* and *Tetragonia implexicoma*. The east islet has the same dominant species as the west islet, with *Acacia sophorae* replacing the casuarinas.

COMMENTS

Pacific Gull castings found on the central islet contained White-faced Storm-Petrel and Common Diving-Petrel remains, which indicates that these species' breeding potential may be inhibited by the presence of Pacific Gulls. Its seabird diversity warrants an upgrading of status.

Schouten Island

(Schouten Island Group, page 410)

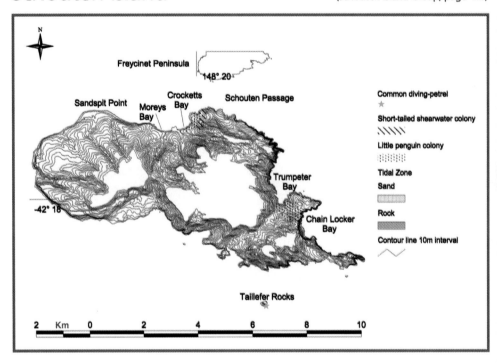

Location: 42°32'S, 148°30'E
Survey date: Jan – June 1977, 6 – 7/9/78 and March 2000
Area: 3439 hectares
Status: National Park

The island is mountainous and rugged with a predominantly cliff coastline, broken by sheltered bays. The highest peak, Mt Storey, is 400 metres high. It is similar in topography and lithology to Freycinet Peninsula, from which it is separated by a one kilometre wide passage. A north – south running fault through the centre of the island divides the granite in the east from the dolerite-capped sedimentary rocks in the west. There are sandstone outcrops beneath the dolerite along parts of the western coast and in two of the steeper western gullies. Abandoned shafts and adits, a tramway and other infrastructure from four phases of coal and tin mining, which occurred between 1842 and 1925, still exist on the island.

BREEDING SEABIRD SPECIES

Little Penguin

Up to 50 burrows were located in March 2000 in the sand above the rocks at the east end of Crocketts Bay in the north of the island and possibly near Trumpeter and Chain Locker Bay at the south-east end of the island.

Short-tailed Shearwater

A colony of about 500 pairs was located east of the penguin colony that is east of Crocketts Bay in the north of the island.

OTHER SEABIRD SPECIES

Common Diving-Petrel – beach washed, Crocketts Bay

Shy Albatross – 2

Australasian Gannet – 4+ in Schouten Passage

Black-faced Cormorant – common

Little Black Cormorant

Pacific Gull – in Schouten Passage

Kelp Gull

Silver Gull

Crested Tern

MAMMALS

Australian Fur Seal – haul-out the eastern side

No reptiles were recorded.

BIRDS

Native:

Brown Quail – calling common towards Sandspit Point

White-faced Heron

White-bellied Sea-Eagle – one nest was located at the back of Sarah Ann Bay in the south and another at Trumpeter Bay on the east coast.

Wedge-tailed Eagle

Brown Falcon — adult female

Tasmanian Native-hen – common, introduced to the island

Masked Lapwing – 2

Brush Bronzewing – 2 seen, others calling

Yellow-tailed Black Cockatoo – 10+

Green Rosella – common

Fan-tailed Cuckoo – calling often

Southern Boobook – calling

Laughing Kookaburra – not common, possibly introduced

Superb Fairy-wren – common

Tasmanian Scrubwren

Tasmanian Thornbill

Yellow Wattlebird – only 1 seen, reported abundant in June and July

Yellow-throated Honeyeater – abundant

Crescent Honeyeater – abundant

Eastern Spinebill – common

Scarlet Robin

Grey Shrike-thrush – not common

Golden Whistler – 2

Grey Fantail

Grey (Clinking) Currawong – often calling

Forest Raven

Beautiful Firetail – 1 seen, reported to be common

Tree Martin – numerous in flocks

Silvereye – common

Peregrine Falcon – adult male at southern end of Milligans Hill

Welcome Swallow

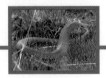

VEGETATION

There is a relatively high incidence of introduced species invading the most accessible parts of the coastline. Pasture species were introduced with sheep in the early 1840s, particularly in the north western corner of the island. To the west of the fault line in the dolerite soils, forests dominated by eucalypts cover most of the land, whereas they comprise less than 40% of the vegetation on the eastern granite soils. Scrub, heath and sedgeland communities dominate most of the eastern half of the island with the sedgeland communities occupying the areas of least drainage. Communities dominated by grasses and herbs are most widespread in the western part of the island and are mostly near the coast and/or associated with human disturbance. These occur where grazing, clearing and frequent burning have destroyed former forest communities which are now reinvading. The herbaceous communities of the rocky coasts are dominated by succulents, which occur in a zonal pattern with *Disphyma australe* being closest to the sea bordered by *Carpobrotus rossii*, stipa tussock-grassland forms and *Poa poiformis* being the most landward.

COMMENTS

During more recent stages of its grazing history, cats were reputedly introduced to control rabbits, which was surprisingly successful. The cats have since disappeared The island is visited by campers, particularly during the summer. The large (approximately 1 km^2) fault-controlled landslip which occurs on the south coast to the west of the fault is considered of representative and outstanding significance for Tasmania (Dixon, 1996).

Diamond Island

(Schouten Island Group, page 410)

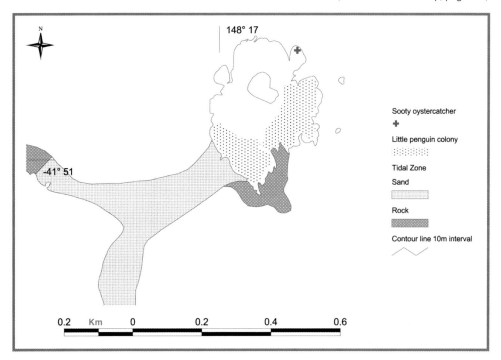

Location: 41°51'S, 148°17'E
Survey date: 11/11/87
Area: 6.76 hectares.
Status: Nature Reserve

It is a low, granite island off Bicheno, sometimes connected at low tide to mainland Tasmania by a sand spit. The shore is mostly smooth, sloping granite and the soil is sandy.

BREEDING SEABIRD SPECIES
Little Penguin

An estimated 200 pairs occur either under matted vegetation or in burrows. They are scattered over the island, with two concentrations, one at the south-west tip and one along the lower eastern areas.

Sooty Oystercatcher

One pair at a nest with 3 eggs on dead *Disphyma crassifolium* was found at the north-eastern tip near bare rocks.

VEGETATION

The dominant vegetation is *Rhagodia candolleana*, which is very matted, forming continuous ground cover, with bracken patches and clumps of cutting grass. Scattered 10 metre casuarinas and 6 metre radiata pines are concentrated in the central to northern part of the island.

COMMENTS

There are several ecotourist ventures to the island, with Little Penguins being the main attraction.

Dogs, feral and domestic cats are potential threats to the penguins. A management plan is currently being developed by the Department of Primary Industry, Water and Environment to ensure that tourist ventures are compatible with the Little Penguin breeding habitat.

No reptiles or other birds were recorded.

MAMMALS
Rabbit

Little Penguins

Governor Island

(Schouten Island Group, page 410)

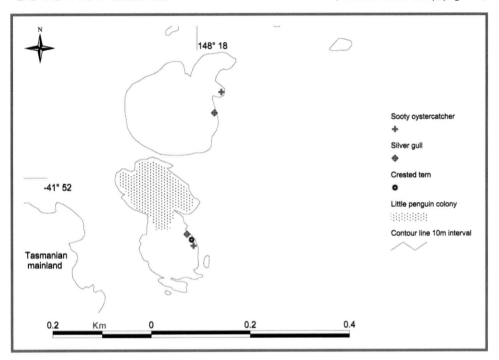

Location: 41°52'S, 148°18'E
Survey date: 11/11/87
Area: 1.94 hectares
Status: Nature Reserve

There are two islands, which are composed of 95% bare smooth granite, with small pockets of vegetation on the west sides.

BREEDING SEABIRD SPECIES
Little Penguin

Up to 20 pairs were nesting under matted vegetation, with a few under rocks, on the south island.

Silver Gull

A total of 130 pairs were nesting, (78 on the north island and 52 on the south island). Most were at empty nests, which had already been used, in narrow gutters of granite, vegetated with *Disphyma crassifolium* and introduced grasses.

Sooty Oystercatcher

One with one egg was found on the south island, nesting on *D. crassifolium* and one pair, with 2 eggs, was found on the north island on bare rock.

Crested Tern

54 nests, with one egg each, were found south of the gull colony on the south island, aligned in vegetated crevices. An additional 50+ unattached birds were roosting at the edge of the south island.

No other fauna was recorded.

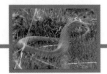

VEGETATION

Vegetation on both islands is dominated by *D. crassifolium*, *Geranium* sp., *Poa poiformis* and introduced grasses.

COMMENTS

The existence of the Crested Tern breeding colony justifies the island's nature reserve status and, as such, it should be protected from inappropriate visitation.

Crested Tern

Picnic Island

(Schouten Island Group, page 410)

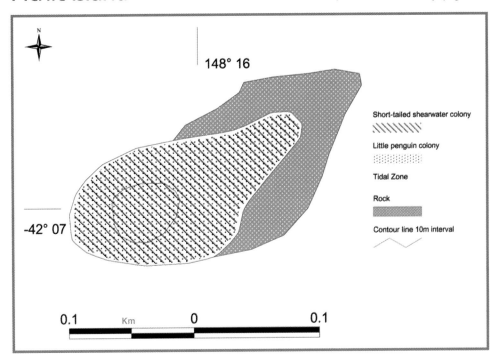

Location: 42°07'S, 148°16'E
Survey date: 12/11/87
Area: 1.01 hectares
Status: Private property

Composed of sandstone over granite, this is a low, rocky islet with a 5 metre cliff on the west side and gently sloping topography elsewhere.

BREEDING SEABIRD SPECIES
Little Penguin

An estimated 150 pairs occur over most of the island in varying densities dependent on soil depth and vegetation cover. They usually nest in association with *Poa poiformis*.

Short-tailed Shearwater

An estimated 160 pairs breed over most of the island, interspersed with Little Penguins, where soil depth is sufficient.

VEGETATION

The vegetation is dominated by *P. poiformis*, introduced weeds and grasses and a few patches of *Tetragonia implexicoma*.

OTHER SEABIRD SPECIES

Black-faced Cormorant – 3

Sooty Oystercatcher – 1

No mammals or other birds were recorded.

REPTILES

Ocellated Skink

COMMENTS

Protective management measures should be put in place to conserve the fragile seabird breeding habitat.

Shearwater raft

Taillefer Rocks

(Schouten Island Group, page 410)

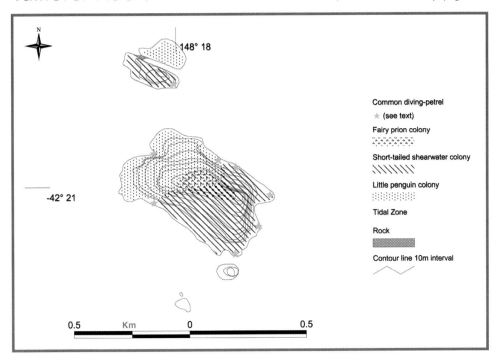

Location: 42°22'S, 148°19'E
Survey date: 12/11/87
Area: 14.83 hectares
Status: National Park

There are three steep, rugged granite rocks in this group, the outer one being by far the largest at 12.59 hectares. There is patchy scrub all over the main island.

BREEDING SEABIRD SPECIES

Little Penguin

120 pairs were mainly nesting in rock crevices and under boulders, due to lack of soil suitable for burrowing.

Short-tailed Shearwater

An estimated 770 pairs nest under matted, dense vegetation, especially *Poa* and under rocks and in crevices screened by *Poa*.

Fairy Prion

On the main island about 100 pairs nest in rock crevices near the summit ridge. On the inner island there were approximately 300 nests in the rock crevices. Most were with eggs.

Common Diving-Petrel

Up to 300 pairs, many with big chicks, were concentrated in low altitude crevices with suitable digging soil.

VEGETATION

Patchy scrub covers the rocks, with Oyster Bay pines being common. The only extensive *Poa* is in the band on the south side of the main rock.

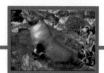

OTHER SEABIRD SPECIES
Black-faced Cormorant – were roosting at the south-west edge of the inner rock.

MAMMALS
Australian Fur Seal – use as a haul-out site in small numbers.

BIRDS
Native:
Crescent Honeyeater

Beautiful Firetail

Forest Raven

Introduced:
Common Blackbird

REPTILES
Metallic Skink

White's Skink

Ocellated Skink

Mountain Dragon

COMMENTS
This is one of very few islands on which Oyster Bay pines occur. It is also regionally significant due to the diversity of seabirds, skinks and in particular, the occurrence of the Mountain Dragon, which is rarely found on offshore islands.

View from the east. Granite with sparse vegetation.

Ile des Phoques

(Schouten Island Group, page 410)

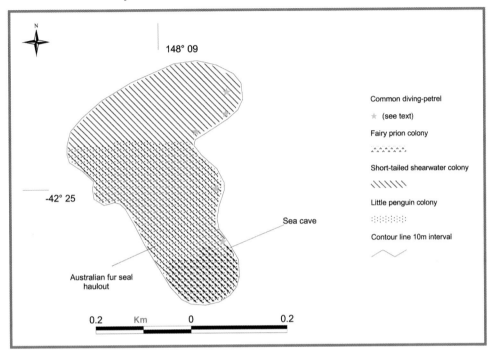

Location: 42°25'S, 148°09'E
Survey date: 28/11/86
Area: 8.05 hectares
Status: Nature Reserve

Sometimes referred to as White Rock by fishermen, Ile des Phoques is an oblong island, orientated east to west. Cliffs along the west shore rise to an undulating plateau over 50 metres high, but are much lower in the east. There is a blowhole on the eastern shore. There are submarine sea caves running through the granite on the south-east, partly associated with large surface caves on the northern and southern sides of the island. Much of the island has sparse soil where burrowing is not possible. There are extensive stone walls on the island.

North-west cliffs.

BREEDING SEABIRD SPECIES

Little Penguin
100 – 150 pairs, with small to full-grown chicks, are unevenly distributed over the island, with access limited to the south-west and western sides. They nest mainly in rock crevices and under boulders due to the scarcity of the soil.

Short-tailed Shearwater
50-100 pairs, many with eggs, were found in crevices or under boulders over the island. In areas where soil was adequate, burrows were dug in loose sandy soil on the flats of the north cliff-top areas.

Fairy Prion
Up to 500 pairs with chicks were found to be breeding. Determining the correct population was difficult due to the inaccessibility of the habitat. Nests were found amongst those of the Common Diving-Petrels, but were concentrated under boulders towards the top of the southern cliff.

Common Diving-Petrel
3000 – 5000 burrows were found. They were vacant with the only sign of occupation being the down from recently-fledged chicks. A small number of fully-feathered chicks were found in the burrows. Many dead (eaten) birds were also found. Half the population nests in burrows in the soil, or under matted vegetation (*Carpobrotus rossii* and *Tetragonia*). The rest nest in crevices and under boulders over the island, with some down the sheer cliff faces on tiny ledges in crevices and burrows. It was difficult to estimate numbers due to the inaccessibility of the habitat.

VEGETATION
The vegetation cover is low and treeless, comprising *Poa poiformis*-dominated tussock grassland with a number of succulents. These include *C. rossii*, *Tetragonia implexicoma* and *Rhagodia candolleana*. Other species include *Pelargonium australe* and *Stylidium graminifolium*.

OTHER SEABIRD SPECIES
Giant Petrel sp – 2 offshore

Fluttering Shearwater – 8 flew by

Shy Albatross – flew by

Australasian Gannet – 2 offshore

Black-faced Cormorant – roosting on central southern side.

MAMMALS
Australian Fur Seal – 61 were hauled out on the south-east side.

BIRDS
Native:
White-bellied Sea-Eagle – a nest was located on the north-east side just below the cliff top.

Peregrine Falcon – 1 and a dead adult

Richard's Pipit

Introduced:
Skylark - 1

No reptiles were recorded.

OTHERS
Grasshoppers – a massive number have had a marked effect on the vegetation.

COMMENTS

It is intersting to speculate on the natural history processes that have shaped this island as it is today. Burrowing habitat is sparse and encrusted guano is evident on exposed rock surfaces over much of the island. The fact that Fur Seals can gain access to the entire plateau area and phosphatic flowstone, formed from seal excrement, veneers granite bedrock at seal haul-out sites, reinforces the theory that in the past seals occurred here in vast numbers. This would also account for the sparsity of vegetation. The flowstone and submarine sea caves and tunnels on the island are considered of outstanding geological significance for the State (Dixon, 1996).

The island is currently used as a site for ecotourism activity which includes diving and seal watching. These should be controlled to ensure the protection of the island as a nature reserve.

'It was then that we began to notice an extremely strong and very disagreeable smell. It grew more and more unpleasant as we got nearer to the little island. Arriving close to its shores, we found it covered with a prodigious number of seals. Their largest, which were of a yellow colour, occupied the upper part, while the small ones, which appeared black, filled the cavities found in the lower part of the rock.' Mr Bailey on board Peron's ship the Le Geographie, February 1802 from Cornell, 1974. (This we interpret as evidence that Isle des Phoques was a breeding colony of Australian Fur Seals.)

Refuge Island

(Schouten Island Group, page 410)

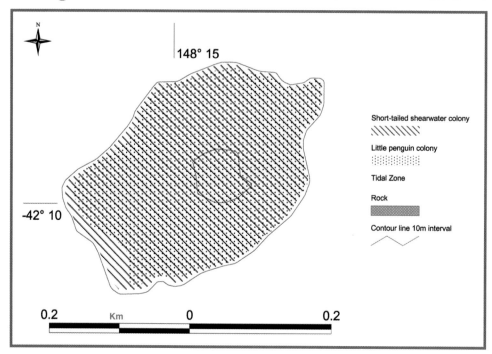

Location: 42°16'S, 148°16'E
Survey date: 12/11/87
Area: 5.76 hectares
Status: National Park

It is a low, flat granite island with a rocky shoreline.

BREEDING SEABIRD SPECIES
Little Penguin

500 pairs were scattered over the island, but tend to be concentrated in pockets, usually in natural depressions in topography where the soil has accumulated. They come ashore at well-worn runways on all shores. Most are in burrows, which are short and shallow, but a few also nest under thick vegetation.

Short-tailed Shearwater

200 pairs nest in low densities scattered throughout the Little Penguin colonies.

No other seabirds were recorded.

BIRDS
Introduced:
Skylark

REPTILES
Ocellated skink

COMMENTS

Management measures should be introduced to ensure that the Little Penguin and shearwater habitats are protected from inappropriate visitation or netting practices nearby.

Little Christmas Island

(Schouten Island Group, page 410)

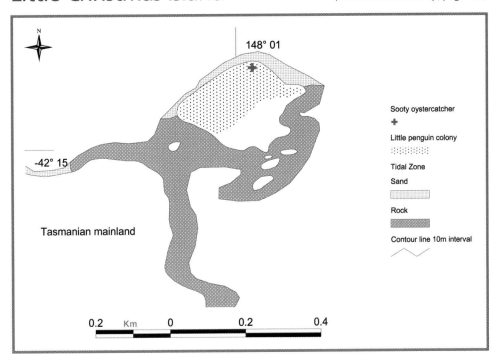

Location: 42°15'S, 148°01'E
Survey date: 12/11/87 and 28/10/99
Area: 2.2 hectares
Status: Non-allocated Crown Land ⇑

This is a small, round flat island with a shingle and dolerite boulder shore, seperated from mainland Tasmania at low tide by a narrow and shallow stretch of water.

BREEDING SEABIRD SPECIES
Little Penguin
243 pairs, most on eggs, some with empty nests, occur in burrows scattered all over the island amongst *Poa poiformis* and under *Tetragonia*, the highest densities on the northern and western sides. Penguins gain access through the coastal vegetation around the island except for the southern tip.

Sooty Oystercatcher
1 pair was sighted on the northern coast of the island between *Tetragonia* patches.

A shallow, narrow stretch of water separates this island from mainland Tasmania.

OTHER SEABIRD SPECIES

Black-faced Cormorant – 7 individuals were recorded on the rocky reef between the island and the mainland.

MAMMALS

Rabbit

No other fauna was recorded.

VEGETATION

Most of the island is covered in introduced grasses and *P. poiformis*. Some boxthorn and *Tetragonia* are present.

COMMENTS

Regionally important for its numbers of Little Penguins, this island is not particularly vulnerable to disturbance, but a status upgrade is desirable. The rabbit population should be eradicated.

Maria Island Group

(Region 6, page 409)

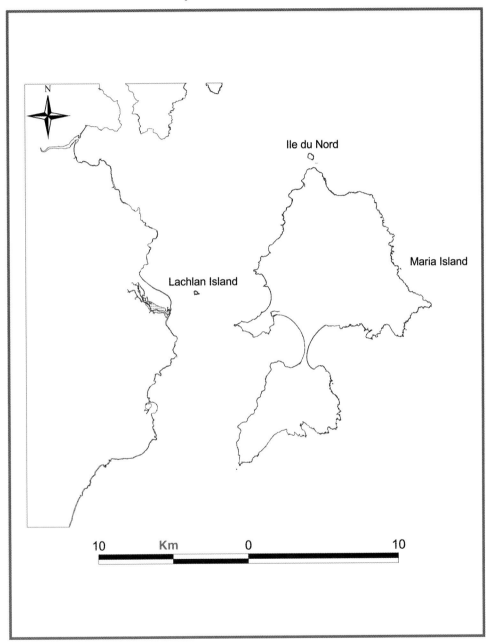

Ile du Nord

(Maria Island Group, page 431)

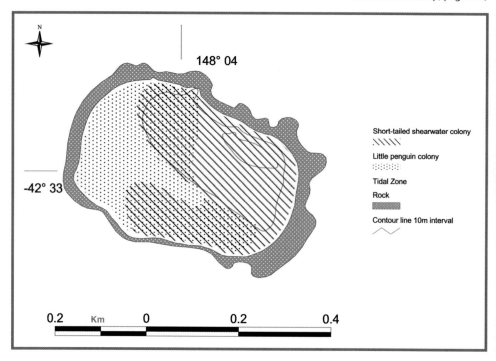

Location: 42°34'S, 148°04'E
Survey date: 19/12/83
Area: 9.65 hectares
Status: National Park

Part of the Maria Island National Park, this is a small oval island with a pebble beach 20 metres wide along the south shore, gently sloping to the north edge where it drops steeply for 30 metres to the rocky shore. There is a light beacon on the west side of the island.

BREEDING SEABIRD SPECIES
Little Penguin

2000 – 3000 pairs, with chicks were found nesting in clefts and burrows around the island but concentrated on the north shore and under dense vegetation on the south-west shore.

Short-tailed Shearwater

An estimated 3150 pairs were surveyed and a number of dead shearwaters were located. The main shearwater colony is concentrated on the north-east end of the island (50 m x 70 m) and is dominated by *Poa* and *Carpobrotus rossii*. Several others are concentrated on the flat inland from the south pebble beach. Other burrows are scattered elsewhere in some places merely under dense matted *Poa*. The soil in the colonies is sandy and loose and collapses easily.

VEGETATION

The vegetation is dominated by *Poa* and introduced grasses.

Looking south with east slope of Maria Island in the background

OTHER SEABIRD SPECIES
White-faced Storm-Petrel – 3 remains were found near the light beacon.

Great Cormorant - 26

Pacific Gull – 1 flew by

Kelp Gull – 1 adult

Silver Gull – 1

MAMMALS
Water Rats – were found amongst the tussock

BIRDS
Masked Lapwing – 2

White-faced Heron

Swamp Harrier – 1

REPTILES
Metallic Skink

COMMENTS
This small island is a significant Little Penguin breeding site for the region and further investigation may reveal White-faced Storm-Petrels breeding here in low numbers. It lies adjacent to the Maria Island Marine Reserve, but is not included within the reserve.

Rock shoreline surrounding the island.

Maria Island

(Maria Island Group, page 431)

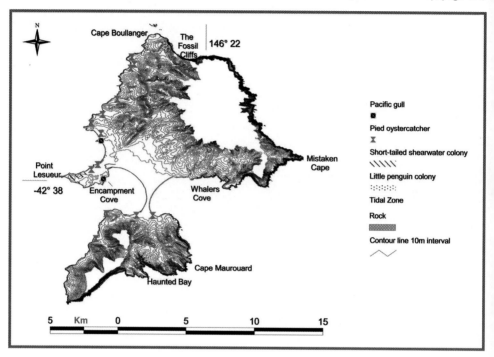

Location	42°40'S, 148°38'E
Survey date:	18/10/99 - 20/10/99
Area:	10,127.10 hectares
Status:	National Park

Maria Island is characterised by two distinct sections joined by a low, narrow isthmus. The Maria Range forms the spine of the northern island, extending from Bishop and Clerk (630 metres) in the north, south to Mt Maria (709 metres) then to Perpendicular Mountain (340 metres). On both the eastern and western sides of the range are steep scree slopes. The eastern coastline consists of an indented line of granite headlands and cliffs which rise to 140 metres at Mistaken Cape. In the north, coastal cliffs rise to 300 metres at Fossil Bay. The western coastline is characterised by dune-barred lagoons behind a series of sandy beaches, interspersed with dolerite and sandstone points. Large stretches of the west coast are either pasture down to the shore or casuarina forest with a thick mat of needles. The major ridge-line of the southern section of Maria Island consists of Big, Middle and Bottom Hills and ends in the dolerite pillars of Cape Peron. The eastern coastline between Barren Head and Cape Bald consists of a series of rounded granite headlands.

BREEDING SEABIRD SPECIES
Little Penguin

775 pairs, on eggs or empty nests, occurred in four small colonies located in the Cape

Boullanger area, the main one of 31 pairs located along the Fossil Cliffs, where they were nesting under boulders and in *Tetragonia* and *Carpobrotus rossii*. Around Darlington Jetty, 14 pairs were found nesting in culverts, under concrete rubble, *Poa poiformis,* barns and other buildings. The largest colony was located at Haunted Bay in the south-east of the southern section of the island, where there are an estimated 720 pairs. Penguins gain access to this area via very steep granite slabs and also via the pebbly beach at the head of the bay. The penguin density was low extending 100 metres into the scrub which consisted of a mixture of *Casuarina*, *Eucalyptus*, bracken, *Leptospermum* and *C. rossii*.

Short-tailed Shearwater

An estimated 5200 pairs inhabit three main colonies, one of about 50 pairs at Whalers Cove, one of about 5000 pairs south of Cape Maurouard and the third of approximately 150 pairs at Point Leseur.

Pacific Gull

One nest was located north of Encampment Cove and an old nest was found on the point in the middle of Booming Bay.

Pied Oystercatcher

13 pairs were sighted all along the west coast of the island, with one pair on the north-east coast of the southern section.

VEGETATION

Fifteen units of vegetation and 566 species of vascular plants have been recorded on Maria Island. The most extensive vegetation type is open-forest of *Eucalyptus obliqua* with *E. globulus* and *E. viminalis* with a shrubby understorey. Of the 566 species, 90 are introduced species.

COMMENTS

One of the major factors affecting the integrity of Maria Island as a national park is the number of introduced plants and animals that have changed and are continuing to change the ecosystem and degrade the natural environment. This degradation started when the island was the focus of mining and agricultural enterprises. Its geology is of global significance, particularly the fossil cliffs at Fossil Bay, which are considered of representative and outstanding significance at an international level (Dixon, 1996).

OTHER SEABIRD SPECIES

Southern Fulmar – 2 were found dead on the coast between Howell Point and Return Point.

Common Diving-Petrel – 5 carcasses were found along the west coast of the island.

Little Pied Cormorant – 2 were roosting on the north-west coast of the south section.

Black-faced Cormorant – 14 were roosting alongside the Great Cormorants.

Great Cormorant – 8 were roosting on points south of Bloodstone Bay and Point Leseur.

Silver Gull

MAMMALS

Native:

Tasmanian Pademelon
Long-nosed Potoroo
Common Wombat
Common Ringtail Possum
Echidna
Water Rat
Swamp rat

Introduced:

Eastern Grey (Forester) Kangaroo
Red-necked (Bennett's) Wallaby
Common Brushtail Possum
Tasmanian Bettong
Eastern Barred Bandicoot
Southern Brown Bandicoot
Rat
Mouse
Cat
Deer

BIRDS

A field survey for the *Maria Island National Park and Ile des Phoques Nature Reserve Management Plan 1998* records 129 species.

REPTILES

Tiger Snake
Copperhead Snake
White-lipped Whip Snake
Blue-tongue Lizard
Metallic Skink
Ocellated Skink
Tasmanian Tree Skink
Three-lined Skink
White's Skink
She-oak Skink

Lachlan Island

(Maria Island Group, page 431)

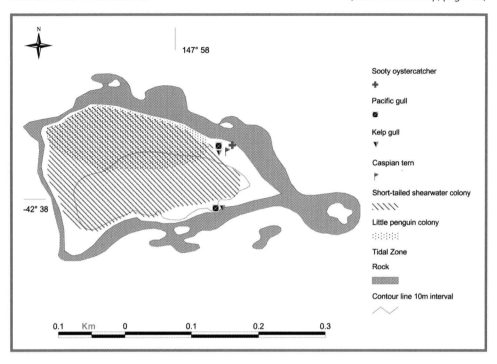

Location: 42°39'S, 147°59'E
Survey: 19/12/83
Area: 2.5 hectares
Status: Non-allocated Crown Land ⇧

This is an elongate island running west to east. It has been severely grazed by rabbits, which has had a marked effect on the seabird breeding habitat. Cliffs on the south shore slope gently to the north, where there is a pebble and rock beach. On the east end is a pebble spit out to an offshore reef.

BREEDING SEABIRD SPECIES
Little Penguin
200 - 300 pairs occur throughout the shearwater colony, particularly in the north of the island.

Short-tailed Shearwater
An estimated 940 pairs in average densities of 0.47/m^2 were scattered everywhere, with the highest densities in the north. They burrow in bare soil as the *Poa* had been ravaged by rabbits.

Pacific Gull/Kelp Gull (indistinguishable)
60 -70 pairs, 53 suspected to be Kelp Gulls, were at nests with eggs, chicks and runners. At least 4 dead runners were seen. The gulls were concentrated at the eastern end, spreading out half-way along the north shore and a quarter of the way along the south shore. Most nests, comprised of dry grass, were along the shoreline. Closer inspection did not take place due to possible disturbance.

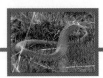

Eastern point. Main gull area.

Sooty Oystercatcher
One pair was defending aggressively on the north-eastern side.

Caspian Tern
1 pair, with one dead chick and one fledgling, was recorded at the north-east end of the island.

Shearwater habitat.

VEGETATION
Poa poiformis is the dominant species. Much of the vegetation has been destroyed by rabbits.

OTHER SEABIRD SPECIES
Black-faced Cormorant – 200+ were roosting on the offshore reef to the east of the island.

MAMMALS
Introduced:
Rabbit – eradication was attempted in the early 1990s, but was not completely successful until the second attempt in the late '90s

No reptiles or other birds were recorded.

COMMENTS
This is an important seabird breeding island, whose status should be upgraded to nature reserve.

Tasman Island Group

(Region 6, page 409)

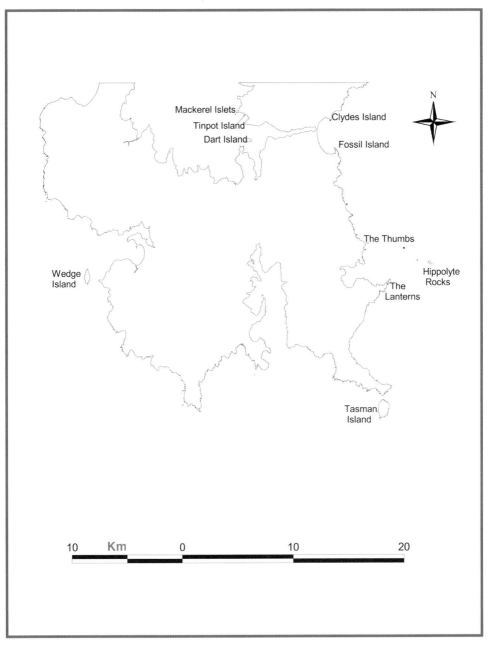

Hippolyte Rocks

(Tasman Island Group, page 439)

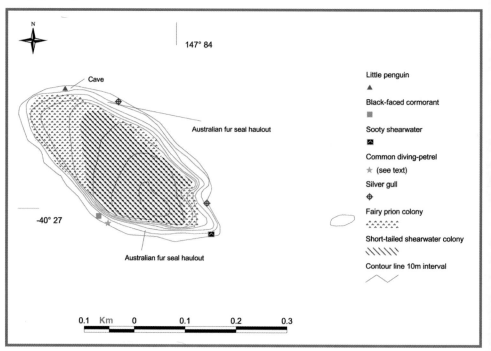

Location: 40°27'S, 148°03'E
Survey date: 26/11/80 and numerous other visits
Area: 5.3 hectares
Status: National Park

The island is composed of granite and has a distinctive, relatively flat top, surrounded by spectacular cliffs ranging in height from 25 m to 65 m, which are intersected by ledges and crevices. The flat top of the island slopes gently to the summit at the south-east end.

BREEDING SEABIRD SPECIES
Little Penguin
Only 1 pair was located in a crevice at the northern end. Access to other suitable breeding sites is prevented by cliffs.

Short-tailed Shearwater
1000 – 2000 burrows are generally restricted to the plateau. At the time of the survey, most of the birds were incubating freshly-laid eggs. At the southern end a few pairs were nesting under boulders. The density of burrows appears to be restricted by shallow soil. The highest densities were restricted to patches of *Poa poiformis*. Some birds were nesting on the surface in the dense *P. poiformis* tussocks.

Sooty shearwater
1 pair was recorded incubating an egg in a deep rock crevice at the southern extreme of the plateau.

Fairy Prion
3000 – 5000 pairs, most with eggs, were recorded all over the island. The nests are in

crevices along the cliffs extending from just above the wave-washed zone to the top. The highest densities were found round the border of the top of the island in crevices and under boulders, especially on the west, south and south-east sides. On the plateau, the burrows are less than 0.5 m in depth and are interspersed with Short-tailed Shearwater burrows. The shallow burrows make the birds accessible to predation by Forest Ravens.

Common Diving-Petrel

One wing was found in a crevice on the south-west side of the island. Birds would have fledged by the time of the visit. There were possibly less than 10 pairs inhabiting crevices.

Silver Gull

An estimated 60 to 80 pairs were breeding on the island. 128 individuals were counted and disused nests and egg shells were present. No chicks or runners were present.

Birds were defending sites on the northern end of the island with another group on the centre of the eastern edge cliffs. Birds were roosting on ledges below the cliff top on the centre of the western cliff.

Black-faced Cormorant

405 pairs were recorded, some at newly-constructed nests and some birds incubating. The majority had 2 or 3 eggs. 194 nests were located on the cliffs at the west-south-west side of the island. These nests extended from the top of the cliffs down to 30 m and were constructed of mainly *P. poiformis*, *C. rossii* and *Senecio* sp. 211 nests were located on the east cliff from the centre to the northern end and two thirds of the way down. These nests were constructed of regurgitate and vegetation. All nests were located on cliff ledges.

VEGETATION

The top is covered in low vegetation dominated by *P. poiformis*, *Tetragonia implexicoma* and *C. rossii*.

Spectacular granite cliffs of the island's west side. Note the gently sloping vegetated habitat of the burrow-nesting birds.

OTHER SEABIRD SPECIES

Cape Petrel – 2 pairs of wings

Australasian Gannet

Arctic Tern – 1 pair of wings and a foot

MAMMALS

Up to 250 Australian Fur Seals haul out on ledges around the island, particularly on the north-east.

BIRDS

Native:

Peregrine Falcon

Forest Raven

REPTILES

Metallic Skink

COMMENTS

The island is accessible by boat from nearby fishing ports and the Fortescue Bay camping ground. Despite a difficult landing and cliff access, Short-tailed Shearwaters are occasionally poached. Any disturbance may cause Black-faced Cormorants to abandon nests exposing the eggs to predation and chicks to premature fledging into the water. The presence of such a large Black-faced Cormorant colony on the island, active every year since 1980, is significant.

The island is becoming a popular site for tourists, with at least two registered charter boats taking tourists to view the seals, birds and spectacular scenery. The island and fauna inhabiting it are of commercial value to tourism in Tasmania, which warrants increased protection via restrictions on access during the breeding season.

Australasian Gannets regularly use this locality as a landing site, indicating a possibility that they may form a new breeding site here, especially given that, in all their other breeding localities, gannet numbers are increasing.

Black-faced Cormorant colony.

The Lanterns (Tasman Island Group, page 439)

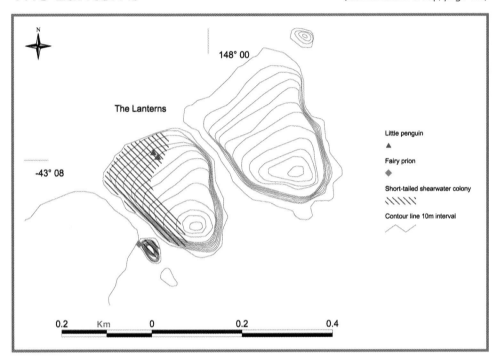

Location: 43°14′S, 148°00′E
Survey date: 26/11/80
Area: 5.35 hectares
Status: National Park

The Lanterns comprise three very steep spires, which are virtually inaccessible.

BREEDING SEABIRD SPECIES

Little Penguin
2 carcasses were found in the shearwater colony, which indicates that they may breed on the rocks.

Short-tailed Shearwater
An estimated 2200 pairs breed on the Lanterns. 92 burrows were located on the steep *Poa*-covered slopes of the westernmost Lantern, extending from the cliff face down to 10 metres. Another smaller colony of about 100 pairs occurs on the north side. The largest colony of approximately 1800 pairs is low down on the north-west side of the middle Lantern and a smaller colony occurs on the cliff top. Between the two colonies, soil depth is either insufficient or vegetation too dense for burrowing.

Fairy Prion
1 burrow was found on the westernmost Lantern.

VEGETATION
It is dominated by *Poa poiformis*.

OTHER SEABIRD SPECIES

Common Diving-Petrel – suspected to breed here

Cape Petrel – a skull was found in the crevice on westernmost lantern

Little Black Cormorant

Black-faced Cormorant

No mammals or reptiles were recorded.

BIRDS
Native:

Peregrine Falcon – a nest was located approximately 5 km south of the Lanterns.

Green Rosella

Tasmanian Scrubwren

Forest Raven

Beautiful Firetail

COMMENTS
This is a spectacular geographical feature.

The Thumbs
(Tasman Island Group, page 439)

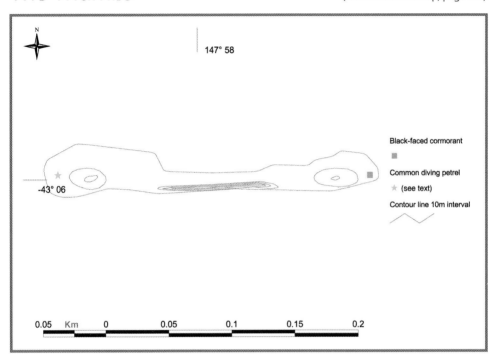

Location: 43°06'S, 147°59'E
Survey date: 21/11/80
Area: 0.45 hectares
Status: National Park

The Thumbs are three spires rising almost vertically from the sea just off the east coast of the Tasman Peninsula.

BREEDING SEABIRD SPECIES
Common Diving-Petrel
An estimated 20 pairs were located on the innermost spire (west). Burrows are in crevices or in shallow soil beneath the roots of *Disphyma crassifolium* and *Poa poiformis*.

Black-faced Cormorant
An old colony exists on the eastern spire, but was not used during 1980.

VEGETATION
Poa poiformis and *D. crassifolium* exist where there is sufficient soil accumulation between the rocks.

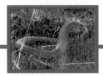

OTHER SEABIRD SPECIES
Australasian Gannet – 2 flew by
Little Black Cormorant – 9
Silver Gull – flew by

MAMMALS
Australian Fur Seals haul out on the western rock slopes.

BIRDS
Native:
White-bellied Sea-Eagle
Forest Raven

REPTILES
4 lizards (unidentified) were seen.

The three spires of The Thumbs.

COMMENTS
A spectacular topographical feature, these spires provide a breeding habitat for the Common Diving-Petrel.

Tasman Island

(Tasman Island Group, page 439)

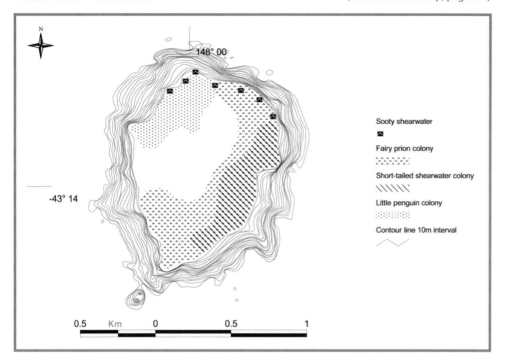

Location: 43°14'S, 148°05'E
Area: 120 hectares.
Survey date: 11 – 15/1/82, 14/6/83 and numerous other visits
Status: National Park

This is an oval island measuring approximately 1.6 km long by 1 km wide and rising to 300 metres at its peak. The average altitude of the plateau is about 280 metres. The rugged shoreline of coastal cliffs rises to meet boulder-strewn slopes, which surround a generally level plateau. Soils, where present, are mostly sandy peat. The cliffs – spectacular, vertical columns of dolerite, particularly on the western side – are considered representative and outstanding for Australia (Dixon, 1996). Deep fissures occur on the north-west side.

BREEDING SEABIRD SPECIES
Little Penguin

300 – 700 pairs, most with full-grown chicks (14/1/82) were breeding among boulders in a small area on the north-west coast.

North-east side, landing and haulageway in centre.

Short-tailed Shearwater

3000 – 7000 burrows were concentrated on the eastern slopes north of the boulder slope amongst *Poa poiformis* and *Tetragonia implexicoma* in which the main Fairy Prion colony was also located.

Fairy Prion

300 000 – 700 000 pairs, many with new chicks, were recorded in what is possibly the largest Fairy Prion colony in Tasmania. The birds breed in all suitable crevices on all cliff faces but mainly on the secondary cliffs. Only a few were found breeding in burrows. The east boulder slopes and all cliffs facing east supported more birds than any other comparable areas on the island.

Tasman Island lighthouse and associated buildings.

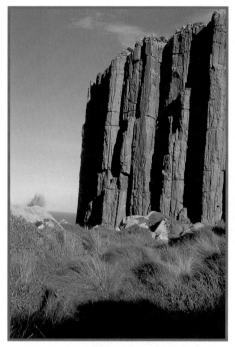

Columnar cliffs surround much of the island.

Sooty Shearwater

Up to 1000 pairs were located in sparsely scattered burrows on the steep slopes, particularly on the island's eastern and north-western sides. They appeared to be segregated from those of the Short-tailed Shearwaters.

VEGETATION

Approximately 100 plants were recorded on the island, the following species being dominant: *P. poiformis*, *T. implexicoma* (coastal and foreshore) *Acacia verticillata*, *Banksia marginata* (scrub and *Leptospermum scoparium* (heaths).

COMMENTS

Tasman Island is a very important breeding site for Fairy Prions, possibly the largest in Australia. Recently proposals have been prepared to

undertake cat eradication, which is considered essential to safeguard the many thousands of prions and other seabirds that are killed annually. Cats were introduced to the island by lighthouse keepers in the 1940s and, throughout much of its occupation by lightkeepers, the island was subjected to livestock grazing. However, since the island was abandoned as a manned light station in the mid 1970s, the native vegetation has been rapidly recolonising the plateau area with extensive areas of shrubs present. Apparently shearwaters once nested to the central east of the plateau itself, but were eliminated by agricultural activity.

OTHER SEABIRD SPECIES
Common Diving-Petrel
Shy Albatross – foraging nearby
Australasian Gannet – foraging nearby
Black-faced Cormorant – foraging in nearby water
Silver Gull

MAMMALS
Cat – A feral cat eradication program had taken place from 1978 – 1983 and the numbers were drastically reduced. In 1982 there were very few Short-tailed Shearwater carcasses, no dead sooty shearwaters and only a few dead Fairy Prions as a result of cat predation, unlike in 1978 when it was reported that one cat lair contained the remains of 91 adult prions, 11 Sooty Shearwaters and one Short-tailed Shearwater. However, the cat population has since recovered to cause immense destruction to the seabirds.

Australian Fur Seals were seen regularly in the sea around the island and were occasionally sleeping on the rocky western shore. In recent years increasing numbers have commenced occupying ledges at the north-east end adjacent to the landing platform.

BIRDS
Native:
White-face Heron
White-bellied Sea-Eagle
Wedge-tailed Eagle
Brown Falcon
Brown Goshawk
Peregrine Falcon
Yellow-tailed Black Cockatoo
New Holland Honeyeater
Crescent Honeyeater
Flame Robin
Forest Raven
Richard's Pipit
Tree Martin
Silvereye

Introduced:
Skylark
House Sparrow
European Goldfinch
Common Blackbird
Common Starling

REPTILES
Metallic Skink
White's Skink
She-oak Skink – common
Ocellated Skink

Wedge Island

(Tasman Island Group, page 439)

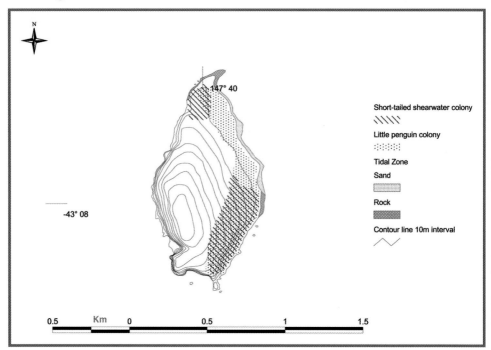

Location: 43°08'S, 147°40'E
Survey date: 5/2/85 and 24/7/85
Area: 42.58 hectares
Status: Non-allocated Crown Land

A teardrop-shaped island, orientated north to south, it is composed of Jurassic dolerite. Its cliffed western and south-eastern coasts and gentle eastern coasts with cobble beaches, give it a wedge cross-profile. It has been fairly heavily grazed by sheep and has resultant highly-modified vegetation.

BREEDING SEABIRD SPECIES
Little Penguin

150 pairs were interspersed with Short-tailed Shearwaters, particularly in the north and to the south-east of the island.

Short-tailed Shearwater

An estimated 20 000 – 25 000 pairs breed here. The main colony is in south-east of the island, with smaller numbers to the north-west in vegetation dominated by *Poa* and *Carpobrotus rossii* with some patches of *Rhagodia*. However much of the island is unsuitable for burrowing.

VEGETATION

The north-west is dominated by *Poa poiformis*, and *C. rossii* with some patches of *Rhagodia candolleana*. In some areas where the sheep have grazed heavily, particularly the north end, the *Poa* has been destroyed, replaced by cutting grass or bare sandy soil. There are a few remnant eucalypts and other scattered trees and shrubs.

OTHER SEABIRD SPECIES
Fairy Prion – flew by
Shy Albatross – fishing offshore
Australasian Gannet – diving offshore
Pacific Gull
Silver Gull
Crested Tern

MAMMALS
Cat – This population was the subject of studies into behaviour and predation rates. Eradication has yet to be attempted.

Sheep – The last were gone by 1986.

No reptiles were recorded.

BIRDS
Native:
White-bellied Sea-Eagle
Swamp Harrier
Brown Falcon
Peregrine Falcon
Tasmanian Native-hen – numerous. An old nest was found in rushes by the swamp.
Forest Raven
Eastern Spinebill
Silvereye– few

Introduced:
European Goldfinch
Skylark
Common Starling

COMMENTS
Grazing and associated burning have emaciated the island. However as grazing no longer occurs there is no justification for regular fires. The island gets visited regularly, particularly for poaching shearwater chicks in March and April.

Apparently cats were introduced to the island in the early 1970s in the belief that they would eliminate rabbits, abundant at the time. They were successful, so it seems, but cat eradication is now necessary.

The island could be used for controlled ecotourism ventures.

Sloping Island Group

(Region 6, page 409)

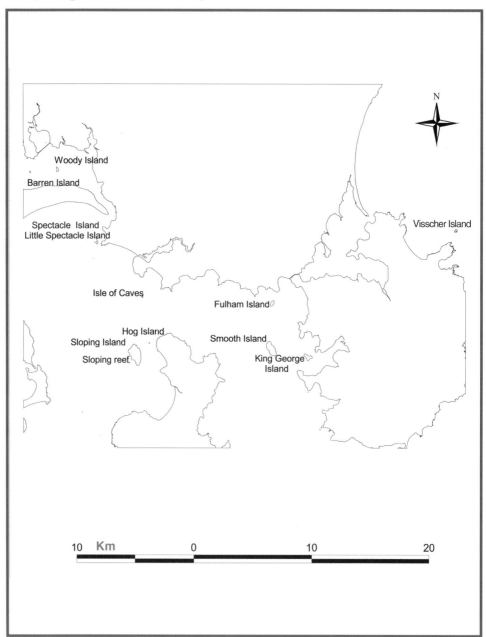

Spectacle Island

(Sloping Island Group, page 452)

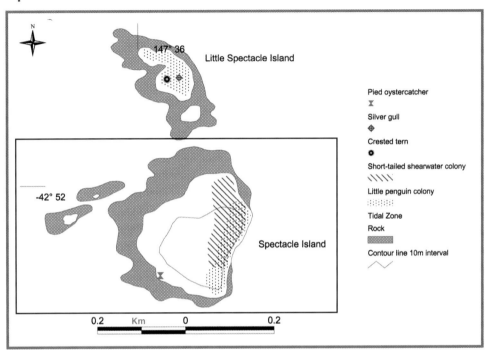

Location: 42°52'S, 147°36'E
Survey date: 23/12/87
Area: 3.5 hectares
Status: Non-allocated Crown Land ⇧

This is a flat-topped island, round to oval shape with steep to cliff sides, especially at the southern end. At the north-east tip is a beach of mussel shells.

BREEDING SEABIRD SPECIES
Little Penguin
600 pairs nest mainly on the eastern side of the island in *Rhagodia* with the occasional burrow found on the western side in the casuarina scrub. The nests are concentrated at the northern end of the island.

Short-tailed Shearwater
An estimated 8000 pairs occur on the eastern side of the island in *Rhagodia* and introduced grasses. They are interspersed with the Little Penguins.

Pied Oystercatcher
1 pair, with 2 half grown chicks, and 12 adults were sighted in the south-west.

VEGETATION
Rhagodia is the dominant vegetation, with scattered *Poa* and *Allocasuarina* sp.

No other fauna was recorded.

COMMENTS

Fortunately this island and its inhabitants are relatively resilient to the high visitor levels that occur here due to its close proximity to habitation and intensive summertime aquatic recreation. Fire is perhaps the greatest threat, a risk associated with visitation. Upgraded status and more appropriate management procedures may mitigate some of the threats.

Little Spectacle Island from Spectacle Island.

Little Spectacle Island

(Sloping Island Group, page 452)

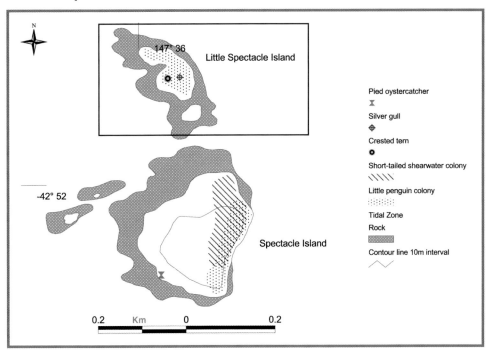

Location: 42°52'S, 147°36'E
Survey date: 23/12/87
Area: 0.62 hectares
Status: Non-allocated Crown Land ↑

It is a long, narrow, sandstone islet running north-west to south-east, steep to cliff-sided, approximately 6 metres high, with a flat top.

BREEDING SEABIRD SPECIES
Little Penguin
Up to 30 pairs occur in low densities all over island. Their burrows are short, due to the nature of the soil.

Silver Gull
500 pairs, at all stages of breeding, were sighted all over the island including on the cliff ledges.

Crested Tern
200 – 250 pairs, most on eggs or with small chicks, but some with large two-thirds grown chicks, were concentrated in a colony at the edge of the southern end. Some were amongst *Rhagodia* but the densest concentrations were in bare areas that were once *Poa*-covered but had been largely destroyed.

VEGETATION
Rhagodia is the dominant vegetation (low bushes) with scattered *Poa* and 4 – 5 metre spindly *Allocasuarina* sp. There has been much vegetation damage by gulls and terns.

OTHER SEABIRD SPECIES

Black-faced Cormorant – 3 on the south end rock

Sooty Oystercatcher – 2

Pied Oystercatcher – 8

Kelp Gull – sub-adult

No other birds, mammals or reptiles were recorded.

COMMENTS

Unlike nearby Spectacle Island, this site is particularly vulnerable to high visitation, causing chaos to surface-nesting species. When disturbed, terns will depart, leaving their eggs and chicks prey to the gulls. Signs are erected annually to caution visitors about landing during the breeding season, but stronger management procedures need to be employed and the status upgraded.

Silver Gull colonies can have an adverse impact on vegetation.

Barren Island

(Sloping Island Group, page 452)

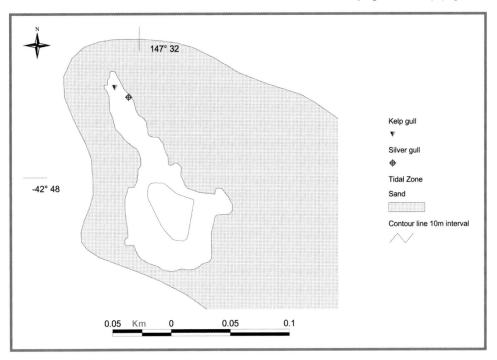

Location: 42°49'S, 148°01'E
Survey date: 21/12/89
Area: 0.53 hectares
Status: Non-allocated Crown Land

It is a triangular-shaped island orientated north to south, composed of wind-eroded sandstone with 3 metre high cliff ledges along the rock shoreline.

BREEDING SEABIRD SPECIES
Silver Gull
An estimated 10 pairs were recorded breeding. 3 had 2 eggs, 1 had 1 egg and 21 runners up to pre-fledgling stage were sighted. The nests were found on the northern shoreline within about half a metre from the water amongst rocks. There were also several nests amongst boxthorn on the flat part of the point.

Kelp Gull
1 pair, with a large downy chick and 1 immature bird, were located on the northern tip, near boxthorn.

VEGETATION
Introduced grasses are the dominant vegetation. African boxthorn is dominant along the northern tip. One radiata pine and a number of seedlings were also recorded.

View to west along north cliffs.

OTHER SEABIRD SPECIES

Black-faced Cormorant – 11 roosting

Pied Oystercatcher – 1 pair, but no sign of breeding and not aggressive.

No mammals were recorded.

BIRDS
Native:
White-faced Heron

Forest Raven

REPTILES
Three-lined Skink – 3 recorded, 2 were pregnant

COMMENT
The situation of this island refuge near human habitation makes it an ideal breeding site for gulls, which breed here in numbers that vary considerably from year to year.

Hog Island

(Sloping Island Group, page 452)

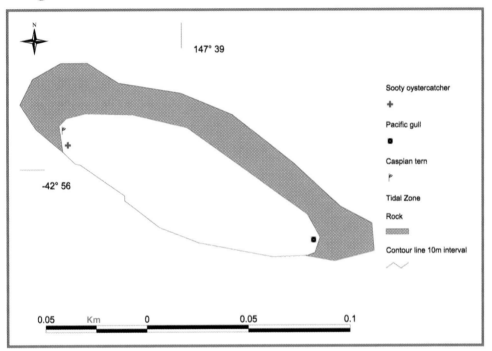

Location: 42°57'S, 147°39'E.
Survey date: 15/10/99
Area: 0.35 hectares
Status: Nature Reserve

It is a low, islet with a rocky shore and a central ridge of 8 to 10 metres running north-west to south-east.

BREEDING SEABIRD SPECIES
Pacific Gull
27 pairs were nesting all over the island, particularly concentrated on the east end.

Kelp Gull
Appoximately 50 pairs are reported to breed here (Bill Wakefield pers. comm.) but none were recorded during this survey.

Sooty Oystercatcher
1 pair was seen on rocks on the western point of the island.

Caspian Tern
1 pair was nesting on rocks at the western point of the island.

VEGETATION
Tetragonia is dominant, with *Lavatera plebeia* and African boxthorn. There is also a patch of *Disphyma crassifolium* at the south-west end.

No other fauna was recorded.

COMMENTS
This may be an important Kelp Gull breeding site.

Sloping Island and Sloping Reef (Sloping Island Group, page 452)

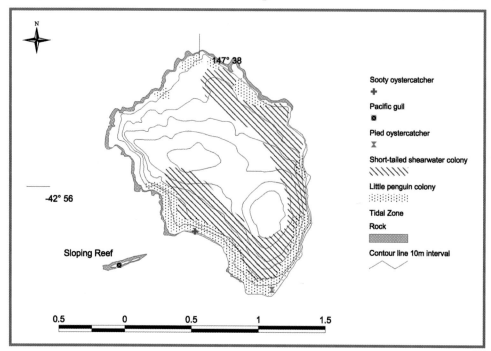

Location: 42°57'S, 147°38'E
Survey date: 15/10/99
Area: 117.19 hectares
Status: Nature Reserve

Sloping Island is a round island with a gently sloping north and north-east coastline and a steep, rocky western coast. Much of the shoreline is rocky with some small cliff gulches in the west. There are two sandy beaches in the south-west, several smaller ones to the north-west and extensive pebble beaches along the east coast. Cliffs rise to two central summits joined by a saddle. A small rocky reef lies to the south-west. There is a house site in the north.

BREEDING SEABIRD SPECIES
Little Penguin

Approximately 2000 pairs nest in colonies mainly located around the southern coastal areas in *Tetragonia*, *Carpobrotus rossii* and some bracken. There are also 3 densely populated colonies at the northern end of the island. Most burrows are interspersed with Short-tailed Shearwater burrows.

Short tailed shearwater

7650 pairs were located in two colonies, one to the north and one to the south ends of the island. The extensive middle section has no suitable burrowing habitat and the north-west area is also steep and rocky.

Pacific Gull
1 pair was sighted on the reef.

Sooty Oystercatcher
1 pair was recorded on the south-west sandy beach

Pied Oystercatcher
1 pair was sighted on the south-west point.

OTHER SEABIRD SPECIES
Black-faced Cormorant – roosting on the reef

Silver Gull

MAMMALS
Native:
Ringtail Possum

Introduced:
Rabbit

BIRDS
Native:
White-faced Heron

Swamp Harrier

Black Currawong

No reptiles were recorded.

VEGETATION
The main species recorded were *Tetragonia*, *Carpobrotus rossii*, bracken, kangaroo apple, *Allocasuarina*, *Leptospermum*, *Poa poiformis* and sedge.

COMMENTS
This island has been severely modified over many years of agricultural use and has not yet recovered. Rabbit populations undergo radical seasonal fluctuations and can cause severe denudation of vegetation, which leads to erosion and burrow collapse in shearwater colonies. Eradication of the noxious weed, serrated tussock, has occurred here for many years. It is possible that other native land mammals occur here as there is adequate habitat to support certain species. Whether ringtail possums are native or not is unknown. This is a favourite shearwater chick poaching site due to the proximity of human habitation and easy access.

Smooth Island

(Sloping Island Group, page 452)

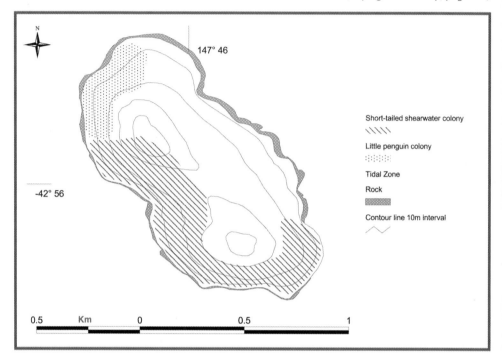

Location: 42°56'S, 147°47'E
Survey date: 2/1/93
Area: 59.31 hectares
Status: Private property

This island is a big grass paddock with thistles, bracken and scattered boxthorn with remnant eucalypts on the eastern shore and *Macrocarpa* at the house ruin site.

BREEDING SEABIRD SPECIES
Little Penguin

There is a scattered concentration of about 20 penguin burrows at the north-west end of the island, in cutting grass.

Short-tailed Shearwater

An estimated 4485 pairs breed on the island. There are two main colonies: one is along the slopes of the west side extending around to the south-east, averaging about 15 metres wide and the other is at the south-west end on the slope in dense *Poa poiformis*.

OTHER SEABIRD SPECIES

Little Pied Cormorant – 6+
Black-faced Cormorant – 10+
Great Cormorant – 15+

BIRDS
Native:
Tasmanian Native-hen
Forest Raven

No reptiles were recorded.

Island sunset

COMMENTS
This island's natural environment had been thoroughly destroyed.

Fulham Island

(Sloping Island Group, page 452)

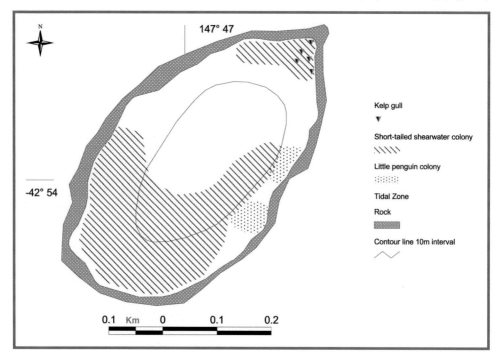

Location: 42°55'S, 147°47'E
Survey date: 2/1/93
Area: 10.14 hectares
Status: Private property

It is a low island rising gently from a rocky shore.

BREEDING SEABIRD SPECIES
Little Penguin
10 burrows were amongst those of the Short-tailed Shearwaters along the east side.

Short-tailed Shearwater
An estimated 9000 pairs occur on the island. The main colony is at the southern end in fine sand and loam amongst introduced grasses, which had formed a dense straw cover concealing the burrows. A smaller isolated colony of about 100 pairs was located on the north-west coast. There were many dead adults, possibly victims of feral cats, but this was not verified.

Kelp Gull
5 pairs, displaying aggressive behaviour, were found at newly-constructed nests in grasses at the northern tip.

VEGETATION
Boxthorns cover 10% of the island, particularly at the south end around the shearwater colony. Outside the shearwater colony, the vegetation is dominated by introduced grasses, bracken and low stunted blackwoods.

OTHER SEABIRD SPECIES

Black-faced Cormorant – 20+ were roosting on northern tip near Kelp Gull

Great Cormorant – at traditional, long-term roost site on northern tip

Silver Gull – 2 were sighted at the northern tip

Caspian Tern – 1 flew by

MAMMALS

Native:
Water Rat – suspected

Introduced:
Cat – suspected

BIRDS

Native:
Forest Raven – 3 nests in *Pinus radiata*

Brown Quail – 6+

Introduced:
House Sparrow – roosting in boxthorn

Common Starling – roosting in boxthorn

COMMENTS

The island's ecology has been severely modified through grazing practices. Due to the island being relatively accessibly there is likely to be shearwater poaching in March and April.

Visscher Island

(Sloping Island Group, page 452)

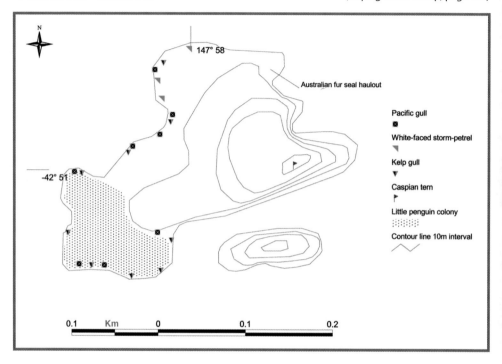

Location: 42°51'S, 147°58'E
Survey date: 21/11/91
Area: 3.38 hectares.
Status: National Park

The island, composed of dolerite bedrock, is cut by three east to west fissures, the southernmost of which is a 5 metre wide gulch separating the main island from a satellite island.

BREEDING SEABIRD SPECIES
Little Penguin
30 pairs, mainly at the south-west end of the island in short burrows, covered by matted vegetation.

White-faced Storm-Petrel
3 pairs were recorded at the north-west end in a stipa patch.

Pacific Gull
20 pairs were widely scattered particularly on the south and west coasts, but differentiating them from Kelp Gulls was difficult during this stage of breeding.

Kelp Gull
Up to 100 pairs were located on the south and west sides.

Caspian Tern
1 pair was recorded at a nest on the summit.

VEGETATION
Vegetation is dominated by stipa and *Poa poiformis* with *Tetragonia* and *Carpobrotus rossii*. Two boxthorns, *Lycium ferocissimum*, were recorded on the eastern half. One was dead. *Atriplex cinerea* was found on the higher ground to the east.

West side.

OTHER SEABIRD SPECIES

Black-faced Cormorant – at several roost sites on the summit, the north-east end and on the rocks on the central west side near the water.

Sooty Oystercatcher – 1 pair

MAMMALS

Native:

Water Rat – suspected

Australian Fur Seal – infrequently small numbers rest ashore here.

BIRDS

Native:

Peregrine Falcon – 2

Crescent Honeyeater – 4+

Eastern Spinebill

Forest Raven – 2

Tree Martin

REPTILES

Metallic Skink

COMMENTS

This is a regionally important breeding site, used consistently by large numbers of surface-nesting species, including Black-faced Cormorants in some seasons. Visitors should be discouraged from landing during breeding seasons.

Caspian Tern, hatched egg.

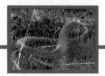

Betsey Island Group

(Region 6, page 409)

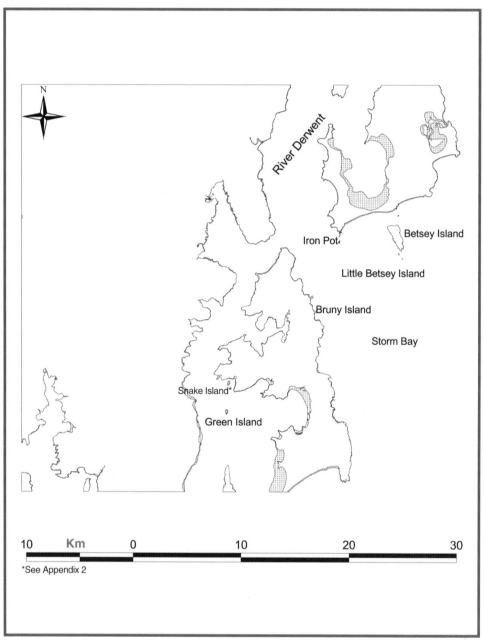

*See Appendix 2

Betsey Island & Little Betsey Island (Betsey Island Group, page 468)

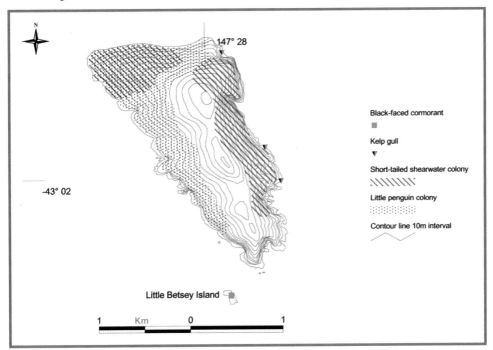

Location: 43°02'S, 147°29'E.
Survey dates: 2/8/85, 21/10/99 and 24/3/99
Area: Betsey: 175.13 hectares
Little Betsey: 0.75 hectares
Status: Nature Reserve

The main island has steep rocky cliffs on the south, east and west sides with a small rocky beach on the northern side. The east coast cliffs are especially precipitous while the west coast has a gentler gradient, providing a more suitable habitat for burrowing birds. Scree slopes dominate the southern end. Open eucalypt woodland covers most of the island, apart from the north and north-west area which has an open sandy terrain which harbours most of the burrowing birds. The main rock type is dolerite with occasional sandstone outcrops occurring on the north-east coast. Little Betsey is a steep-sided rock with a platform on the top. There are remains of buildings from when the island was farmed.

BREEDING SEABIRD SPECIES
Little Penguin

An estimated 15 048 pairs breed on the island, with burrows interspersed with those of shearwaters over most of the west coast, the highest densities occurring in the northern end where *Tetragonia* is dominant. There were many

obvious penguin runways up rocks and through *Tetragonia* on the west coast.

Short-tailed Shearwater
An estimated 150 000 shearwater burrows occur extensively around the island extending from *Tetragonia*-dominant areas to those vegetated by Kangaroo apple and bracken. They are especially ocncentrated in the western half from the north through to the south ends.

Kelp Gull
3 pairs and 11 individuals were recorded at nest sites on the eastern cliffs.

Black-faced Cormorant
232 pairs were recorded roosting on Little Betsey (24/3/99). This locality may sometimes be used for breeding.

VEGETATION
The northern end is dominated by Tasmanian blue gum (*Eucalyptus globulus)* forest, which continues to the edge of the cliffs to the east, while the southern end is sedgeland. The western end has large areas of succulent saltmarsh, dominated by *Tetragonia implexicoma*. Problem weeds include boxthorn and Cape Leeuwin wattle.

Ralphs Bay and Betsey Island. Bob Cliffords Incat parked accidently on top of the nearby reef.

No other seabird species were recorded.

MAMMALS

Silver-haired Rabbit – These were introduced in 1825, when J. King was granted a lease of the island to breed them in a venture which required selling the skins to China. The venture failed, but the rabbits thrived.

BIRDS

Native:

Brown Quail – common

White-faced Heron

White-bellied Sea-Eagle – 1 pair was seen nesting in a tree on the north-west side on the edge of open eucalypt forest.

Swamp Harrier

Brown Goshawk

Wedge-tailed Eagle

Brown Falcon

Spotted Pardalote

Noisy Miner

Yellow-throated Honeyeater

Black-headed Honeyeater

New Holland Honeyeater

Grey Butcherbird

Forest Raven

Nankeen Kestrel

Silvereye

REPTILES

White's Skink

She-oak skink

COMMENTS

Severe erosion caused by rabbits, reported on 2/8/85, was having an adverse impact on the south-east shearwater colony. Damage by rabbits has occurred extensively in the north and north-west, precipitating major burrow collapse in the sandy conditions. Although there were signs of rabbit-induced damage in the north-west in 1999, rabbit numbers appeared to be fairly low. European rabbit fleas were introduced to the island in the mid 1980s to facilitate more effective impact of the myxoma virus which, is presumed to be transported here by flying vectors such as mosquitoes.

Weed invasion, especially of Cape Leeuwin wattle, is a particular problem affecting the island's ecology.

The abundance of Little Penguins here is an important consideration in the management of set nets used legally by recreational and professional fishers. Large numbers of penguins have been known to be caught when such fishing equipment is set along this island's shoreline. These nets should be banned from use during the dawn and dusk periods when the penguins venture to and from the island.

A management plan to reinforce the islands' nature reserve status should be developed and should include weed and feral pest management strategies, as well as issues of visitor access.

Iron Pot

(Betsey Island Group, page 468)

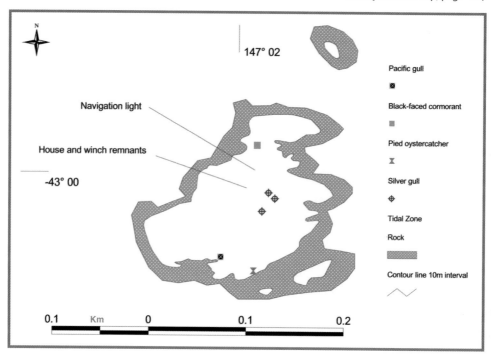

Location: 43°01'S, 147°25'E
Survey date: 21/11/91 and 12/11/99
Area: 1.27 hectares
Status: Non-allocated Crown Land ⇑

This is a small, flat rock, with extensive bare sandstone rock platform ledges around it. It has a working navigation light, ruined house and an old winch, remnants of its permanent occupation as a light station.

BREEDING SEABIRD SPECIES
Little Penguin
1 pair was located under the guttering of the house ruin (12.11.99).

African Boxthorn poses a threat to nearby seabirds but also provides security form predators.

Pacific Gull
1 pair, with 2 eggs, was located just south of the lighthouse.

Silver Gull
359 pairs, quite a few dead from being caught in boxthorn bushes, were nesting all over the island in amongst vegetation and old house ruins. 53 had no eggs, 60 had 1 egg, 124 had 2 eggs, 52 had 3 eggs and 70 had chicks.

Pied Oystercatcher
One pair was seen to the east of the Pacific Gull's nest.

Black-faced Cormorant
20 pairs, with eggs and/or new chicks, were nesting to the north of the lighthouse.

VEGETATION
The rock is dominated by boxthorn, which is hazardous to the gulls. Lupins are also abundant. *Disphyma crassifolium* and *Sarcocornia* were sparsely scattered in the shore crevices.

Western side of Iron Pot.

Remnants of light keeper's dwelling and landing facilities.

No other fauna was recorded.

COMMENTS
This is a regularly visited island, particularly in summer when seabirds are breeding. Its status should be upgraded to protect the Black-faced Cormorants that are known to breed at this site regularly in large numbers. Landing should be prohibited from September to April.

'Ten years later some terrifying experiences were recalled of a raging gale, unprecendented since the establishment of the light. Waves broke over the top of the lighthouse, 35 metres above sea level, carrying with them all movable structures and breaking down a rubble wall.' Mercury, 10/11/82

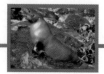

Green Island

(Betsey Island Group, page 468)

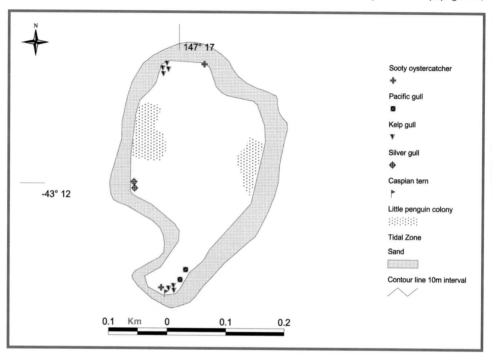

Location: 43°12'S, 147°17'E
Survey date: 5/1/88
Area: 4.17 hectares
Status: Nature Reserve

A teardrop shape, Green Island has rocky shores interspersed with rock pebble beaches, where introduced grasses proliferate.

BREEDING SEABIRD SPECIES
Little Penguin

20 pairs, most with chicks, were located in old rabbit burrows all along the west and east shores. They seem to be confined to the edges of the island and do not extend into the bracken.

Pacific Gull

2 pairs, with 2 eggs each, were sighted on the southern tip of the island.

Silver Gull

47 pairs, 37 with empty nests, the rest with eggs were found at a colony on the mid west coast nesting on shoreline rocks and in association with *Correa* sp. Some nests were built in the *Correa* shrubs one metre above the ground.

Kelp Gull

150 pairs and 185 runners were recorded. Most nests (123) were empty, but others were with eggs and/or chicks. There were two distinct colonies, one on the north-west coast with 77 pairs and one on the south coast with 73 pairs. Nests were scattered in the grassed areas and amongst thistles, with a few on the pebble beaches amongst seaweed.

Sooty Oystercatcher

2 pairs were located. One pair at the south end was defending and one pair with a two-thirds grown chick was at the north end.

Caspian Tern

2 pairs, one with an egg and one with a half-grown chick, were located at the southern end in the Kelp Gull colony.

No other seabirds were recorded.

VEGETATION

Around most of the shoreline, introduced grasses dominate. Dense thistles take over in some locations. A few *Correa* sp. shrubs are scattered amongst the rocks. The central area is dominated by *Acaena* and waist-high bracken, with assorted weed beneath.

MAMMALS

Rabbit – common

BIRDS

Native:
White-fronted Chat

REPTILES

Metallic Skink

COMMENTS

Green Island is representative of an island on which a relatively high diversity of seabirds breed in an area close to regular human activity. It should be more appropriately managed, with visits prohibited from September to April.

Partridge Island Group

(Region 6, page 409)

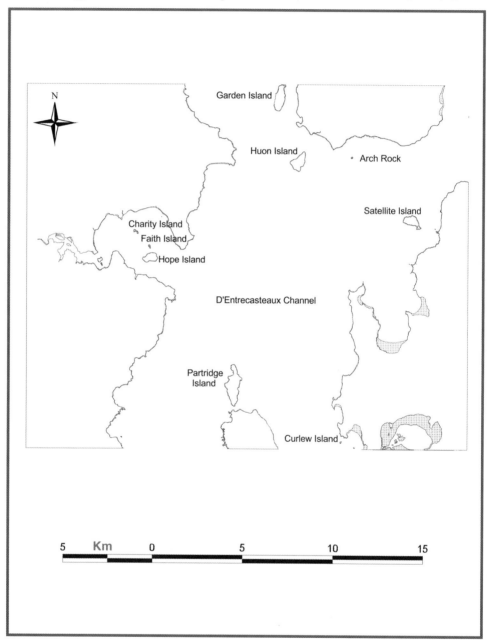

Huon Island

(Partridge Island Group, page 476)

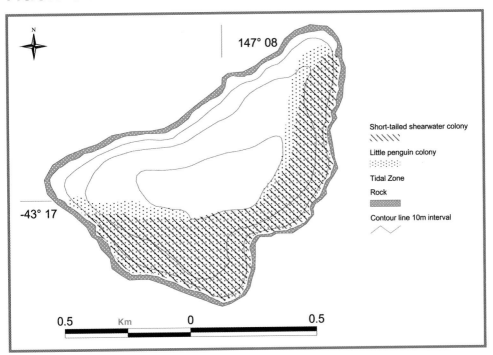

Location: 43°17'S, 147°08'E
Survey date: 4/1/88
Area: 47.01ha
Status: Multiple Land Tenure: Private property and Non-allocated Crown Land

The island is partially inhabited, with neglected, gently undulating pasture, dense bracken and large scattered white gums, mostly on the north-west side. The shores are rocky dolerite, from which the island slopes steeply for 15 – 20 metres before more flat ground is encountered. Infrastructure includes a jetty at the north-east end and several houses.

BREEDING SEABIRD SPECIES

Little Penguin

200 pairs, with big downy or fledged chicks, were found all along the eastern shore in rock crevices at the shoreline and in burrows.

Short-tailed Shearwater

An estimated 1000 pairs breed on the island. The main colony is on the east side on a gentle slope to the rocky shore with the vegetation dominated by *Tetragonia* and scattered *Poa*. There are also scattered burrows in dense bracken on the island slopes. Burrows are generally within 30 metres of shore rocks on slopes.

No other bird species were recorded.

MAMMALS
Rabbit – old warrens were located.

REPTILES
Metallic Skink

VEGETATION
Much of the island is covered with introduced grasses. White gums and bracken dominate the north-west.

COMMENTS
The island has, in the past, been subjected to intensive agricultural activities, having substantial and extensive alluvial soil deposits. Shearwaters are suspected to be a relatively recent arrival and certainly prior to land clearance, habitat would have been entirely unsuitable.

Arch Rock

(Partridge Island Group, page 476)

Location: 43°17′S, 147°11′E
Survey date: 4/1/88
Area: 0.44 hectares
Status: Non-allocated Crown Land ⇑

Lying adjacent to Ninepin Point Marine Reserve, it is an elongate sandstone islet gently sloping to the west with 5 metre high cliffs on the east side. A cave forms an archway through its centre. Marine weathering highlights the fossils. There is no suitable soil for burrowing species.

BREEDING SEABIRD SPECIES
Pacific Gull
A nest was found on the south-east tip on the edge of the *Poa* on bare rock.

Kelp Gull
2 nests were located on the south-east tip on the edge of the *poa* on bare rock 6 metres from the Pacific Gull's nest.

VEGETATION
Poa poiformis is the dominant vegetation.

OTHER SEABIRD SPECIES

Black-faced Cormorant – a roost site was located on the central east cliff and only one bird was present during the visit.

Crested Tern – 2 were diving off the northern tip.

No mammals or other birds were recorded.

REPTILES

Metallic Skink

COMMENTS

Geomorphically, Arch Rock is considered to be of outstanding significance for the local region (Dixon 1996), warranting an increase in its status to conservation area.

Charity Island

(Partridge Island Group, page 476)

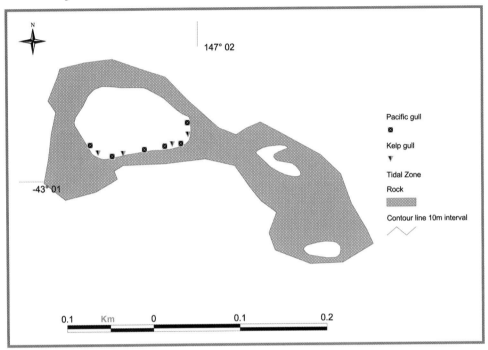

Location: 43°01'S, 147°02'E
Survey date: 21/21/87
Area: 0.6 hectares
Status: Non-allocated Crown Land

It is a low, flat, round island with a rocky shoreline. Its partner islands Faith and Hope have no breeding seabirds, the latter being devoid of native vegetation apart from a few eucalypts. Faith Island, which is covered in scrub, had Black-faced Cormorants, and Pacific and Kelp Gulls present, but they were not acting territorially. Two gravestones on Faith Island, one marked with the date July 18, 1862, the other with November 1, 1872, were concealed by scrub.

BREEDING SEABIRD SPECIES
Pacific Gull
6 pairs, one with chicks, the others with newly-constructed nests, were located along the edge of the vegetation on the southern shore.

Kelp Gull
4 pairs were defending on the rocky shoreline near the Pacific Gulls' nests.

VEGETATION
There is scrub dominated by eucalypts over the island.

OTHER SEABIRD SPECIES
Pied Oystercatcher

No other fauna was recorded.

COMMENTS
The island's vegetation is very characteristic of that of the adjacent Tasmanian mainland.

Partridge Island

(Partridge Island Group, page 476)

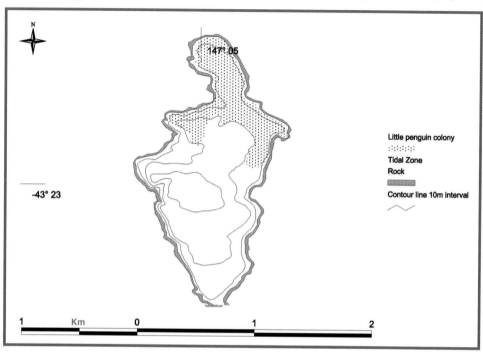

Location: 43°24'S, 147°05'E
Survey date: 17/5/88
Area: 102.58 hectares
Status: National Park

It has a highly modified landscape, being easily accessible by boat from Bruny Island and mainland Tasmania. The island is mostly flat to the east, with a flat interior sloping to a west coast escarpment at around 25 metres, which drops steeply to the dolerite rock shoreline. To the south end, vegetation is very dense and often impenetrable but this reverts to more open eucalypt woodland to the north. Generally the soils are shallow and inadequate for burrow-nesting species, with grazing and burning having been regular activities. There is a substantial jetty and a timber shelter.

BREEDING SEABIRD SPECIES
Little Penguin

50 pairs were recorded in short burrows and cavities at the north end up to 50 metres from sea level with a few at the south end.

VEGETATION

The vegetation is dominated by open, dry eucalypt forest with grass understorey, similar to nearby Bruny Island.

OTHER SEABIRD SPECIES

Short-tailed Shearwater – several skeletons were found after the fire in 1988, but there are no obvious signs that they breed here.

MAMMALS

Rabbit – were once on the island, but they have now been eradicated.

Native mammals are likely to occur as habitat is suitable and the adjacent mainland is very close.

BIRDS
Native:
Brown Quail
White-bellied Sea-Eagle
Brown Goshawk
Peregrine Falcon
Green Rosella
Forty-spotted Pardalote
Tasmanian Scrubwren
Strong-billed Honeyeater
Grey Shrike-thrush
Black-headed Honeyeater
Forest Raven
Introduced:
Common Blackbird

REPTILES
Blue-tongue Lizard

COMMENTS

The island was burnt in 1988 and has been subjected to grazing in the past. Being a popular boating destination, it is always likely to be subjected to fires and, whilst they are not necessarily of much consequence to the seabirds, the effects on vegetation and species such as the Forty-spotted Pardalote, an endangered species, is an important consideration that requires management.

Curlew Island

(Partridge Island Group, page 476)

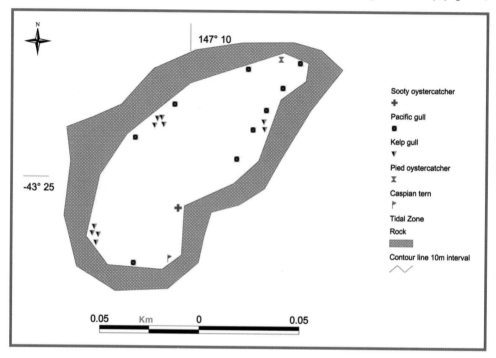

Location: 43°26'S, 147°10'E
Survey date: 16/11/91
Area: 0.415 hectares
Status: Non-allocated Crown Land ↑

An elongate island orientated north-west to south-east, it is flat, only half a metre above sea level and has a rocky shoreline surrounding it. There is no burrowing habitat on the island.

BREEDING SEABIRD SPECIES
Pacific Gull
40 pairs, although difficult to ascertain amongst the Kelp Gulls, were recorded all around island. They were generally at new nests.

Kelp Gull
20 pairs were estimated to be breeding.

Sooty Oystercatcher
1 pair, at a nest with 2 eggs, was located at the south-east amongst *Carpobrotus rossii*.

Pied Oystercatcher
1 pair was sighted at the north end.

Caspian Tern
1 pair was located at the south end of the island.

VEGETATION
It is dominated by introduced grasses, *Lavatera plebeia* and *C. rossii* with a thin stipa band around the shoreline. In from the immediate shore, *Sarcocornia quinqueflora* is dominant. There is a small patch of stunted blackwoods on the north-east side with *Rhagodia* adjacent.

OTHER SEABIRD SPECIES

Black-faced Cormorant – 18 roosting

Little Black Cormorant – 3 roosting

Great Cormorant – 10 roosting on the east side

Silver Gull – 24 adults present

No mammals or other birds were recorded.

REPTILES

Metallic Skink

COMMENTS

It is locally important as a gull nesting site and a status upgrade should be considered.

Actaeon Island Group

(Region 6, page 409)

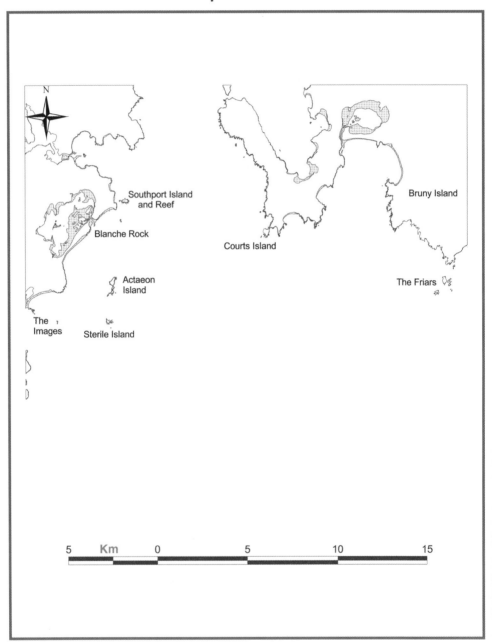

Courts Island

(Actaeon Island Group, page 486)

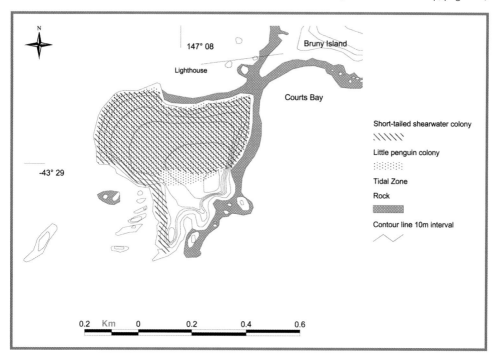

Location: 43°30'S, 147°08E
Survey date: 5/12/91
Area: 15.83 hectares.
Status: National Park

Part of the South Bruny Island National Park and connected by a spit at low tide, this island is divided into two distinct sections. The northern section provides good burrowing habitat. The southern section is characterised by severely wind-pruned scrub, bare open patches, rock, no soil at all and small dolerite boulders. The southern slopes are steep to sheer dolerite, with a boulder beach on the mid south side.

BREEDING SEABIRD SPECIES
Little Penguin
150 – 200 burrows are scattered all over shearwater areas, with well-worn tracks up the steep north-west side. Some birds also nest on the upper shore boulder slope and on slopes beneath *Rhagodia* and *Tetragonia*.

Short-tailed Shearwater
Breeding in densities of $0.94/m^2$, 180,825 pairs occur in the northern end of the island in well-vegetated areas of *Tetragonia*, *Rhagodia*, *Carpobrotus rossii* and grasses. Burrows are very deep and long. The small colony of 50+ on the south side of the escarpment is dominated by *C. rossii*.

Sooty Shearwater
Although this species was not found on the island during this survey, it is known to breed here.

OTHER SEABIRD SPECIES
Sooty Oystercatcher – feeding on spit rock area

MAMMALS
Cat – very common around the lighthouse

Goat – possibly visit the island: faeces were seen and one dead goat was found at the base of the shore-side track.

BIRDS
Native:
Tawny-crowned Honeyeater

Forest Raven

Little Grassbird

VEGETATION
The northern section is dominated by *Tetragonia*, *Rhagodia*, *C. rossii* and grasses. The southern section is characterised by severely wind-pruned scrub dominated by *Banksia* sp.

COMMENTS
Adjacent to a popular tourist site, it would benefit from being fenced and used as a controlled ecotourist destination.

View to the south over Courts Island from nearby Cape Bruny.

The Friars

(Actaeon Island Group, page 486)

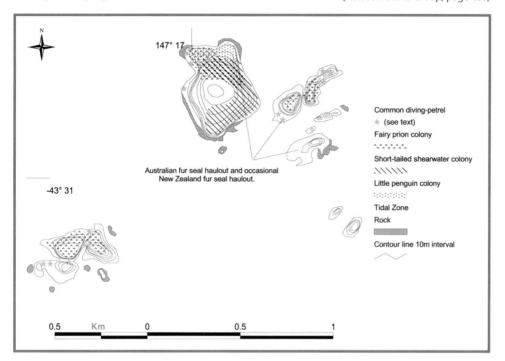

Location: 43°32'S, 147°17'E
Survey date: 18/11/91
Area: North rock: 10.28 hectares, north-east rock: 1.55 hectares, south rock: 5.13 hectares
Status: National Park

The Friars are four steep dolerite rocks off South Bruny Island, forming part of the national park. The northern rock has a long narrow ridge running north to south with the summit at the south end. It has a high ridge on the east side which is mostly sheer. The west side is steep to very steep. The north-east rock is elongated running north to south and pyramid shaped with the summit at the south end. It is steep to sheer all around except for the north face which is very steep. The southern rock slopes to the summit at the north end, then flattens out onto a south-east gently sloping expanse of densely matted Poa and Carpobrotus rossii. The south, east and west slopes are very steep. Evidence suggests that this rock was burnt in the early 1980s.

BREEDING SEABIRD SPECIES
Little Penguin
50 pairs were found on the northern end of the main rock, where there is access extending up to the summit.

Short-tailed Shearwater
An estimated 1870 pairs were located in a colony on the south-east gently sloping expanse of dense,

East side looking north to Bruny Island with mainland Tasmania in background.

matted *Poa* co-dominant with *C. rossii* on the northern end of the summit of the main rock.

Fairy Prion
An estimated 600 pairs breed on the northern rock, approximately 1000 on the north-east rock and over 20 000 on the southern rock in the same areas as the diving- petrels.

Common Diving-Petrel
An estimated 1800 pairs were located on the northern rock, with the densest concentrations on the lower slopes, where vegetation is more sparse. Burrows are shallow. On the north-eastern rock 1440 pairs occur in areas dominated by *Carpobrotus rossii* and 40% *Senecio* sp. On the southern rock, an estimated 20 000 pairs occur, but concentrations are densest in *C. rossii* and *Senecio* sp. on the southern and eastern slopes. Most surveyed had large chicks.

VEGETATION
The dominant species are *C. rossii* and *Senecio* sp., with patches of *P. poiformis*.

OTHER SEABIRD SPECIES
Black-faced Cormorant

MAMMALS
Australian Fur Seal – use the rocks as a regular haul-out site.

New Zealand Fur Seal – may also haul out here at times.

BIRDS
Native:
Peregrine Falcon

Forest Raven – 15+

REPTILES
Metallic Skink

COMMENTS
Whilst fur seals here, as at many localities in the region, are persecuted, these islands are otherwise secure from disturbance, as access ashore is often difficult. It is, however, a site for seal-watching tours.

Southport Island & Southport Island Reef
(Actaeon Island Group, page 486)

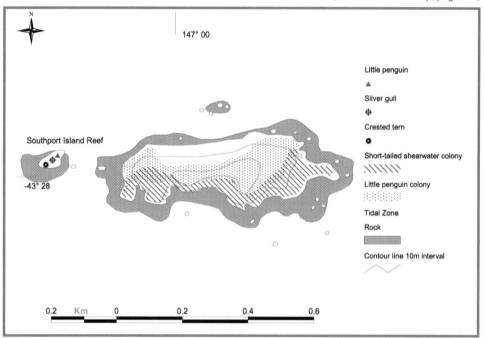

Location: 43°28'S, 147°00'E
Survey date: 4/12/80
Area: Island: 6.9 hectares,
Reef: 0.2 hectares.
Status: Non-allocated Crown Land ⇑

The island is steep to the water on the south side with more gentle slopes on the north side. There is a small amount of flat ground along the central ridge, which runs north to south. The reef located 50 metres off the south-west end of the island is flat, rocky and low with no suitable burrowing habitat.

BREEDING SEABIRD SPECIES
Little Penguin
Interspersed with Short-tailed Shearwater colonies throughout the island, an estimated 200 burrows occur in variable densities, with high densities limited to areas of *Poa* and *Tetragonia*. Many were with chicks.

One pair was found breeding beneath *Poa poiformis* on the reef.

Short-tailed Shearwater
An estimated 21 000 pairs occur in average densities of $0.46/m^2$ predominantly in areas dominated by *Poa* and *Tetragonia*. There were none in the scrub patches and bracken or blackberries at the northern end. Most were on eggs in fresh, cleaned burrows.

Silver Gull
210 pairs, 46 with 1 egg, 108 with 2 eggs, 34 with 3 eggs, 5 with 1 chick, 13 with 2 chicks and 4 with 3 chicks were counted on the reef. Nests are scattered all over, mainly out in the open, though

some were located under bushes and *Poa* and in crevices very close to the waterline.

Crested Tern
137 pairs, 127 nests with 1 egg, 6 with 2 eggs and 4 with newly-hatched chicks, were located in a concentrated group west of centre of the reef, mainly on bare ground with small *Poa* clumps and *Senecio* sp.

VEGETATION
Coastal heath, dominated by *Acacia melanoxylon, A. verticillata, Banksia marginata, Leptospermum scoparium, Melaleuca squarrosa* and *Westringia brevifolia* is the main vegetation community, with *Poa* and *Tetragonia* dominating the Little Penguin and Short-tailed Shearwater habitat. At the time of the survey, blackberries were widespread and believed to be a problem.

Surface nesting seabirds like gulls and terns are disturbed by visitors

OTHER SEABIRD SPECIES
Australasian Gannet – flew by
Black-faced Cormorant – 2 on reef
Sooty Oystercatcher – 1 on island
Pacific Gull – 1 on island

BIRDS
Native:
Crescent Honeyeater
Tree Martin – 2 on reef
Little Grassbird
Silvereye

Whilst no mammals were recorded, habitat here is adequate to support a native rodent population.

REPTILES
Metallic Skink – found on the island.

COMMENTS
The island is currently being used for the *ex situ* planting of the endangered heath *Epacris stuartii* because its natural habitat, Southport Bluff, is threatened by the fungus *Phytophthera cinnamomi*. The plant's translocation to the island is an attempt to provide an environment similar to its natural habitat, but in which *Phytophthera* is less likely to occur. The status of the island should be raised to afford some protection to the breeding seabirds and the relocated threatened species.

Blanche Rock

(Actaeon Island Group, page 486)

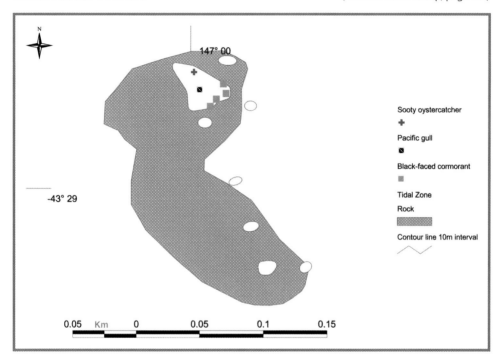

Location: 43°29'S, 147°00'E
Survey date: 4/12/80
Area: 0.07 hectare
Status: National Park

It is an oval-shaped dolerite island with a relatively flat top and surrounded by sheer sides with crevices. The southern side drops more gently to the sea and is bisected by a gulch. There is no habitat suitable for burrowing birds.

BREEDING SEABIRD SPECIES
Pacific Gull
One nest with 3 chicks was located on the summit. 3 other birds were seen.

Sooty Oystercatcher
One pair at a nest was recorded on the northern end of the island.

Black-faced Cormorant
4 old nests from the recent breeding season, constructed from seaweed, were located on the eastern side.

VEGETATION
The small flat areas on the summit are vegetated with *Tetragonia implexicoma*, *Carpobrotus rossii*, *Chenopodium glaucum*, *Solanum vescum* and *Senecio lautus*, as are some crevices.

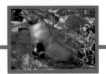

OTHER SEABIRD SPECIES
Silver Gull

No other fauna was recorded.

COMMENTS
The island is accessible by boat from nearby fishing ports but it is unlikely that people land on it.

East side, mainland Tasmania in background.

Actaeon Island

(Actaeon Island Group, page 486)

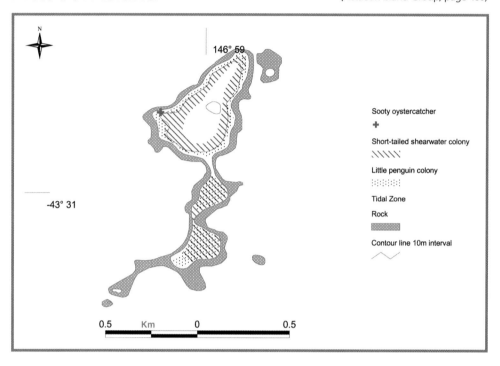

Location: 43°32'S, 146°59'E
Survey date: 4/12/80
Area: 15.65 hectares
Status: Game Reserve ⇑

The island has three sections, which are joined by a narrow neck of land formed from dolerite beach rocks. The main northern section rises sharply for up to 4 metres from a rocky shore then slopes very gently to a navigation beacon at the highest point, 14 metres above sea level. The two southern sections are flat and only a few metres above sea level. The southernmost section is completely surrounded by wave-worn cobbles, which form a narrow tombolo, some 100 metres long, linking the two bedrock parts of the island. Ridges and terraces on the tombolo and a general lack of vegetation, indicate it is continually wave-washed. This geomorphic phenomenon is considered of outstanding significance on a State level (Dixon, 1996).

BREEDING SEABIRD SPECIES
Little Penguin

An estimated 1000 pairs occur, with many on eggs or with small chicks. Burrows are scattered throughout the Short-tailed Shearwater colonies but are more common along the western and eastern shores of the southern half of the northern section. Birds were also found in rock crevices on the western coast.

Short-tailed Shearwater

25 000 pairs are estimated to be breeding on the island. Dense burrow concentrations occur in

small areas of the two southern sections of the island, with lesser concentrations occurring over most areas where soil is sufficient for burrowing. In the northern section, burrows are in a strip of up to 30 metres wide extending around the edge of the vegetated area. In some places burrows extend a further 10 metres into very dense, chest-high bracken. However, in the vegetated area, the density of the root systems prevents burrowing.

Sooty Oystercatcher

One pair, with 2 eggs, was found on the eastern rocky shoreline of the main part of the island.

VEGETATION

Vegetation includes areas of *Poa poiformis* tussock grassland, extensive patches of *Pteridium esculatum*, *Melaleuca squarrosa* and *Leptospermum scoparium*. *Carpobrotus rossii* and *Tetragonia implexicoma* are found in the littoral areas. The vegetation of the island has been frequently burnt.

Looking south over Actaeon Island to Sterile Island.

OTHER SEABIRD SPECIES

Southern Giant-Petrel – 1 feeding on a whale carcass

Northern Giant-Petrel – 17 feeding on whale carcass

Shy Albatross – derelict

Black-faced Cormorant

Great Cormorant

Pacific Gull – 2 pairs (not breeding)

MAMMALS

Rabbit – have been on the island for many years, but their numbers in 1980 appeared to be extremely low

Rat – a species was observed but its identity is uncertain. It is possibly an indigenous species as the island's habitat is suitable.

Seals – occasionally haul out

Southern Right Whale – dead on beach

REPTILES

Metallic Skink – numerous

Thick vegetation of main section, south over narrow boulder formation.

COMMENTS

Because the vegetation has been burnt over many years, the habitat suitable for burrowing has been significantly modified. Short-tailed Shearwater chicks, probably thousands are harvested annually, previously legally, now illegally. It is the island's use for this activity that has determined its status, which is inappropriate, given its population of Little Penguins. Its status should be upgraded to conservation area. Topographically it is an attractive locality, readily accessible by small boat.

Sterile Island

(Actaeon Island Group, page 486)

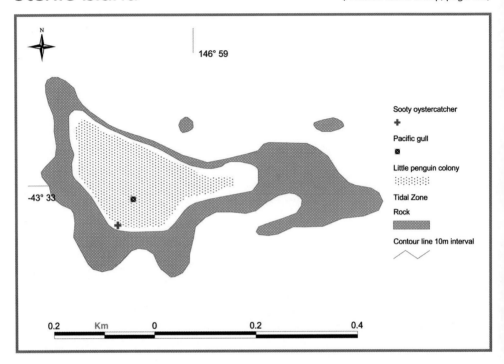

Location: 43°33'S, 146°59'E
Survey date: 4/12/80
Area: 3.68 hectares
Status: Game Reserve

The island is very flat, with a slight depression towards the centre, encompassed by an 8.5 metre levee-like rim, presumably formed by storm waves. Small round boulders form the beaches surrounding the island. There is very little soil even at the centre, which is covered only by a thin layer. It is considered of representative and outstanding geomorphic significance on a State level (Dixon, 1996).

BREEDING SEABIRD SPECIES
Little Penguin

An estimated 500 pairs, with chicks, were nesting all over the island under *Poa* and *Tetragonia*, anywhere that afforded sufficient concealment. The densest concentrations were along the north shore.

Pacific Gull

One pair and a juvenile were recorded.

Sooty Oystercatcher

One pair, with one egg, was located at the edge of the vegetation and beach rocks on the southern side.

VEGETATION

The centre of the island is dominated by *Poa poiformis* which is surrounded by a strip of shrubs, mainly *Olearia phlogopappa*. There is an outer zone dominated by *Senecio* sp which meets the stony beach. Creeping mats of *Tetragonia implexicoma* occur on the beach.

North shore, obelisk and Tasmanian mainland in background.

OTHER SEABIRD SPECIES
Silver Gull – 100+ not breeding, but displaying territorial behaviour

MAMMALS
Rabbit – although none were seen during this survey, they are believed to exist on the island.

BIRDS
Native:
Forest Raven

REPTILES
Metallic Skink

COMMENTS
Rocky reefs and rough seas make access to the island difficult, so it is little visited. Forest Raven, which visit islands in the area in large numbers, prey on penguin chicks and eggs here due to the exposed breeding conditions. A stone obelisk used for navigation stands at the west end. Whilst the assignation of an upgraded status would have little conservation impact on this island, the status of game reserve is entirely inappropriate.

The Images

(Actaeon Island Group, page 486)

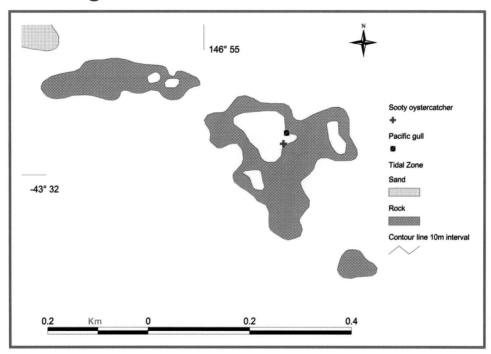

Location: 43°32'S, 146°55'E
Survey date: 7/1/88
Area: 0.53 hectares
Status: Conservation Area ⇩

The Images is a group of low, bare, rocky reefs running east to west with the central section dominated by dense stunted blackwoods, matted with extensive blackberries. There is insufficient soil for burrowing birds. The shore is rocky with patches of rock pebble beaches.

BREEDING SEABIRD SPECIES
Pacific Gull
One pair was sighted 3 metres from the Sooty Oystercatcher's nest on bare ground by *Poa* clumps on the mid east coast of the island.

Sooty Oystercatcher
One pair, acting defensively, was located 3 metres from the Pacific Gull.

OTHER SEABIRD SPECIES

Black-faced Cormorant – 5 were roosting on the outer bare rocks and one flew by.

No mammals or other birds were recorded.

REPTILES
Metallic Skink

COMMENTS
The conservation area status appears to be inappropriate.

South and West Islands – Region 7

(Regions, page 44)

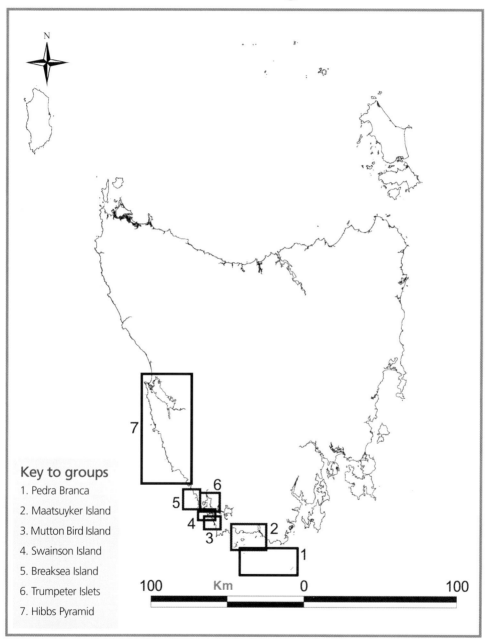

Key to groups
1. Pedra Branca
2. Maatsuyker Island
3. Mutton Bird Island
4. Swainson Island
5. Breaksea Island
6. Trumpeter Islets
7. Hibbs Pyramid

Pedra Branca Group

(Region 7, page 501)

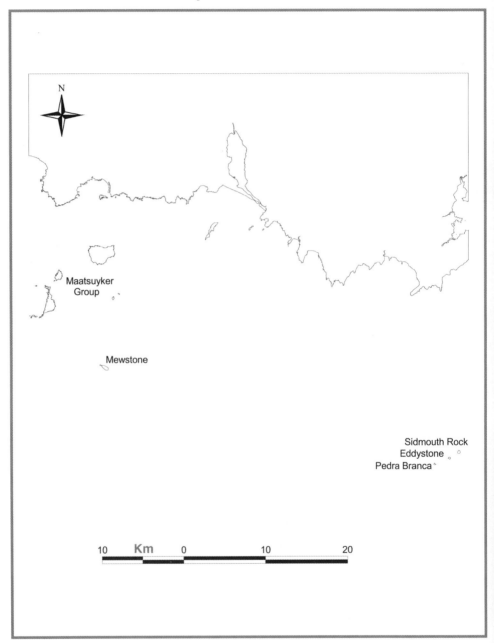

Mewstone

(Pedra Branca Group, page 502)

Location: 43°44'S, 146°22'E
Survey date: 31/12/77, 2/11/85, 4/11/86, 5/11/90, and other visits
Area: 13.1 hectares
Status: National Park

This is an oval-shaped, granite island with steep cliffs rising to a small flat summit of approximately 150 metres at the south-east end. A ridge consisting of loose boulders and crevices runs from the summit towards the north-west and slopes down to a saddle two thirds of the way along the island before rising to the lower north-west summit. The eastern and western slopes rise to meet at the ridge, the western slope being the steeper of the two. Plants ony occur in crevices where soil has accumulated.

Shy Albatross at summit of Mewstone, Maatsuyker Island and south coast of mainland Tasmania in background.

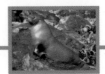

BREEDING SEABIRD SPECIES

Fairy Prion
An estimated 20 000 pairs breed in crevices and under boulders along the summit ridge, where such habitat is most abundant. They also occur over the whole island where crevice habitat occurs.

Silver Gull
A few breed on the lower south-facing cliff ledges and at least 40 individuals were recorded roosting on the island.

Black-faced Cormorant
20 pairs were counted. 10 nests and 15 roosting birds were located on a big offshore rockstack at the south end of the island. A small colony that could not be counted (possibly up to 10 pairs) was also located on the centre of the east side.

Shy Albatross
The most recent census in 1997 revealed approximately 7500 pairs. Breeding birds occur all over the island with the greater numbers on the western slopes, the summit ridge and the south-east slopes. On the eastern slope the nests occur in scattered groups.

VEGETATION
Senecio lautus, Carpobrotus rossii, Poa poiformis, Asplenium obtusatum and *Salicornia quinqueflora* were the only plant species found.

Shy Albatross on the wing

OTHER SEABIRD SPECIES

Australasian Gannet – occasional bird landed

Buller's Albatross – a single individual has been seen with the Shy Albatrosses over several seasons.

MAMMALS

Australian Fur Seal – up to 50 were hauled out on small ledges around the eastern and southern sides of the island and on the south-east rock.

BIRDS

Native:

Peregrine Falcon

Forest Raven

REPTILES

Tasmanian Tree Skink

COMMENTS

The topography of the island restricts human access to it. This is the largest of only three Shy Albatross breeding colonies in Australia. The World Heritage status and general lack of accessibility combine to give the island adequate protection. The entire island is composed of Muscovite granite, which occurs on none of the adjacent islands or coast, making it an outstanding feature for the local region (Dixon, 1996).

Tasman sighted Mewstone in 1642 and described it as 'a small island like a lion'. It was named by Captain Furneaux in 1773 and in the Admiralty sailing directions, was recorded as 'swarming with birds'.

Pedra Branca

(Pedra Branca Group, page 502)

Eastern cliff face

Location: 43°52'S, 146°58'E
Survey date: 2/8/90 and numerous other visits
Area: 2.5 hectares
Status: National Park

This rock measures 270 metres long by 100 metres wide with a maximum height of 60 metres. The east and west slopes rise steeply to meet a central ridge, which runs in a north to south direction. The eastern face commences as a rock platform sloping gently up to meet the base of a 25 metre high sheer rock face. Above the sheer face, the rock becomes less steep, thus making conditions suitable for seabirds to nest on cliff ledges.

BREEDING SEABIRD SPECIES
Fairy Prion
Approximately 10 pairs nest in the south-east crevices where the skinks are most abundant.

Pacific Gull
One pair was sighted at a nest at the northern end of the rock, but usually this species only visits.

Silver Gull
52 pairs were primarily nesting on the east and south-east slopes on the rock ledges and overhangs where skinks are most abundant. There is also a small colony on the north-west coast.

Adult gannets on a lower roosting ledge.

Kelp gull
One pair was nesting south of the Silver Gull colony on the north-west ledge, but in most years, this species is a regular visitor in small numbers.

Black-faced Cormorant
5 pairs were located at new nests. Two nests are situated on the rock ledge on the western side of the northern tip about 25 metres above sea level. Another one is on the south-east coast and two are located at the north side of the saddle at the rock's summit, all regularly-used sites.

Australasian Gannet
6000-8000 pairs breed on Pedra Branca. The major colony covers the rock's east side from about 18 metres above sea level to the top of the central ridge. The other major colony is in the north of the island. Both are interspersed with Shy Albatross nesting sites.

Australian Fur Seals asleep and playing along side Pedra Branca on a quiet sunny day.

Shy Albatross

250 pairs breed on the island. The main sites are in the south-east and central east with up to 10 in the north. Nests are interspersed with those of gannets.

VEGETATION

Sarcocornia quinqueflora is the only plant species found on the island; it is sparse and confined to rock cracks.

Pedra Branca skinks feed on fish remains when seabirds provide them with the opportunity.

Marine life flourishes here despite constant pounding by the sea.

OTHER SEABIRD SPECIES

Short-tailed Shearwater – thousands were seen feeding on dense schools of fish close to the rock

Common Diving-Petrel – 20+ come ashore to sleep, occupying rock crevices, but do not breed here.

White-fronted Tern – 3 flew by

Giant Petrel sp – 1 offshore. They regularly visit, preying on gannet fledglings in March and April.

Great Skua – 1 flew by

MAMMALS

Australian Fur Seals use this island as a regular haul-out site, with up to 500 present, when sea conditions permit.

New Zealand Fur Seal – visits occasionally

BIRDS

Native:

Silvereye – 3 on eastern side

Cattle Egret – on the south-east

Satin Flycatcher – 1

REPTILES

Pedra Branca Skink – a population of currently around 400 individuals in 4 discrete colonies. All inhabit a maze of crevices and tunnels formed like catacombs in the weathering rock, giving the skinks the protection that is critical for their survival.

COMMENTS

Pedra Branca is the site of a long-term population monitoring program of birds, seals and skinks and extensive ecological studies of their marine prey. A marine debris survey of the island has also been undertaken over the past 10 years. Silver Gulls are thought to be responsible for the declining numbers of the Pedra Branca Skink which, because of its habitat being restricted to this one island, is listed as vulnerable under the Commonwealth *Endangered Species Protection Act 1992* and the Tasmanian *Threatened Species Protection Act 1995*. The Australasian Gannet populations have been steadily increasing.

Unregulated visits to the island are prohibited due to the vulnerability of the resident species to disturbance.

Here on Pedra Branca Shy Albatross and Australasian Gannet occupy adjacent nest sites.

Eddystone Rock & Sidmouth Rock (Pedra Branca Group, page 502)

Eddystone Rock

Location: 43°50'S, 147°01'E
Survey date: 2/8/90
Area: 4 metres (east – west) by 5 metres (north – south)
Status: National Park

Several kilometres to the east of Pedra Branca lie Eddystone Rock and Sidmouth Rock. Sidmouth Rock, approximately 90 metres in diameter, is constantly wave-washed and therefore supports no terrestrial wildlife. Eddystone Rock, rising like a tower to a summit of 30 metres, was named by Captain Cook in 1777, owing to its resemblance to Eddystone Lighthouse.

Australasian Gannet colony.

BREEDING SEABIRD SPECIES
Fairy Prion
Two active nests were located in crevices on the summit.

Black-faced Cormorant
A nest was found on the north ledges and observations from Pedra Branca in subsequent years indicates that this speices breeds here regularly but in very low numbers.

Australasian Gannet
Approximately 100 pairs nest on the summit plateau.

MAMMALS
Australian and New Zealand Fur Seals – 50+ were hauled out on the lower ledges. They occupy the exposed rock reef to the west of the main tower of rock, but can do so only when sea conditions fall below moderate swell height.

No other bird species were recorded and although a thorough search of all suitable habitat was conducted, no skinks were found.

VEGETATION
These rocks cannot sustain plant life

Eddystone Rock

COMMENTS
Eddystone Rock is accessible by helicopter through the winter when seabirds are not nesting. A plastic piton at the summit suggests a previous visit, possibly by navy personnel.

Maatsuyker Island Group

(Region 7, page 501)

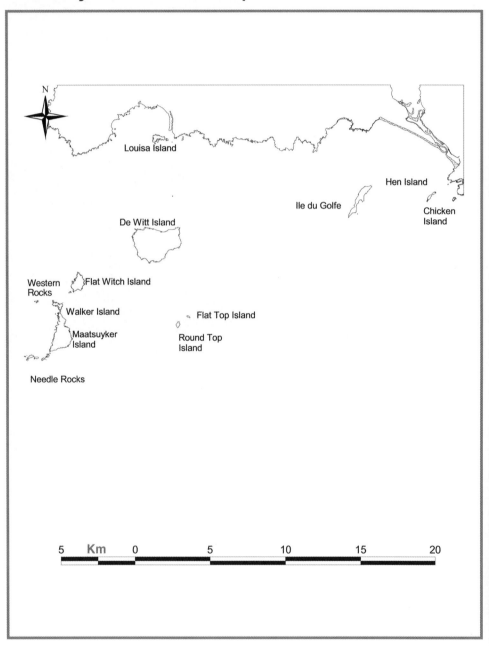

Chicken Island

(Maatsuyker Island Group, page 512)

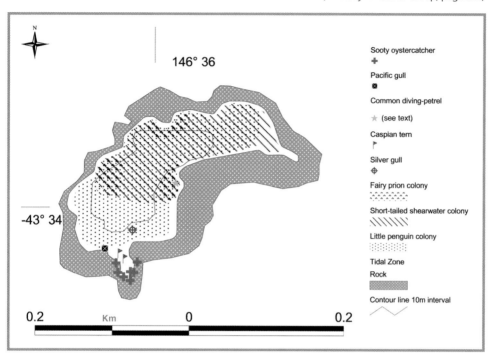

Location: 43°34'S, 146°36'E
Survey date: 12/12/87
Area: 1.95 hectares
Status: National Park

This is a low, flat islet rising to only 15 metres with a rocky shoreline. The west side is dominated by 10 metre cliffs while the east side slopes gently to the sea with 5 metre rock falls to the rock shoreline. It has loose, sandy soil.

BREEDING SEABIRD SPECIES
Little Penguin

288 pairs are interspersed with shearwater colonies in low density. They come ashore almost everywhere.

Little Penguin tracks worn in limestone rock.

Short-tailed Shearwater

2500 burrows cover most of the island apart from the rocky south end. They are predominantly in 60% *Carpobrotus rossii* with *Senecio* sp. and occasional *Poa* clumps with 10% bare, sandy soil. The loose soil makes burrowing conditions very fragile.

Fairy Prion

250 pairs occur in three distinct colonies. On the mid east coast, they are interspersed with Common Diving-Petrels at the cliff edge in very unstable soil, slumping badly in some sections. Here burrows are most concentrated and interconnected. Vegetation is dominated by sparse *Sarcocornia quinqueflora*, new-growth *Senecio* sp. and *Tetragonia*. The other major colony on the north-west coast is in very steep *C. rossii* and scattered stipa, which is not suitable for shearwaters, thus providing a non-competitive environment.

Common Diving-Petrel

100 pairs were recorded on the mid east coast, interspersed with the Fairy Prion colony.

Pacific Gull

1 pair was located 3 metres to the north of the Caspian terns' nests at the south end.

Silver Gull

14 pairs, with eggs and chicks, were counted on the south east coast amidst scattered stipa and *C. rossii*.

Sooty Oystercatcher

7 pairs, most with eggs, were recorded at the southern end on the bare rocks.

Looking towards Hen Island.

Caspian Tern
2 pairs, defending with 2 small downy runners, were found at the south end on flat bare rock with shell grit in cracks.

VEGETATION
Sarcocornia quniqueflora and *Senecio lautus* are the dominant species with scattered patches of *C. rossii*, *Tetragonia*, *Poa* and stipa.

No other fauna was recorded.

COMMENTS
This is an important island with a high seabird diversity. The east coast has good habitat for seals to haul out. It should be monitored to ascertain when this occurs. The absence of reptiles should be verified.

Hen Island

(Maatsuyker Island Group, page 512)

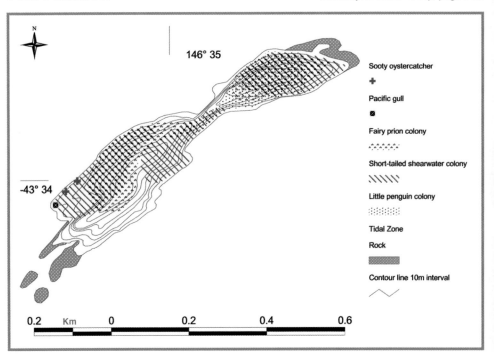

Location: 43°34'S, 146°35'E
Survey date: 12/12/87
Area: 7.6 hectares
Status: National Park

It is a long, narrow, north-east to south-west running island in two main sections with a razor-back ridge that is joined by a lower ridge. The north-east is rimmed by 30° slopes leading to sheer cliffs. The south-west is steeper with 40° slopes extending from the ridge top to mostly bare steep cliffs up to 20 metres and a rugged rocky shoreline. The northern peak rises to 60 metres and the southern to 75 metres. The south-west slope is dominated by *Senecio* sp which had been stripped by caterpillars at the time of the survey, leaving the soil exposed and the 30° – 40° slope vulnerable to erosion.

BREEDING SEABIRD SPECIES
Little Penguin

90 pairs occur in two colonies of 20 pairs each on the east coast with a further colony of 50 pairs living mainly in rock crevices near their landing site on the north-west tip.

Short-tailed Shearwater

11 627 pairs were interspersed with Fairy Prions along the ridge slopes. The northern colonies are dominated by *Carpobrotus rossii*, *Senecio* sp and matted *Tetragonia* over shrubs, which form a low canopy. The east colonies are in a mixture of *C. rossii*, *Rhagodia* and *Tetragonia* and the south-west colony is dominated by *Senecio* with scattered *Poa* tussocks.

Fairy Prion
27 328 pairs occur over the whole island, interspersed with Short-tailed Shearwaters.

Pacific Gull
One pair was defending vegetated slopes near the Sooty Oystercatchers' nests in the south-west section.

Sooty Oystercatcher
2 pairs, displaying defensive behaviour, were found on vegetated slopes in the south-west section.

East slopes to south.

VEGETATION
The main species are *C. rossii*, *Senecio*, *Tetragonia*, *Rhagodia* and *Poa*.

No mammals or other seabirds were recorded.

REPTILES
Metallic Skink

COMMENTS
This is a pristine island with interesting topography and high seabird diversity and population size relevant to its size. A comparison between Hen and its close neighbour, Chicken Island, reveals quite substantial differences in topography and soils. It is also interesting to note that Hen Island has skinks and no Common Diving-Petrels, whereas Chicken Island has Common Diving-Petrels and no skinks.

East slopes north to Tasmanian mainland.

Ile du Golfe

(Maatsuyker Island Group, page 512)

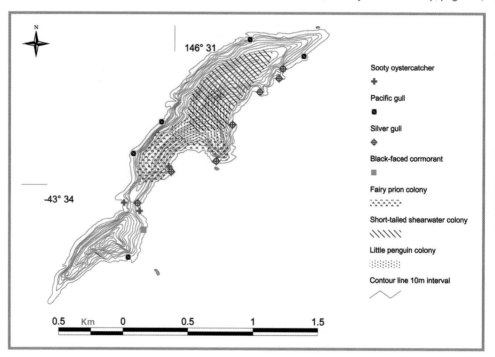

Location: 43°34'S, 146°32'E
Survey date: 28/12/80 - 2/1/81 and 12/12/87
Area: 68.16 hectares
Status: National Park

It is a long, narrow island, with a central razorback ridge, running north-east to south-west. From the north-east, the ridge rises steeply before ascending much more gradually to the northern peak of 150 metres, then drops away to a saddle before again rising steeply to a southern peak of 120 metres. It descends and drops away sharply to a bare rocky outcrop at the island's south-western tip. It is the only completely limestone island in Tasmania.

BREEDING SEABIRD SPECIES

Little Penguin

40-60 pairs, most with downy chicks and some with eggs (28/12/80), occur in light densities along the east coast, where they come ashore at numerous places.

Short-tailed Shearwater

An estimated 133 760 pairs, most with eggs (28/12/80) breed on the island.

The greatest concentrations occur in the northern section, with highest densities on the gentler slopes amongst vegetation dominated by *Senecio* sp. (80%) with scattered *Poa*.

East slopes, mainland in distance to south.

Fairy Prion
An estimated 356 400 most with chicks (28/12/80), occur all over the island, where vegetation is present. The only places virtually devoid of burrows are the densest shearwater colonies and where bracken fern is especially dense. Many Fairy Prions are under vegetation and not in burrows while some also nest in rocky clefts along the shores. They tend to favour the steeper slopes. The Forest Raven is very common along the island's ridge top, taking shearwater eggs and apparently causing the deaths of some Fairy Prions. However, the prions' main predator is the Pacific Gull.

Pacific Gull
5 pairs, with eggs and chicks (12/12/87), were scattered around the coast.

Silver Gull
67 pairs at nests with eggs and small chicks (12/12/87) were recorded on the island.

The main colony of 30 nests was located in the large eastern gulch with smaller ones all along the eastern coast and a few scattered nests on the west coast. All gull colonies are within 20 metres of the sea in areas of broken rock. Silver Gulls usually nest among bare rocks on the higher vegetated ledges using dry *Poa* as their main nest material.

Sooty Oystercatcher
3 pairs were located in sheltered bays around the coast.

Black-faced Cormorant
22 pairs were recorded breeding on cliff ledges on the south-east coast 20 metres up from the shoreline, just below the vegetation line.

No other seabirds were recorded.

MAMMALS
Dolphin – 2000+ were sighted travelling west.
Swamp Antechinus

BIRDS
Native:
Swamp Harrier
Peregrine Falcon
Tasmanian Thornbill
Tasmanian Scrubwren
Silvereye
Forest Raven – 30+
Introduced:
Common Blackbird

REPTILES
Tasmanian Tree Skink
Metallic Skink
Three-lined Skink

VEGETATION
Very dense vegetation covers most of the island, with 1.2 metre high bracken and fireweed in many places preventing easy penetration. Vegetation in the more sheltered parts is dominated by *Leptospermum scoparium* which forms a 5 – 6 metre canopy. Other woody shrubs, particularly *Drimys lanceolata*, *Pittosporum bicolor* and *Olearia phlogopappa* often provide a canopy for a wide variety of ferns.

On the more exposed western slopes *Correa backhousiana*, *Acacia verticillata*, *Solanum vescum*, *Dianella tasmanica* and *Pelargonium australe* are dominant.

Pteridium esculentum grows densely in places along the northern half of the ridge top. The most abundant and widespread species on the island are *Poa poiformis* and *Senecio lautus,* with *Rhagodia candolleana* also being plentiful.

COMMENTS
Its limestone features, outstanding topography and abundance of seabirds, particularly Short-tailed Shearwaters and Fairy Prions, make this a highly significant island. It has suitable habitat for Fur Seals.

North-west slopes looking towards south coast of mainland Tasmania.

Louisa Island

(Maatsuyker Island Group, page 512)

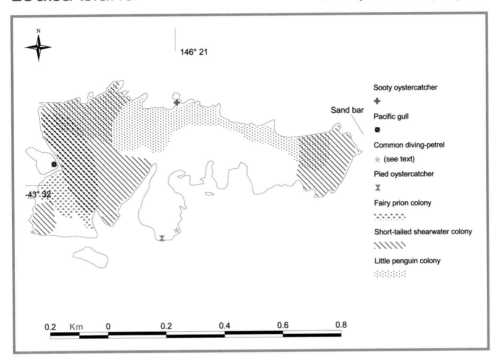

Location: 43°32'S, 146°21'E
Survey date: 8 – 12/11/77 and 25/4/85
Area: 23.04 hectares
Status: National Park

Louisa Island is an irregular-shaped island approximately 800 metres long and about 500 metres at the widest part, with the summit just over 80 metres high towards the western end. On the southern side of the island there is a small deep hollow, which was formerly an extensive sea cave, now with a collapsed roof section. The island is joined to mainland Tasmania at Louisa Bay by a sand spit about 250 metres in length, which can provide access at low tide. The soil is peaty and fibrous, creating excellent burrowing habitat.

BREEDING SEABIRD SPECIES

Little Penguin

Up to 650 pairs breed on the island. During the 1977 survey eggs were being incubated or pairs had very young chicks. Burrows are mainly over the eastern half of the island, but also extend across to the north-western parts. There is also a colony recorded on the north-east tip of the islet to the west of Louisa Island.

Short-tailed Shearwater

An estimated 205 969 pairs breed over most of the island but in fewer numbers in the lightly-forested parts, except near the fringes. Colonies are generally dominated by *Poa poiformis*. They are also on the islet to the east of Louisa Island on

Eastern end and Tasmanian south coast.

the summit patch and north-east slopes in *Carpobrotus rossii* and *Rhagodia candolleana*.

Fairy Prion
An estimated 400 pairs breed predominantly on the western slopes in dense *Poa*.

Common Diving-Petrel
An estimated 1600 pairs breed on the lower slopes before the bare cliffs under scrub on the island off Louisa Point. The habitat for diving-petrels seems to be extensive, with *Rhagodia* and *Poa* providing a dense cover.

Pacific Gull
One pair was sighted on the ledge of the small islet that is about 100 metres to the west of Louisa Island.

Sooty Oystercatcher
One pair was sighted on the mid northern coast.

Pied Oystercatcher
1 pair was recorded on Spit Beach on the mid southern coast.

VEGETATION
The vegetation is typical of the coastal communities of south-west Tasmania, but with the effects of fire apparent in many areas densely covered by *Pteridium esculentum*. The central parts are lightly forested, with *Eucalyptus nitida* and *Eucalyptus ovata* the dominant species. *Leptospermum scoparium* and *Melaleuca squarrosa* form much of the understorey along with the larger fern species including *Dicksonia antarctica*. In the more exposed areas *Acacia verticillata*, *Banksia marginata*, and the sedges and tussocks dominate. *R. candolleana*, *Tetragonia implexicoma* and *C. rossii* are the most common species near the shoreline.

OTHER SEABIRD SPECIES
Little Pied Cormorant – 10+
Black-faced Cormorant – 4
Silver Gull

MAMMALS
Native:
Tasmanian Pademelon
Long-nosed Potoroo

BIRDS
Native:
Pacific Black Duck – 10

White-faced Heron
Peregrine Falcon - flying up river
Brush Bronzewing
Yellow-throated Honeyeater
Crescent Honeyeater
Black Currawong
Forest Raven

REPTILES
Tasmanian Tree Skink

COMMENTS
This is a pretty and diverse island, which is perhaps more vulnerable to visits than most in the region because it has a reasonable anchorage in suitable weather on the northern side. Its geology and diversity of fauna and flora make it an island of outstanding significance.

Dense vegetation, mainly giant ferns.

Louisa Island and sand spit adjoining Tasmanian mainland at the Louisa River.

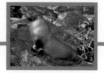

De Witt Island

(Maatsuyker Island Group, page 512)

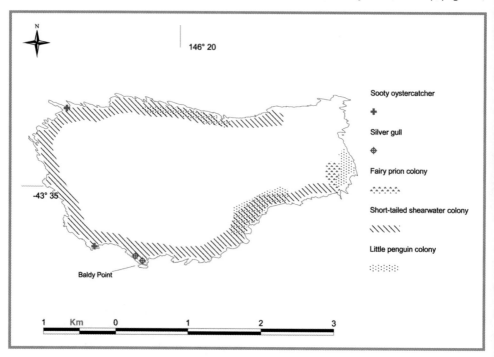

Location: 43°36'S, 146°22'E
Survey date: 20/1/77 – 7/2/77, 13/12/87, 14/12/87 and 15/3/91
Area: 515.7 hectares
Status: National Park

The largest island of the Maatsuyker group, its shore is characterised by precipitous cliffs interrupted by bays, limestone caves and gulches. It rises to a summit of 354 metres on the southern side. The flattest parts are located close to the shore on the northern side, where there are two small inlets suitable for landings. Three streams, one running out of each of the landing inlets and the third over a cliff on the island's western side, provide permanent water sources.

BREEDING SEABIRD SPECIES
Little Penguin

500 pairs breed on the island. The main colonies are located about 300 metres in from the two northern landing inlets, in the forest where the ground is relatively flat. There are also smaller colonies on the east and south coasts. Many in the south nest under boulders 10 metres in from the shoreline.

Short-tailed Shearwater

11 141 pairs breed in colonies, which are well distributed around the periphery of the island, tending to be in a strip up to 60 metres wide, except on the eastern side. The largest colonies occur around the two landing inlets. The highest

Extensive forest areas often with open understorey.

densities are on the open forest floor beneath eucalypts, avoiding nearby dense bracken areas, as is usually the case for this species.

Petrel – species unknown
400 burrows were found along the southern summit ridge in very low densities on the steep slopes. The 118 burrows inspected were empty, possibly due to predation from the eastern Swamp Rat, *Rattus lutreolus*.

Fairy Prion
50 pairs breed in the south-east, just west of the Little Penguin colony of 20 pairs.

Silver Gull
2 pairs, with 3 eggs each were recorded breeding. Their nests, made from *Poa*, were located on rock ledges on the offshore rock south of Baldy Point. 9 adults, in total, were sighted.

Sooty Oystercatcher
One nest with 2 eggs was located in the north-west gulch and another with two eggs was found in the southern cove (13/12/87).

VEGETATION
Eucalyptus nitida, *E. ovata* and *E. obliqua* are the main species. Other species occurring in the forest and in relatively sheltered places include *Eucryphia lucida*, *Nothofagus cunninghamii*, *Leptospermum scoparium*, *Banksia marginata*, *Cyathodes juniperina*, *Melaleuca squarrosa* and *Pimelea drupacea*. A dense growth of Austral bracken, *Pteridium esculentum*, covers the area near the shoreline on the northern side and is gradually spreading into parts of the forest.

South-east coast rock formation in vicinity of limestone caves.

MAMMALS
Native:
Swamp Rat – very numerous
Long-nosed Potoroo
Tasmanian Pademelon

BIRDS
Native:
White-bellied Sea-Eagle
Peregrine Falcon
Green Rosella – common
Swift Parrot – 10+ feeding in heavily flowering *Eucalyptus ovata* and leatherwoods
Spotted Pardalote
Striated Pardalote
Tasmanian Scrubwren – many
Brown Thornbill – many
Strong-billed Honeyeater – common
Crescent Honeyeater – common
New Holland Honeyeater – common
Pink Robin – common, especially in forested parts
Olive Whistler – common
Grey Fantail
Black Currawong
Forest Raven
Tree Martin
Silvereye

Introduced:
Common Blackbird

REPTILES
Tasmanian Tree Skink
Metallic Skink

On the more exposed slopes and cliffs and in the immediate vicinity of the shoreline, species include *Acacia verticllata, Westringia brevifola, Epacris myrtifolia, Correa backhousiana, Pomaderris apetala, Rhagodia candolleana, Carpobrotus rossii Tetragonia implexicoma* and *Poa poiformis.*

COMMENTS
The large population of Swamp Rats apparently caused considerable interference to the breeding success of Little Penguins and Short-tailed Shearwaters in 1975 and 1976. Geologically, the island is considered of outstanding significance for the State. A zone of folding, in conglomerate-siltstone succession, transects the island, well-exposed in the west and south-east coasts. There is also a series of closed and partially-closed depressions, associated fissure caves and openings, stepped profiles and boulder pile from a past collapse. This is probably associated with large scale subsidence of the top of the 250 metre cliff due to stress release caused by ongoing erosion or perhaps triggered by massive past collapse (Dixon, 1996). The island has a long history of human use, including logging and some occupation. An assessment of the human impact and further seabird and mammal surveys would greatly enhance the relatively small reservoir of knowledge of this large, important island.

Flat Witch Island

(Maatsuyker Island Group, page 512)

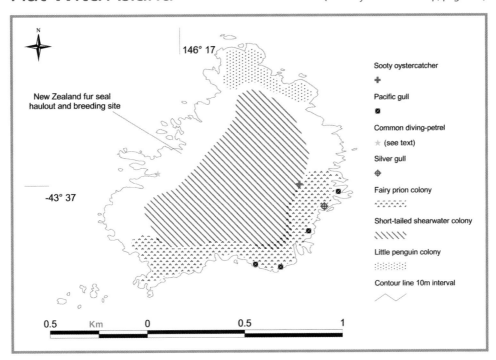

Location: 43°37'S, 146°17'E
Survey date: 28/12/78 - 8/1/79 and 15/12/87
Area: 64.32 hectares
Status: National Park

Also known as Little Witch, it has fairly steep vegetated slopes and sheer cliffs around its coast but is gently sloping across the top, with the summit exceeding 100 metres towards the southern end. Soil across the island is extensive but not deep.

BREEDING SEABIRD SPECIES
Little Penguin

400 pairs nest mainly in the north and north-east of the island with low numbers well inland and a few scattered along the western slopes. None are on the eastern or southern sides as there is no access here from the water.

Short-tailed Shearwater

An estimated 499 696 pairs breed extensively over most of the island, with only the steepest slopes being totally devoid of burrows. Densities are highest in *Poa*-dominant areas. In the scrub areas, there is good burrow density where there is open canopy and fewer burrows in closed canopy areas. On 15/12/87, eggs were being incubated.

Fairy Prion

An estimated 10 000 pairs breed on the island. Burrows are most common on the steep slopes

around the shores, but they also extend inland to about 200 metres. They are densest in a 30 metre wide strip on the east coast round to the south-west on very steep slopes in predominantly *Carpobrotus rossii*, scattered *Senecio* sp. and bat's wing fern. On 15/12/87, birds had eggs and small chicks.

Common Diving-Petrel

Up to 100 pairs were located nesting in the central west on lower *C. rossii* slopes with a few burrows interspersed with those of the Fairy Prions.

Pacific Gull

4 nests were located on sheer cliffs mainly on the south and east coasts.

Silver Gull

40 pairs were defending their nests, which were located on sheer cliffs on the mid east coast.

Sooty Oystercatcher

One nest was located on the mid east coast on edge of *Poa*.

VEGETATION

In the relatively sheltered parts, vegetation is dominated by the woody shrubs *Leptospermum*

From Maatsuyker Island to Flat Witch and the south coast of Tasmania.

No other seabirds were recorded.

MAMMALS
Swamp Antechinus
Australian Fur Seal
New Zealand Fur Seal

BIRDS
Native:
White-bellied Sea-Eagle
Peregrine Falcon
Green Rosella
Black Currawong – many
Forest Raven – 50+
Beautiful Firetail
Silvereye

REPTILES
Tasmanian Tree Skink

scoparium and *Drimys lanceolata*. Ground cover beneath the canopy is generally sparse, consisting mainly of *Drymophila cyanocarpa* and ferns, which also occupy steep gulches. On the more exposed slopes, including the cliffs and ridges, *Senecio lautus* and *Poa poiformis* grow very extensively with *Cyathodes abietina, Rhagodia candolleana, C. rossii Tetragonia implexicoma, Apium prostratum* and *Sarcocornia quinqueflora*.

COMMENTS
It is generally an undisturbed island, due to having poor access and an unrealiable anchorage, as is largely the case in the region. Since the survey, New Zealand Fur Seals have taken up residence here with 5+ pups born annually. The island has limited habitat, which will restrict expansion to an estimated maximum of 50. There is a haul-out site for both Australian and New Zealand Fur Seals in the south.

Western Rocks

(Maatsuyker Island Group, page 512)

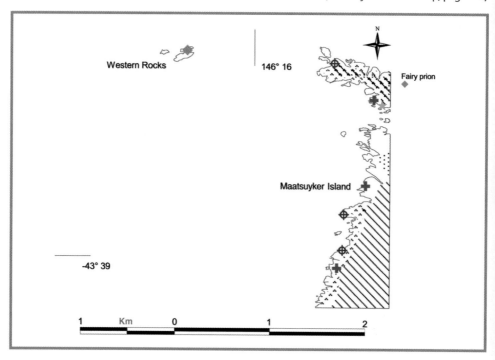

Location: 43°38'S, 146°15'E
Survey date: 15/12/87
Area: 0.29 hectares
Status: National Park

Part of the Maatsuyker group, these are steep rocks, which are completely wave-washed in heavy storms. The main rock is bare apart from a very small area of succulent vegetation on the north-west face below the summit.

BREEDING SEABIRD SPECIES
Fairy Prion

Up to 20 pairs were located in a small area of vegeation on the north-west face just below the summit.

COMMENTS

These are commonly known as Black Rocks.

Maatsuyker Island

(Maatsuyker Island Group, page 512)

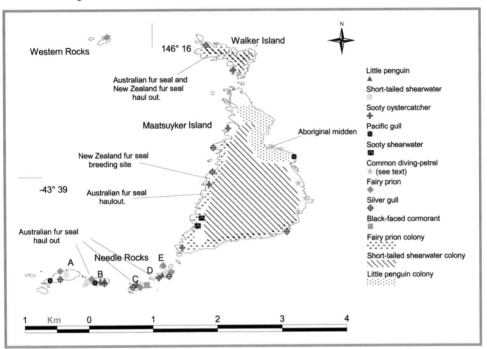

Location: 43°39'S, 146°16'E
Survey date: 13/12/85, 15/12/87, 6/1/88
and numerous other visits
Area: 186.25 hectares
Status: National Park

This is a jagged island characterised by steep cliffs, small bays, gulches and subterranean caverns. The west coast has very steep slopes which are often inaccessible. Its lighthouse is an important landmark. An Aboriginal midden is a striking example of the maritime activities of the Tasmanian Aborigines. Human-made structures include a haulage way, houses, sheds, vehicle tracks, dams and lighthouse keepers' name plaques.

BREEDING SEABIRD SPECIES

Little Penguin

Up to 700 nests are concentrated on the north-east slopes. Access is via a gulch on the mid north-east coast.

Short-tailed Shearwater

An estimated 800,000 pairs breed over much of the island, but nesting density varies substantially through open *Poa* areas and beneath *Melaleuca* scrub, which forms a thick canopy.

Fairy Prion

Up to 5000 pairs are scattered all around the perimeter of the island on steep slopes where

vegetation is dominated by succulents such as *Tetragonia* and *Carpobrotus rossii*.

Common Diving-Petrel
An estimated 10 000 pairs breed around the edge of the island, with concentrations along the western side on the very steep slopes frequently extending well into the dense scrub, which is dominated by *Correa* sp.

Pacific Gull
One pair was sighted on a headland on the north-east coast.

Silver Gull
Two colonies of 4 pairs each were recorded on the mid west coast and one colony of 6 pairs was located on the south-east coast.

Sooty Oystercatcher
3 pairs were sighted on the western shore.

Sooty Shearwater
2 pairs were found 100 metres into the tea tree scrub above the north-east landing gulch amongst Short-tailed Shearwaters. There is a possibility that this species is more abundant, but locating them amongst vast numbers of shearwaters is difficult.

Small numbers of southern elephant seals occur at several locations at Maatsuyker Island.

Soft-plumaged Petrel
This rare species is yet to be confirmed to be breeding on the island, but observations over many years have revealed small numbers courting on the east to north-east slopes.

VEGETATION
The vegetation is dominated by *Leptospermum scoparium*, which covers most parts of the island and occurs in all but the most exposed situations. *Melaleuca squarrosa* and *Banksia marginata* are frequently found in association with *Leptospermum*, forming a dense canopy, stunted and considerably wind-pruned in the more exposed areas, but reaching approximately 6 metres in height in some relatively sheltered situations around the higher slopes. In the more sheltered parts, species include *Drimys lanceolata*, *Cyathodes juniperina*, *Pimelea drupacea*, *Acacia verticillata*, *Pittosporum bicolor* and several ferns. In exposed areas, including the cliffs and in the immediate vicinity of the shoreline the dominant species are *Westringia brevifolia*, *Helichrysum paralium*, *Correa backhousiana*, *Rhagodia*

East side landing gulch and jetty (since demolished).

candolleana, Senecio pinnatifolius, Carpobrotus rossii, Tetragonia implexicoma and the tussock grass *Poa poiformis*.

MAMMALS

Australian Fur Seal – a haul-out site only

New Zealand Fur Seal – breeding, with 130 – 140 pups counted annually

Southern Elephant Seal – 2 males were recorded at the landing. Both were moulting. One was estimated as 4 – 5 years old and the other as 6 – 8 years old. They occupy three main sites and small numbers use the area regularly. Pups are known to have been born here on at least three occasions.

Swamp Antechinus

BIRDS

Native:

White-bellied Sea-Eagle

Brown Falcon

Peregrine Falcon

Nankeen Kestrel

Lewin's Rail

Brush Bronzewing

Green Rosella

Orange-bellied parrot

Southern Boobook

Tasmanian Scrubwren

Tasmanian Thornbill

Crescent Honeyeater

New Holland Honeyeater

Eastern Spinebill

Pink Robin

Black Currawong – common

Forest Raven – common

Welcome Swallow

Silvereye – very common

Introduced:

House Sparrow

Common Blackbird – nest with nestling

European Goldfinch

REPTILES

Metallic Skink

Three-lined Skink

Tasmanian Tree Skink

COMMENTS

This is an island of great historical significance, having the longest continual occupation of any island in Tasmania, with Aboriginal seafarers visiting possibly for thousands of years followed by a series of resident lighthouse keepers. The absence of lighthouse personnel since June 1998 has left the island more vulnerable to visitation and subsequent disturbance. Its management plan proposes the removal of weeds, predominantly blackberries and the retrieval and maintenance of its historical artefacts to prevent them from decay and destruction. It is geomorphololgically representative and outstanding for Australia due to unusual soils which have been influenced by activities (burrowing and nutrient addition) of the Short-tailed Sheawaters combined with geology (Precambrian metasediments) (Dixon, 1996).

Walker Island

(Maatsuyker Island Group, page 512)

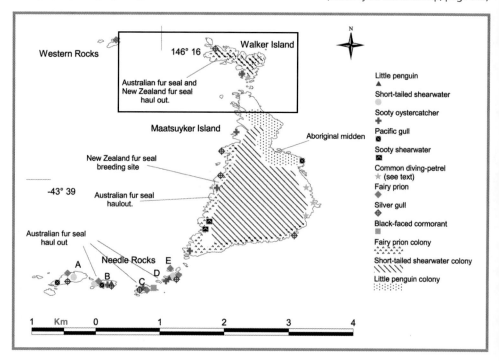

Location: 43°38'S, 146°16'E
Survey date: 12/1/78, 12/12/87 and numerous other visits
Area: 15.3 hectares
Status: National Park

This is a steep-sided island with a few breaks in the coastline on the northern side, particularly near the north-west corner. Away from the shores, it slopes more gently to the summit of 70 metres near the south-eastern end. There are several deep gulches and a blowhole on the western side. A tunnel penetrates the island in the narrowest part. Soil cover is extensive but shallow.

BREEDING SEABIRD SPECIES
Little Penguin
200 – 300 pairs, most with fledglings (12/1/78), are very lightly scattered throughout the island, with the main concentrations around the northern landing inlet and in the vicinity of the north-western corner of the island, extending across to the higher slopes.

Short-tailed Shearwater
An estimated 146 286 pairs nest over the whole island apart from on cliffs or very steep slopes. Densities are greatest in areas dominated by a mixture of *Senecio lautus* and *Poa poiformis* (50%) and *Carpobrotus rossii* (50%).

Fairy Prion
3000 pairs occur all over the island apart from the densest shearwater areas. The main concentrations are along the east and mid west coasts on the very steep vegetated slopes, where the soil is either insufficient or too precarious for shearwaters.

Common Diving-Petrel
An estimated 500 pairs were interspersed with Fairy Prions, with the main concentration on the very steep vegetated slopes of the east end, where the soil is either insufficient or too precarious for shearwaters.

Silver Gull
16 pairs were counted, most with eggs and chicks. One nest in the west side blowhole gulch colony had 5 eggs. One colony of 8 pairs was on the edge of the gulch to the west of the blowhole, one colony of 4 pairs was on the north-west tip and one colony of 4 pairs was just north of the Fairy Prion colony in the mid west.

Sooty Oystercatcher
1 pair was sighted on the south-west coast.

VEGETATION
The vegetation is severely stunted and wind-pruned. Woody shrubs are found mostly over the eastern half of the island but other species including succulents are more widespread and abundant. Species recorded were *Leptospermum scoparium, Olearia phlogopappa, O. persoonioides, Correa backhousia, Cyathodes abietina, Westringia brevifolia, Helichrysum paralium, Pelargonium australe, Asplenium obtusatum, Apium prostratum, Lepidium foliosum, Rhagodia candolleana, S. lautus, Tetragonia implexicoma, Salicornia quinqueflora, Carpobrotus rossii, Poa poiformis* and stipa.

From the northern tip of Maatsuyker Island to Walker Island and Tasmanian mainland south coast in distance.

East slopes

OTHER SEABIRD SPECIES
Pacific Gull – (12/1/78)

MAMMALS
Australian Fur Seal – small numbers have infrequently occupied the shoreline as a haul-out. Since they were first recorded in 1994, numbers have been increasing over the past few years. Approximately 1000 seals were observed on Walker Island in March 2000, most hauled-out on the high central section.

A New Zealand Fur Seal – harem was present in 1995, with one pup born.

REPTILES
Tasmanian Tree Skink

COMMENTS
The Australian Fur Seals' occupation of the central ridge of the island has caused the decline of the Short-tailed Shearwater and Fairy Prion populations in this area. Continued surveys are essential to monitor the change in the size and structure of the seal colonies and the reduction in breeding habitat for burrowing seabirds.

Flat Top Island

(Maatsuyker Island Group, page 512)

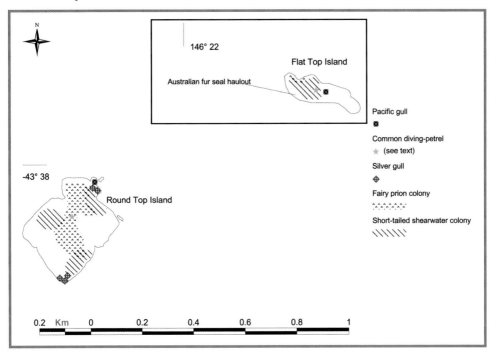

Location: 43°38'S, 146°23'E
Survey date: 28/12/80, 3/11/83 and March 1998
Area: 1.58 hectares
Status: National Park

Approximately 240 metres long by 120 metres wide, Flat Top Island has sheer cliffs rising to about 60 metres from the sea on all sides to a relatively flat surface across the top. It is the smallest island in the group and has numerous sea caves, at various altitudes, running through it. Soil cover is mostly confined to the flattish area on top, where it is extensive but not deep.

BREEDING SEABIRD SPECIES
Short-tailed Shearwater

Up to 400 pairs breed extensively all over the plateau and down the northern slopes in crevices, burrows and under rocks. They are restricted to these areas due to lack of soil depth in other parts of island. Birds were incubating eggs during the survey of 28/12/80.

Fairy Prion habitat on the plateau top.

537

From the east, Flat Top Island with Round Top Island immediately behind.

Fairy Prion

Up to 5000 pairs are estimated to be nesting across the top of the island and in a very small area just down from the top on the northern side. Their burrows are interspersed with those of Short-tailed Shearwaters. During the survey of 3/11/83 most had eggs and on 28/12/80 most had downy chicks.

Common Diving-Petrel

1000 pairs, most with newly-hatched chicks, were located all over the island where soil is sufficient.

Pacific Gull

One pair was sighted.

VEGETATION

The vegetation is dominated by *Carpobrotus rossii* while other species include *Asplenium obtusatum*, *Rhagodia candolleana*, *Poa poiformis*, and *Sarcocornia quinqueflora*.

East ledge fur seal haul out.

OTHER SEABIRD SPECIES
Grey Petrel – a carcass was found at the west end of the island's plateau

Shy Albatross – at sea nearby

Black-faced Cormorant – this species breeds here intermittently, with 26 near-fledging chicks present in March 1998 on the high west end cliff ledges at an altitude of approximately 240 metres.

Silver Gull – 4

MAMMALS
Australian Fur Seals haul out in the sea caves and the cliff ledges on the east side.

BIRDS
Introduced:
European Goldfinch

REPTILES
Tasmanian Tree Skink – numerous

OTHERS
East coast land snail abundant

COMMENTS
This is a pristine, topographically distinctive island with geological features considered to be of outstanding significance for the State (Dixon, 1996) and a large seal haul-out. Its relative inaccessibility gives it security from disturbance, although the fur seals have, at times, been harrassed.

Sea caves cutting through the island, negotiable in a 4m dinghy in calm seas.

Round Top Island

(Maatsuyker Island Group, page 512)

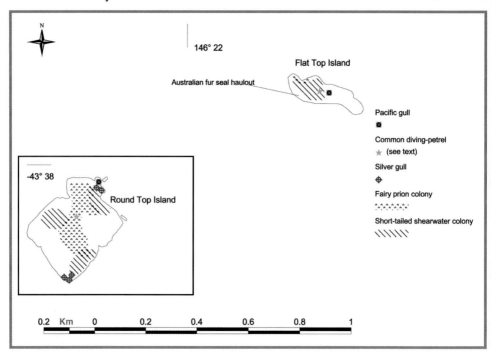

Location: 43°39'S, 146°22'E
Survey date: 11/12/87
Area: 6.25 hectares.
Status: National Park

Part of the Maatsuyker group, it is an oblong island with steep to sheer cliffs rising to a flat to gently undulating, 165 metre summit. The eastern side has less steep cliffs, where vegetation is more extensive. Soil cover is generally shallow over the whole of the island.

BREEDING SEABIRD SPECIES
Short-tailed Shearwater

An estimated 8927 pairs occur in three relatively distinct colonies, one in the north-east, one in the south-east and one in the summit area. The densest colony is in the summit area in dense *Poa* and *Carpobrotus rossii* patches with occasional scattered shrub species.

Fairy Prion

9327 pairs occur throughout most of the island, mainly in *Poa*, *Senecio* and *C. rossii*, interspersed with Short-tailed Shearwaters and Common Diving-Petrels.

Limestone band rock formation.

Common Diving-Petrel
An estimated 8660 pairs breed throughout most of the island, mainly in *Poa*, *Senecio* and *C. rossii*, interspersed with Fairy Prions.

Pacific Gull
1 pair was recorded on a ledge covered with *C. rossii* on the north-east end.

Silver Gull
13 pairs were located on boulders at the end of the south-east coast where *Senecio*, *Poa* and *C. rossii* predominates. 2 pairs were on the north-east end ledge. Most pairs had eggs and/or very small chicks.

VEGETATION
The vegetation is dominated by *C. rossi*, *Cyathodes juniperina*, *Senecio lautus*, *Rhagodia candolleana*, *Poa poiformis* and *Asplenium obtusatum*, with patches of *Leptospermum*.

No mammals or other birds were recorded.

REPTILES
Tasmanian Tree Skink – many were sighted feeding off insects in a flowering tea tree up to 2 metres off the ground.

Metallic Skink

COMMENTS
Access is very difficult, so it is well-protected from human disturbance. Its distinctive topography, seabird diversity, skink population and pristine environment make this an important island.

Southern side.

Needle Rocks

(Maatsuyker Island Group, page 512)

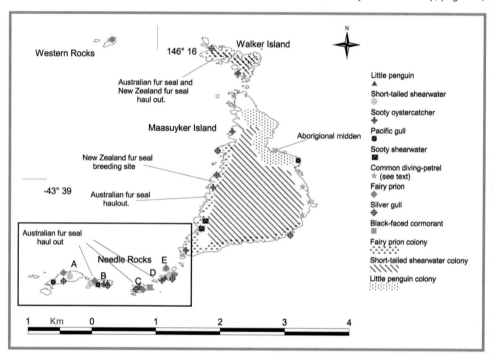

Location:	43°40'S, 146°15'E
Survey date:	11/12/87
Area:	5 rocks with seabirds:
	A = Main rock and most westerly - 4.78 hectares
	B = Next rock to the east - 2.25 hectares
	C = Next rock to the east - 1.77 hectares
	D = Next rock to the east (Seal Rock) - 1.37 hectares
	E = Eastern most rock - 0.34 hectares
Status:	National Park

Needle Rocks are a group of 5 main rocks to the south of Maatsuyker Island with very steep to sheer cliffs ranging from 30° to vertical slopes. There is very little vegetation or soil for burrowing on any of the islands. The largest rock, the westernmost, is the highest with a summit of 90 metres and 45°-50° vertical cliffs, while the smallest most easterly rock is pyramid-shaped with very steep sides all around and virtually no burrowing habitat.

BREEDING SEABIRD SPECIES
Little Penguin

40 pairs were recorded on rock B. A colony of 30 was located at the south-east tip 20 metres up in the *Carpobrotus rossii* and under rocks. On rock D, 10 were estimated to be nesting in rock crevices predominantly on the south-east side, generally in *C. rossii*-dominated areas.

The rugged beauty of Needle Rocks.

Short-tailed Shearwater

860 pairs were estimated to breed on rocks A, B and D, generally on very steep slopes in rock crevices or under *C. rossii*.

Fairy Prion

A large population of 10 640 pairs breeds here, often in rock crevices or burrows in bare soil patches on rock ledges competing with shearwaters for space on very steep slopes. Many prion burrows showed signs of interference from shearwaters and some had been taken over.

Pacific Gull

The only breeding pair was located on the mid north side on steep slopes of rock B. Another pair was located on rock A, but there was no sign of breeding. An old nest was located on rock E on the eastern side of the summit.

Silver Gull

On rocks B, C and D, 15 pairs were breeding on the boulder slopes. On rock D a colony of 14 nests was located around the boulders of the summit and under boulder ledges. A further 10 were seen on rock A, but there was no sign of breeding.

Sooty Oystercatcher

3 pairs were recorded, 2 nests with single eggs and one with a chick. One pair was located on rock B on the mid north side only 5 metres away from the Pacific Gull's nest. One pair was located on rock C on the north-east end on the bare rock area above the seal haulout and another pair was located on the west end of rock D on the bare rock gravel patch.

Black-faced Cormorant

100-150 nests were located on the north-east cliff ledges to a height of 70 metres of rock C and at the south-east end of rock D. (13/12/85)

View of Needle Rocks from within the lighthouse prisms.

MAMMALS

Australian Fur Seals – up to 1200 haul out here regularly.

Rock A — small number on the north-west rock ledges

Rock B – ashore on rock ledges

Rock C – many ashore

Rock D – many ashore

Rock E – 100-200 present on lower rock ledges, even though they had no shade.

BIRDS

Forest Raven – 3 on the west coast of the main rock

REPTILES

None on rock A

Metallic Skinks were common on all other rocks.

VEGETATION

Most of the rock slopes were dominated by *C. rossii*, with scattered *Senecio*, *Poa*, *Tetragonia* and *Sarcocornia*.

COMMENTS

Access ashore is difficult, which protects the rocks from disturbance. They are an outstanding geographical feature.

Mutton Bird Island Group

(Region 7, page 501)

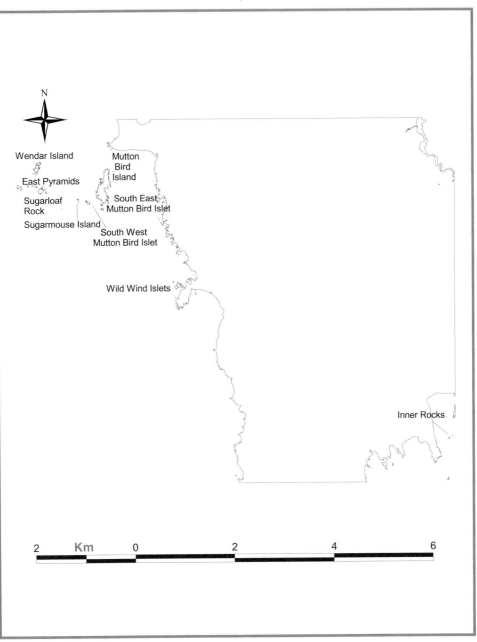

Inner Rocks, New Harbour

(Mutton Bird Island Group, page 545)

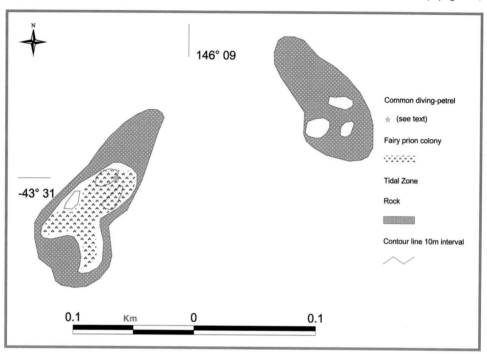

Location: 43°31'S, 146°09'E
Survey date: 17/12/81
Area: 0.23 hectares
Status: National Park

There are three rock spires in this group, with the two northern ones joined at low tide and the third separated by a 2 metre gulch. The summits of all three are on the north face.

BREEDING SEABIRD SPECIES
Fairy Prion

The peaty ground beneath the dense fern is honeycombed with prion and diving-petrel burrows. Prions were also nesting beneath the thick patches of button grass and in tunnels under the grass. An estimated 100 pairs, with eggs, were recorded.

Common Diving-Petrel

Up to 100 pairs, most with fully-fledged chicks, were recorded. Burrows were difficult to investigate due to the peaty nature of the soil and very narrow tunnels.

VEGETATION

The vegetation is dominated by fern species, extremely dense in patches, and *Carpobrotus rossii*. Vegetation is concentrated towards the summit on the north face of all the spires. There are also several patches of button grass.

Seabirds nest on these steep rocks.

OTHER SEABIRD SPECIES

Black-faced Cormorant – roosting on the western spire

Pacific Gull – a midden site of this species was found.

REPTILES

Metallic Skink – on the western spire

COMMENTS

This island is an example of unique burrowing habitat with its combination of topography and peat soils.

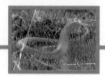

Wild Wind Islets

(Mutton Bird Island Group, page 545)

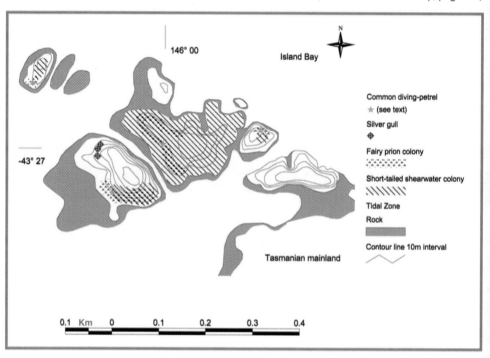

Location: 43°27'S, 146°00'E
Survey date: 27/11/89
Area: 3.95 hectares total
Status: Non-allocated Crown Land

This is a group of five small islets just to the west of South West Cape. Most of the islets have very irregular, steep rock and cliff shorelines, but all, apart from the most south-westerly, provide habitat for burrowing birds. The main island with an area of 2.05 hectares has the largest diversity and population of species.

BREEDING SEABIRD SPECIES
Short-tailed Shearwater
The main island's population of 20 300 pairs was located mainly on the upper steep eastern slopes in a mixture of *Poa*, *Carpobrotus rossii* and *Senecio* sp. The smaller islets have very small colonies of less than 200 pairs, as their terrain is generally very steep and unsuitable shearwater habitat.

Fairy Prion
An estimated 3000 burrows were scattered over four of the islets. They were generally interspersed with Common Diving-Petrels and most common on the lower 10 metre slopes where soils are shallow and the slopes very steep and unsuitable for shearwaters.

Common Diving-Petrel
An estimated 3000 pairs were generally interspersed with those of the Fairy Prions on the steeper cliffs dominated by *Carpobrotus rossii*.

Seperated by narrow waterways these fine islets are exposed to prevailing weather.

Silver Gull
3 pairs were found only on the island directly to the south of the main island on the north-west tip cliff ledge.

OTHER SEABIRD SPECIES
Black-faced Cormorant – 6 were roosting on the main island and 6 on the island directly to the south of the main island.

No mammals were recorded.

BIRDS
White-bellied Sea-Eagle – a nest was located amongst the scrub on the most south-westerly islet.

REPTILES
Metallic Skink – were common amongst bird burrows on the main island

VEGETATION
The vegetation on most of the islets is predominantly *Poa* and *C. rossii* with *Senecio* sp. and small patches of scattered shrubs, which constitute less than 5% of the total vegetaton cover. The most south-westerly islet is quite different, with little soil. Its vegetation is similar to the closest mainland Tasmania shores.

COMMENTS
A soil slip had occurred on the main island in the middle of the shearwater colony on the eastern slopes. The islands are not recorded on the Nomenclature list, but the name Wild Wind Islets has been suggested to honour the Tasmanian trading ketch skippered by Neil Smith, which was used for transport to conduct surveys. This is a pristine group of islands, which are of representative value for their local region.

South East Mutton Bird Islet (Mutton Bird Island Group, page 545)

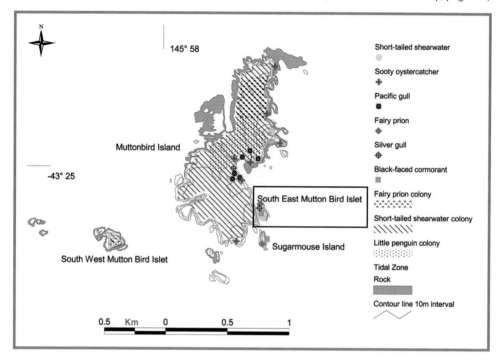

Location: 43°25'S, 145°58'E
Survey date: 2/12/81
Area: 0.52 hectares
Status: National Park

A predominantly steep to sheer-sided small islet off the south-east of Mutton Bird Island, it has a north to south running ridge and a lot of bare rock. Gentle slopes occur on the west coast, before steeply rising to a summit of 22 metres.

BREEDING SEABIRD SPECIES
Short-tailed Shearwater
250 pairs breed mostly on the west side where the slope is more gentle and there is sufficient peat soil depth. The dominant vegetation is *Carpobrotus rossii* and *Senecio* sp.

Fairy Prion
Up to 1000 pairs are estimated to breed all over the island amongst rocks and in burrows, primarily interspersed with Short-tailed Shearwaters.

Silver Gull
6 pairs and 35 individuals were nesting amongst *Poa* on very steep, bare cliff ledges on the east side towards the northern end.

Black-faced Cormorant
23 pairs, one with an empty nest, 9 with 2 eggs, 13 with 3 eggs were recorded on steep bare cliff ledges near the southern end of the island on the east side.

VEGETATION

Poa dominates the steep cliff ledges on the northeast side, while *Senecio* sp and *C. rossii* dominate the western, more gentle slopes.

No other fauna was recorded.

COMMENTS

This is a small island with a high seabird diversity.

South West Mutton Bird Islet (Mutton Bird Island Group, page 545)

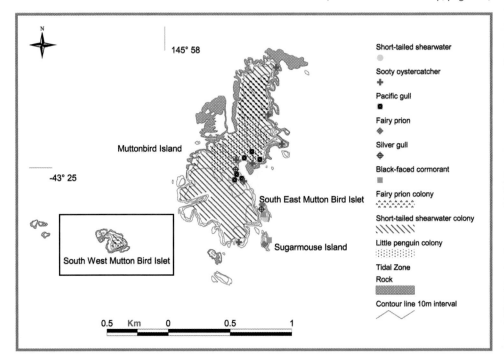

Location: 43°25'S, 145°58'E
Survey date: 2/12/81
Area: 0.52 hectares
Status: National Park

This is a dome shaped island rising to a height of 28 metres with a central ridge running north to south. Much of the east coast is a gentler slope except for the last 20 metres, when it becomes a bare rock cliff dropping to the sea. The western and southern sides are very sheer.

BREEDING SEABIRD SPECIES
Short-tailed Shearwater

Up to 1000 pairs are confined to the east side, as the west side is barren cliffs and ledges. They burrow in loose peat soils in association with *Carpobrotus rossii*, where burrows are in densities of $0.48/m^2$ and *Senecio* sp., where burrow densities rise to $0.7/m^2$ but cover a smaller area.

Fairy Prion

Up to 200 pairs are lightly scattered on the east side, mainly in burrows with a few in rock crevices. No other fauna was recorded.

VEGETATION

The east coast burrowing habitat is dominated by *Carpobrotus rossii* and *S. lautus*, while the barren west side is dominated by *Poa*.

COMMENTS

Although the island is accessible to Little Penguins, none appear to be here.

Mutton Bird Island

(Mutton Bird Island Group, page 545)

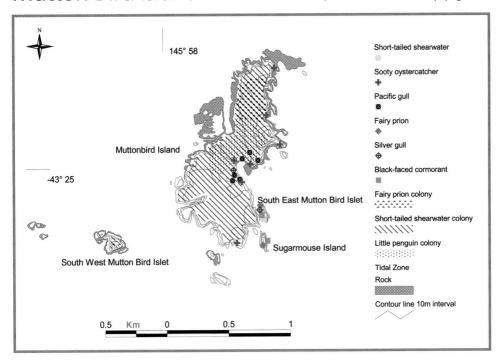

Location: 43°25'S, 145°57'E
Survey date: 2/12/81 and 7 – 20/1/77
Area: 43.7 hectares total
Status: National Park

Sometimes referred to as Flat Island, Mutton Bird Island is a large irregular shaped island running north to south. It is relatively flat with the highest slopes occurring on the eastern side where the summit reaches just over 40 metres. Its eastern shores are gently sloping with areas of pebble beach, whereas the western shores are generally more cliffy with gulches and occasional pebble beaches. The gulch on the northern side is a striking feature. Soil cover is extensive but not deep. Several small streams flow under wet conditions, probably originating in the swamps found on the higher slopes in the south-east. Iron fastenings in the rocks of the west coast harbour possibly indicate past visitation by bird or guano collectors.

BREEDING SEABIRD SPECIES

Little Penguin

An estimated 3000 pairs, with eggs and chicks (2/12/81), were breeding over most of the island interspersed with Short-tailed Shearwaters. The greatest concentrations occur on the western landing inlet.

Short-tailed Shearwater

An estimated 529 521 pairs breed extensively and relatively densely over almost the entire island.

East side.

Fairy Prion

An estimated 2500 pairs on eggs (2/12/81) were nesting mainly at the edge of the soil cover in the immediate vicinity of the shore. However, they also extend across the island, particularly in the area between the western and eastern landing inlets.

Pacific Gull

6 pairs were observed feeding on the eastern shoreline and near the Silver Gull colony but did not appear to be breeding (2/12/81).

Silver Gull

12 pairs were nesting primarily on the slopes at the eastern landing inlet.

Sooty Oystercatcher

7 pairs on nests were located on all shores but mostly along the east coast.

VEGETATION

The vegetation is dominated by the perennial herb *Senecio pinnatifolius*, which covers most of the island while co-dominant species are the fern *Histiopteris incisa*, tussock grass *Poa poiformis* and pigface *Carpobrotus rossii*.

Woody shrubs including *Drimys lanceolata*, *Correa backhousiana*, *Leptospermum scoparium* and *Westringia brevifolia* occur only in small areas and are most common around the higher slopes. Even in these parts, the vegetation is usually less than 4 metres in height. Ferns and sedges also occur in patches.

Little Penguin tracks.

OTHER SEABIRD SPECIES
Great Cormorant (20/1/77)

No mammals were recorded.

BIRDS
Native:
White-faced Heron – common
White-bellied Sea-Eagle – 1 immature (2/12/81)
Swamp Harrier – 2
Brown Falcon
Striated Fieldwren (Calomanthus)
Sulphur-crested Cockatoo – 2 feeding in the scrub
Black Currawong – common
Forest Raven – breeding on east cliff amongst *Poa*
Silvereye – very common

Introduced:
Common Starling –10

REPTILES
Metallic Skink
Tasmanian Tree Skink

COMMENTS
This is an ideal location for breeding seabirds. As there is almost no canopy, landing is easy for the shearwaters, while the extensive areas of gentle slopes provide easy access for Little Penguins. It has probably more birds nesting per hectare than any other island in the area, but unfortunately once ashore, all parts of the island are readily accessible, making the birds particularly vulnerable to human disturbance. Fires and the killing of birds for crayfish bait have been problems in the past. Reports of seals hauling out are unverified to date.

'The landing of shearwaters on Mutton Bird Island is a truly incredible sight…On January 8, 1977 sunset occurred at 2100 hours at which time small flocks of shearwaters were observed flying close to the water a short distance out at sea in the west. These flocks rapidly became larger as the light faded, before soaring high into the air at about 2115 hours, by which time numbers were building up at an incredible rate forming huge rafts over the water. The first birds commenced landing at 2120 hours, and within minutes an endless stream converged on the island, circling it several times before coming in for their final run. The main landing period was from 2130 to 2300 hours, during which time countless thousands of shearwaters practically blotted out the entire sky for as far as the eye could see and made any human travel about the island virtually impossible…This was typical of all other evenings during the visit.' (White, 1980)

Sugarmouse Island

(Mutton Bird Island Group, page 545)

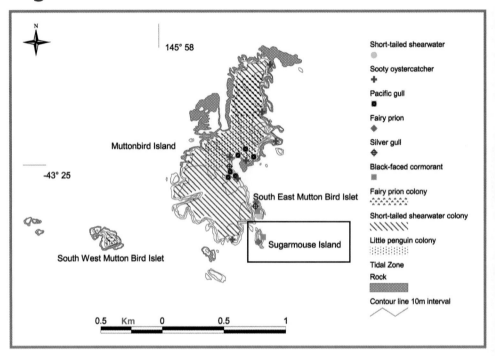

Location: 43°26'S, 145°58'E
Survey date: 2/12/81
Area: 0.54 hectares
Status: Non-allocated Crown Land

This is a very small islet off the south-east corner of Mutton Bird Island with a razorback ridge running north to south with steep to sheer cliffs all around. The soil is sandy and loose.

BREEDING SEABIRD SPECIES
Fairy Prion
An estimated 2000 pairs inhabit honeycombed nests in all suitable areas especially on the east side of the island and the top of the ridge.

Black-faced Cormorant
22 pairs were located on the south end and east side near the ridge top on bare rock ledges in nests constructed of vegetation and seaweed. 14 had empty nests, 4 nests had one egg, 2 had 2 eggs and 2 had 3 eggs.

No other fauna was recorded.

East Pyramids

(Mutton Bird Island Group, page 545)

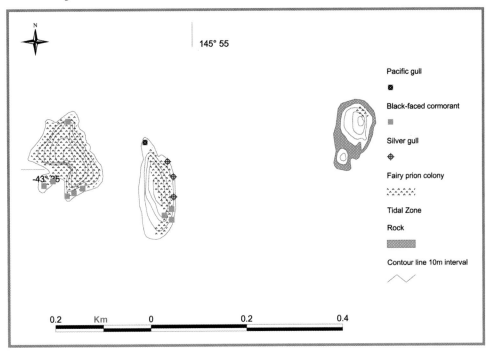

Location: 43°25'S, 145°55'E
Survey date: 2/12/81
Area: Total = 6.69 hectares
West Rock = 3.27 hectares,
Central Rock = 2.97 hectares
East Rock = 0.45 hectares
Status: National Park

East Pyramids are composed of three rocks referred to as west, central and east rocks each with little soil and few opportunities for burrowing birds. West Rock is steep sided with a central razorback ridge running north to south. Central Rock has a similar structure with a north to south razorback ridge and sheer sides, but its southern end drops gently. The East Rock has steep to sheer cliffs on the east side, and a north to south ridge with more gentle slopes on the west side, becoming sheer half way down.

BREEDING SEABIRD SPECIES
Fairy Prion

An estimated 100 pairs are concentrated in an area approximately 10 metres by 13 metres on the west side of the westernmost rock, with several burrows up to one metre long, also on the east side below the ridge beneath *Carpobrotus rossii*. On the central rock 50 – 100 pairs nest mainly in crevices or in short burrows. On the East Rock up to 50 pairs nest in rock crevices, with very few in burrows, found only where soil has accumulated in gullies, especially at the north and east end.

Pacific Gull

One pair was seen acting territorially on the Central Rock. Castings on the East Rock contained Fairy Prion remains.

West Rock.

Silver Gull

3 pairs were defending the bare ledges of the Central Rock about 15 – 20 metres above sea level on the east side. One fresh nest was found to the north of the highest point on the east side, 15 metres from the top ridge.

Black-faced Cormorant

On the west rock, 5 old nests were found towards the south end and there was evidence of another colony at the north end of the summit ridge. On the central rock 39 nests were found in an area 20 metres by 10 metres, just below the Silver Gull nest. Nests were freshly constructed from tussock and pigface, but evidenced from the many broken eggs and some nests not having been laid in, it appears that breeding was not successful. 5 were seen sitting on a a low rock near the water. There was no sign of nests on the east rock.

VEGETATION

The dominant vegetation on the west rock is C. rossii and Poa. On the central rock vegetation is similar, but more sparse, dominating the eastern side. The east rock has sparsely scattered vegetation dominated by C. rossii, Poa and fern species, with concentrations around the summit and along the north spur.

OTHER SEABIRD SPECIES

Sooty Oystercatcher – 3 landed on central rock, but were not breeding.

Common Diving-Petrel – wing found on westernmost rock, so may breed here.

Silver Gull – flew by west rock

No other fauna was recorded.

COMMENTS

The Fairy Prions on these islets readily abandoned their nests when humans walked by, demonstrating that they are particularly vulnerable to disturbance.

Sugarloaf Rock

(Mutton Bird Island Group, page 545)

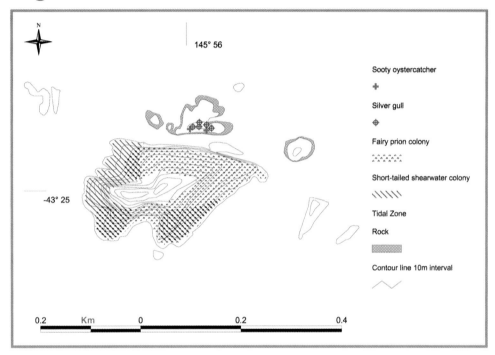

Location: 43°25'S, 145°56'E
Survey date: 2/12/81
Area: 3.56 hectares
Status: National Park

Rising to a summit of 103 metres, Sugarloaf Rock is very steep, with most parts of the eastern slope being bare rock, making these areas unsuitable for burrows. The main ridge runs from east to west. A small rock of 0.27 hectares lies to the north.

BREEDING SEABIRD SPECIES
Short-tailed Shearwater

An estimated 15 000 pairs breed predominantly on the west side, half way up, in an area 30 metres by 30 metres, where soil depth is adequate. Burrows are ususally very short beneath *Carpobrotus rossii* and the soil is loose and peaty, held together by plant roots. Average densities range from $1.17/m^2$ in the north-west to $0.32/m^2$ in the south-east. Shearwaters breed amongst the Fairy Prions in this area and seem to oust them from burrows, as evidenced by prion eggs lying at the entrances of enlarged burrows.

Fairy Prion

An estimated 2000 burrows were found all over the island in rock crevices extending to the edge of the vegetation in peaty loose soil held together by plant roots.

Sooty Oystercatcher

A nest with two eggs was located next to a *Poa* tussock on the summit of the rocky islet to the north of Sugarloaf Rock.

Silver Gull
5 pairs were defending and a nest was located with two eggs on the north-east corner on the summit of the north rock.

OTHER SEABIRD SPECIES
Pacific Gull – one pair

MAMMALS
Australian Fur Seal – haul-out of about 12 was located on the small rocky islet to the north of Sugarloaf Rock.

BIRDS
Black Currawong

No reptiles were recorded.

VEGETATION
C. rossii is dominant on the west side and largely confined to the western slope, with *Senecio* sp. co-dominant on the east side.

COMMENTS
Sugarloaf Rock is an island with impressive topography and high numbers of seabirds, for its land area.

Wendar Island

(Mutton Bird Island Group, page 545)

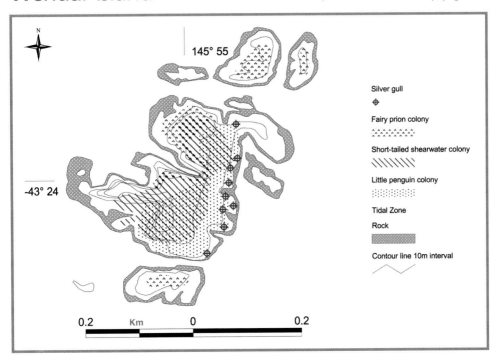

Location: 43°24'S, 145°55'E
Survey date: 2/12/81
Area: 5.8 hectares
Status: National Park

This is a small group of islands which consists of one main island with five small nearby islets. They are predominantly bare rock with sheer cliffs with only enough soil depth for short Fairy Prion burrows on three of them. Apart from the southern barren coast, the main island has greater soil depth, which provides adequate habitat for Little Penguins and Short-tailed Shearwaters.

BREEDING SEABIRD SPECIES
Little Penguin
An estimated 500 pairs, most with eggs and chicks, occur all over the main island, with the highest concentrations on the central east coast. They are commonly interspersed with Short-tailed Shearwaters.

Short-tailed Shearwater
Between 3000 and 4000 pairs burrow over the main island, with the highest concentrations being on the central east coast. They are interspersed with Little Penguins.

Fairy Prion
An estimated 1500 pairs, most on eggs, were located. The largest colony of 1000 pairs is on the main island, where burrows are scattered over the island with the higher densities towards the west side on the edge of *Poa* and rocks. On the smaller islets their burrows are short and often under *Carpobrotus rossii* or *Poa*.

East side

Silver Gull

10 pairs, most with 1 or 2 eggs, were recorded. One pair was located at the northern end of the main island. 8 pairs were nesting half way down the eastern side on steep slopes.

VEGETATION

Poa poiformis and *C. rossii* are the co-dominant species on the smaller islets, and are interspersed with scrub on the main island.

COMMENTS

This is a small, pristine, locally important island because of its diversity of seabirds.

OTHER SEABIRD SPECIES

Sooty Oystercatcher – 3 pairs, 2 on the main island and one on the northernmost islet, were defending but not nesting.

Pacific Gull – 1 pair was seen defending the southern shore on the northernmost islet, but no nest was found.

No mammals or reptiles were recorded.

BIRDS

Black Currawong – with 2 three-quarter grown chicks in a nest on tussock on the western cliff ledge.

Swainson Island Group

(Region 7, page 501)

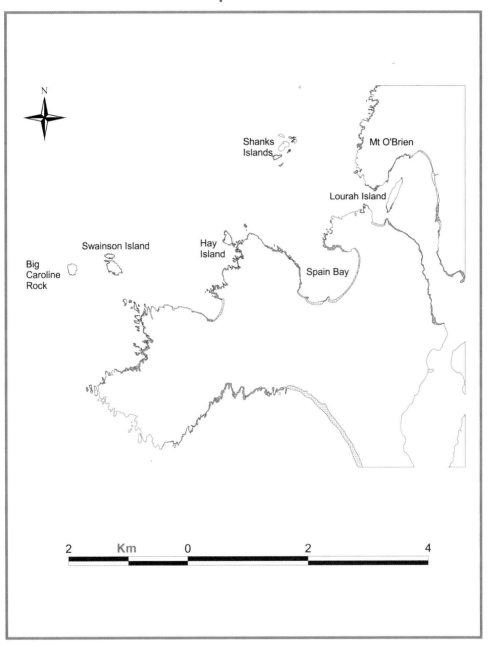

Big Caroline Rock

(Swainson Island Group, page 563)

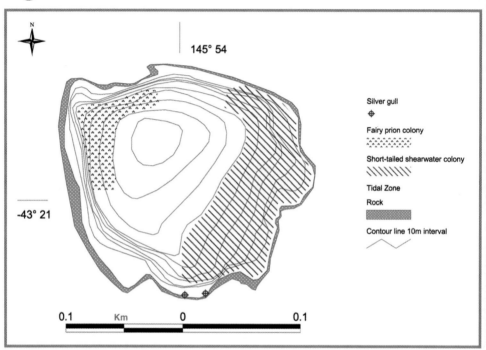

Location: 43°22'S, 145°55'E
Survey date: 2/12/81
Area: 2.19 hectares
Status: National Park

This is a steep, roughly oval-shaped rock rising to a height of 86 metres creating an impressive spectacle at the entrance of Port Davey. Cliffs on the western side rise vertically to within 20 metres of the summit whereas on the eastern side cliffs give way to steep vegetated slopes, which extend to the summit.

BREEDING SEABIRD SPECIES
Short-tailed Shearwater

15 000 pairs breed on the rock. Burrows are densest in areas where *Senecio lautus* is dominant. Insufficient depth of soil prevents burrowing in the lower 10 – 15 metres of the vegetated area. Elsewhere, the soil is loose, soft and peaty. Burrows occur in an area of about 1 hectare with a burrow density of $1.47/m^2$.

Fairy Prion

1000-2000 pairs were found throughout most of the vegetated area except where shearwater burrows are concentrated. Prion burrows are most numerous near the summit and just below it on the western side. At the time of the survey birds were incubating eggs.

Silver Gull

2 pairs were defending inaccessible cliff ledges on the central eastern side.

At the entrance to Port Davey is impressive Caroline Rock. (south-east side)

VEGETATION

Thirteen species of plants were recorded. *Senecio lautus* and *Carpobrotus rossii* are dominant. Shrubs including *Tetragonia implexicoma*, *Correa backhousiana* and *Leptospermum scoparium* are mainly confined to the summit and less steep areas.

No mammals were recorded.

BIRDS

Striated Fieldwren (Calamanthus) – 1

Black Currawong – 1 pair

REPTILES

Tasmanian Tree Skink

North-west side.

COMMENTS

With difficult access, this island should remain undisturbed. Its small size, pristine environment, vegetation diversity and the presence of skinks gives it representative value in the south-west region.

Swainson Island

(Swainson Island Group, page 563)

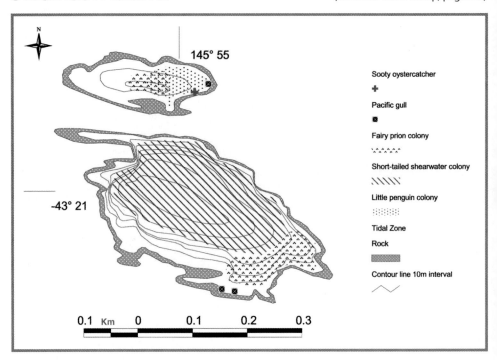

Location: 43°22'S, 145°55'E
Area: 4.14 hectares
Survey date: 3/12/81
Status: National Park

The main island of 3.42 hectares has a central ridge running north-west to south-east with very steep to sheer cliffs on the north-east side. The small north islet of 0.72 hectares has a north to south running ridge with sheer cliffs on the west side and relatively gentle slopes on the east.

BREEDING SEABIRD SPECIES

Little Penguin

100 pairs, with eggs or downy chicks, were located in depressions beneath the *Poa* on the eastern side of the small islet.

Short-tailed Shearwater

An estimated 38 000 pairs were recorded on the main island. Burrows were very dense amongst the *Senecio* sp. which is co-dominant with *Carpobrotus rossii* on the upper slopes. Birds were also nesting on the surface beneath thicker scrub patches around the central ridge.

Fairy Prion

3050 pairs breed on the two islets. An estimated 3000 pairs breed on the main island with the densest concentrations on the east coast lower slopes away from the shearwaters. A maximum of 50 pairs breed on the steeper east coast slopes of the small islet in rock crevices and burrows where there is peaty soil.

East side.

Pacific Gull
One nest, with two eggs, was located on the eastern shore of the small islet at the north end beneath the *Poa* just above the bare rock line. Another two were in *Poa* on the main island.

Sooty Oystercatcher
1 pair at a nest with 2 eggs was located on the small islet 25 metres to the south of the Pacific Gull nest.

VEGETATION
Poa poiformis is dominant on the small islet while *Senecio* is co-dominant with *C. rossii* on the main island. The main island also has scrub patches with *Correa* sp. dominant.

COMMENTS
A small diverse island with undisturbed seabirds breeding habitat.

OTHER SEABIRD SPECIES
Silver Gull – 3 were defending on the small islet but there was no sign of nesting.

No mammals were recorded.

BIRDS
Native:
Brown Thornbill
Black Currawong
Forest Raven
Little Grassbird
Introduced:
European Greenfinch
European Goldfinch

REPTILES
Tasmanian Tree Skink

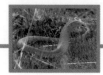

Hay Island

(Swainson Island Group, page 563)

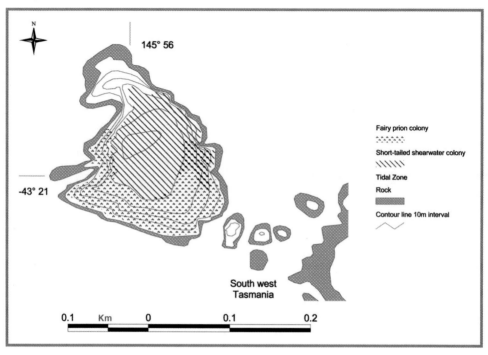

Location: 43°22'S, 145°57'E
Survey date: 3/12/81
Area: 1.85 hectares
Status: National Park

The island is surrounded by steep to sheer cliffs with an area of almost flat ground forming a saddle between the two little peaks at both the north and south ends. The summit is 78 metres high.

BREEDING SEABIRD SPECIES
Short-tailed Shearwater

An estimated 7500 pairs were recorded. The greatest concentration occurs in the saddle between the two peaks which has good soil depth and peaty soil on gentle slopes. Densities are highest amongst the *Senecio* and more lightly scattered amongst *Carpobrotus rossii*. There are no burrows at the northern end where there is only bare rock.

Fairy Prion

1000-2000 pairs nest on the lower slopes all over the island apart from the northern end. Burrows have been destroyed by shearwaters in those areas, which are co-inhabited.

No mammals, reptiles or other seabirds were recorded.

BIRDS
Native:
Peregrine Falcon – female circling the island
Tasmanian Scrubwren
Brown Thornbill
Forest Raven
Silvereye

VEGETATION
The vegetation is dominated by *Senecio* and *C. rossii*.

COMMENTS
This is a small, pristine island with a low diverity of flora and fauna.

Shanks Islands

(Swainson Island Group, page 563)

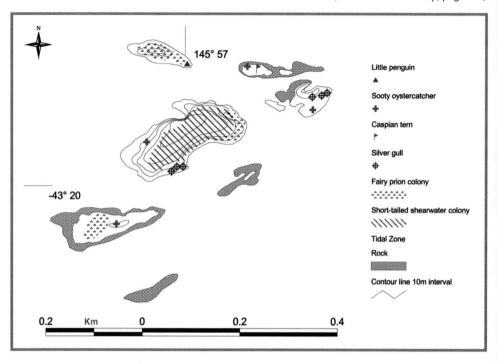

Location: 43°21′S, 145°58′E
Survey date: 3/12/81
Area: 5 islands: total of 2.72 hectares
Status: National Park

This is a group of five small rocky islands, with steep cliffs and central ridges. Mostly they are bare rock with *Poa poiformis* and *Carpobrotus rossii* being the dominant vegetation. The main island, of 1.6 hectares, is the only island which has sufficient soil for burrowing seabirds.

BREEDING SEABIRD SPECIES
Little Penguin
2 pairs were observed on the northernmost island low down on the east coast in rock clefts.

Short-tailed Shearwater
An estimated 8700 pairs breed on the main island. Burrows were located along the east coast from 20 metres above the *C. rossii*, extending to the summit along the north-south ridge amongst thick scrub. Only in a few places do burrows continue down the west side, as it is rocky with thick scrub.

Fairy Prion
Up to 5500 pairs breed on all the islands except the easternmost. They nest in *C. rossii* on lower slopes. They are scattered throughout the main island with an estimated 3000 – 5000 pairs, concentrated on steep *C. rossii* lower slopes away from the shearwaters.

Silver Gull
6 pairs, with eggs, were sighted. 3 pairs were located on the easternmost island with 7 empty nests. 3 pairs were also located at the south end of the eastern side of the main island, 5 metres above water at the edge of the vegetation.

Sooty Oystercatcher
4 pairs, with 2 eggs each, were observed. One pair was located on each of the islands apart from the westernmost one. Their habitats ranged from rock clefts to thick *Poa* to bare ground.

Caspian Tern
One pair, with 2 eggs, was on the north-eastern islet in the open amongst *C. rossii*.

VEGETATION
The vegetation is predominantly *C. rossii* with *Poa poiformis* and scrub and some ferns.

No mammals or reptiles were recorded.

BIRDS
Native:
Forest Raven – nesting on a cliff ledge on the western side of the main island

COMMENTS
The high diversity of seabird species makes these islands regionally important.

Main island.

Lourah Island

(Swainson Island Group, page 563)

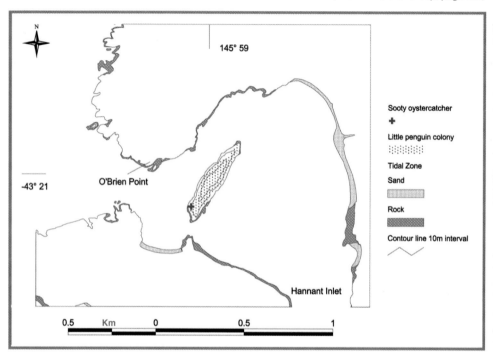

Location:	43°21'S, 145°59'E
Survey date:	14/12/81
Area:	4.86 hectares
Status:	National Park

It is a long, narrow island which rises gently from the shore to a central ridge. The west coast is bare rock.

BREEDING SEABIRD SPECIES
Little Penguin
A colony, approximately 50 metres long by 25 metres wide, containing 200 – 300 pairs extends from a smaller number amongst boulders to larger numbers in scrub. The area slopes gently, then rises more steeply to a ridge. Burrows are in sandy peat to peat soil where fern species is the dominant ground cover. The penguins' access is from a small beach.

Sooty Oystercatcher
1 pair, with 2 eggs, was recorded at the south-west end of the island 16 metres above sea level on gravel that had accumulated in a rock cleft.

OTHER SEABIRD SPECIES
Pied Oystercatcher – 1 pair flew by

Pacific Gull – 1 flew by

VEGETATION
It tends to be typical of that of nearby mainland Tasmania. The west coast is mainly devoid of

vegetation and windswept near the summit. It then progresses into stunted heath, before myrtaceous shrubland commences just beneath the summit. At the north end, the top of the island is flat and covered in butttongrass. Along the eastern shore the vegetation is very thick with *Melaleuca*, *Leptospermum* and *Eucalyptus*. Fern species dominate the sandy peat soils.

BIRDS
Native:
Black Swan – 2

COMMENTS
This island has no faunal significance.

Breaksea Island Group

(Region 7, page 501)

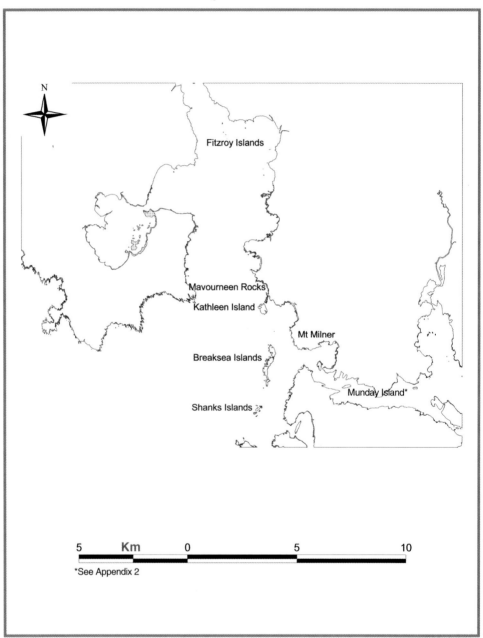

*See Appendix 2

Breaksea Islands

(Breaksea Island Group, page 574)

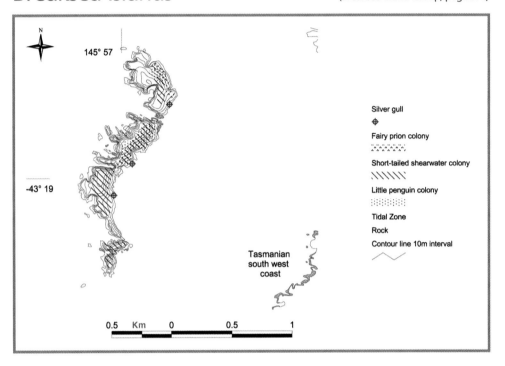

Location: 43°20'S, 145°58'E
Survey date: 6/12/81
Area: 16.25 hectares
Status: National Park

There are two islands in this group – north and main Breaksea Island. North Breaksea has a central ridge running north to south, covered in scrub. The east coast of the island has several main access points. The west coast is rocky with steep cliffs. The main Breaksea Island is in two sections separated by sheer cliffs joined at the base by a sea platform. It also has a central ridge running north to south with scrub covering the top and becoming dense on the summit. The western slopes are rugged with rock and scrub and very little soil.

BREEDING SEABIRD SPECIES
Little Penguin

A total of approximately 400 pairs breed on both islands. Up to 200 pairs, many with small chicks or eggs, breed on the main island. They are found amongst a small patch of rock rubble on the central east coast just above the water, where access is gained via a short 5 – 10 metre rock face. There are few burrows in the south. On the north island there is a small colony of about 200 pairs amidst the *Poa* on the north-east coast and the *Senecio* on the central east coast just inland from the penguins' main access point. Eggs and chicks were present during the survey.

North-west cliffs.

Short-tailed Shearwater

An estimated 3000 – 5000 pairs were located in varying densities over both islands, mainly amidst *Senecio* sp. and *Carpobrotus rossii*, where the soil is sandy and peaty. Burrows are in high densities in the strips of *Poa* on the north-east of both islands, otherwise densities vary according to soil availability and type. Burrowing is more successful in sandy soils with average burrow lengths of approximately 1 metre whereas burrows in the peat soils are generally shallow and unstable. Birds also nest on the cliffs in clefts and crevices beneath slabs of rock and overhanging *C. rossii*. During the survey, many were incubating eggs and some had chicks.

Fairy Prion

20 pairs, on eggs, were located in a small area of the east coast of North Breaksea Island just above the sheer rock line in peaty soil.

Silver Gull

Three separate colonies were found on the west side of the main island, about 30-40 metres apart, with nest sites associated with *Poa* or *C. rossii*. The northern colony had 20 pairs, most with eggs, the middle colony 15 pairs, half with eggs and the southern colony had 13 pairs, most with no eggs.

VEGETATION

Senecio sp is dominant on the north island, with a strip of *Poa* extending inland from the central east

OTHER SEABIRD SPECIES

Sooty Oystercatcher – 3 pairs were on the shore of the north island but were not breeding. One pair was present on the main island but also not breeding.

Pacific Gull – 3 pairs were observed on the north-west point of the main Breaksea Island on rocky coast and several flew by, but there was no evidence of breeding.

MAMMALS

Introduced:

Rabbit – on the main island, where they have denuded an area of vegetation close to the central ridge, causing localised severe erosion.

BIRDS

Native:

White-bellied Sea-Eagle – a juvenile was seen on the main island.

Green Rosella

Crescent Honeyeater

Superb Fairy-wren

Eastern Spinebill

Black Currawong

Forest Raven

Tree Martin – 4

REPTILES

Tasmanian Tree Skink – on both islands

coast. Tea tree and pepper bush cover the top of the ridge on the main island, while the southern end is rocky with low, thick scrub. *Poa* and *Senecio* are dominant on the eastern slopes and a 25 – 30 metre strip of *Poa* extends inland from the north-east coast.

COMMENTS

An assessment of the impact of rabbits should be carried out and an eradication program implemented. Accredited tourist ventures to the islands could be encouraged.

Sea cave on the main island.

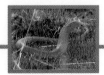

Kathleen Island

(Breaksea Island Group, page 574)

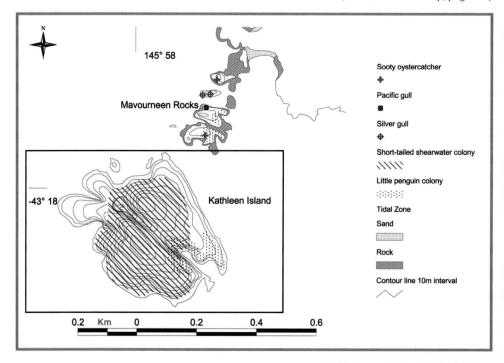

Location: 43°31'S, 145°59'E
Survey date: 5/12/81
Area: 11.35 hectares
Status: National Park

This is an irregular-shaped island with steep cliffs particularly on the southern side, rising to a summit of 72 metres. A deep gulch almost splits the island.

BREEDING SEABIRD SPECIES
Little Penguin

100 – 200 pairs were recorded in the main southeast cove. They were breeding in burrows from the water line up to at least near the saddle, possibly further. Most were on eggs but a small percentage had downy or large chicks.

Intensive burrowing by shearwaters has resulted in loss of vegetation and erosion.

Short-tailed Shearwater

An estimated 66,919 burrows were located beneath scrub and rainforest in sandy soil and in peat gullies. They were substantial, strong burrows with no ground vegetation to protect them. The south-west colony ends at the cliff and is dominated by *Senecio* sp. and *Carpobrotus rossii*, except for the last 10 metres.

VEGETATION

Thick scrub and rainforest exists over much of the island. On the south-west side, inland to 10 metres, *Senecio* sp. and *C. rossii* are co-dominant.

BIRDS

Native:
Green Rosella
Superb Fairy-wren
Scrubtit
Tasmanian Scrubwren
Tasmanian Thornbill
Crescent Honeyeater
New Holland Honeyeater
Black Currawong – nesting in the tea tree

REPTILES
Metallic Skink

COMMENTS

It is a pristine island with examples of erosion caused by burrowing seabirds.

South-east side.

North-west side.

Fitzroy Islands

(Breaksea Island Group, page 574)

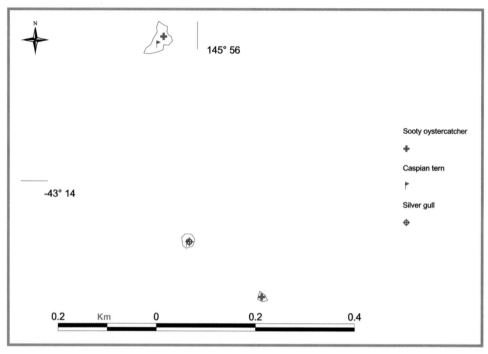

Location: 43°14'S, 145°56'E
Survey date: 12/12/81
Area: 0.18 hectares
Status: National Park

These are a group of four small rocky islets, one without seabirds, as it is wave-washed. The northern islet is the largest at 0.12 hectares, the middle and southern islets are virtually low rocky reefs with areas of 0.04 and 0.02 hectares respectively.

BREEDING SEABIRD SPECIES
Silver Gull
4 nests with 1 egg, 4 with 2 eggs, 5 with 3 eggs and 4 empty nests were found amongst *Sarcocornia*, *Poa* and *Tetragonia*.

Sooty Oystercatcher
One pair was located breeding on the southern islet and another 2 pairs were present on the northern islet but did not appear to be breeding.

Caspian Tern
One pair with a half-grown chick was nesting at the north-eastern end of the northern islet amongst *Poa*. The other pair with a fully-fledged chick was nesting on the north end of the middle islet.

VEGETATION
Poa, *Sarcoconia* and *Tetragaonia* dominate the coastal areas with patches of scrub on the eastern slopes.

OTHER SEABIRD SPECIES

Black-faced Cormorant – 2 were sitting on the shore of the middle islet.

Pacific Gull – 1 pair was flying over the middle islet, but there were no signs of breeding.

No other fauna was recorded.

COMMENTS

Small islets which have little faunal significance.

South coast, rowing ashore

Mavourneen Rocks

(Breaksea Island Group, page 574)

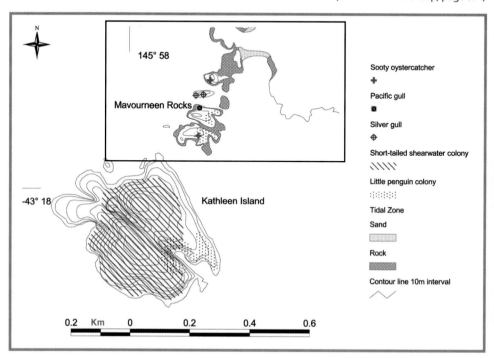

Location: 43°18'S, 145°58'E
Survey date: 6/12/81
Area: 0.88 hectares total
Status: National Park

This is a group of four small rocks, which all have sheer bare cliffs on their northern, western and southern coasts. Their eastern coasts are covered in scrub. The northernmost rock which is closest to shore has no breeding seabirds. The two southernmost rocks closest to Kathleen Island are joined at low tide and have the largest area of 0.68 hectares.

BREEDING SEABIRD SPECIES

Little Penguin

Up to 20 pairs were located all along the east coast slopes of the southernmost rocks under scrub to almost the summit. Several pairs were also in depressions under the canopy of thick vegetation. All birds were on eggs, most newly laid.

Pacific Gull

An old nest was located on the third rock south of the mainland.

Silver Gull
2 pairs were recorded in *Poa* near the summit.

Sooty Oystercatcher
One pair was on the second rock from the mainland shore by *Poa*, 10 metres above the water. The other pair was on the southernmost rock again by *Poa*, at the edge of the rocks in the south-east corner. Each pair had 2 eggs.

VEGETATION
Scrub, which forms a thick canopy, dominates the east coast slopes and *Poa* exists in patches on the edge of the rocky shores.

No other fauna was recorded.

COMMENTS
Small islets which have little faunal significance.

Trumpeter Islets Group

(Region 7, page 501)

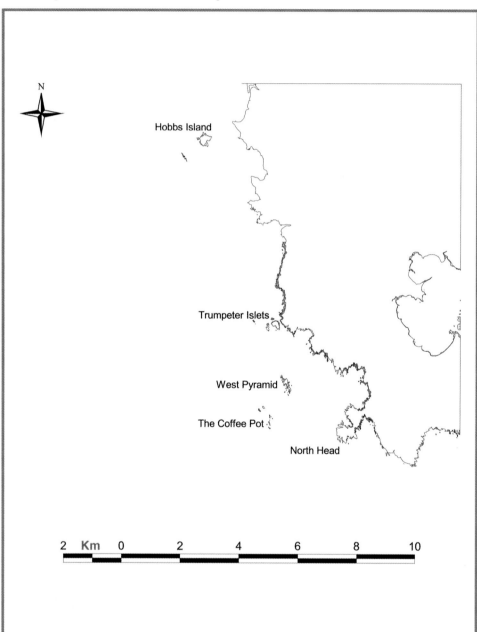

The Coffee Pot

(Trumpeter Islets Group, page 584)

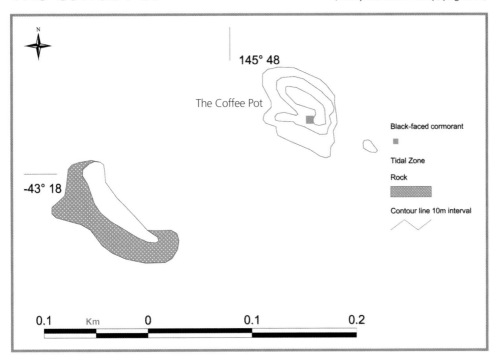

Location: 43°18'S, 145°48'E
Survey date: 4/12/81
Area: 0.31 hectares
Status: National Park

This rock is steep to sheer sided and shaped, as the name suggests, like a coffee pot. Its spout, a rock spire, on the southern end is half the height of the summit. There is very little soil present.

BREEDING SEABIRD SPECIES
Black-faced Cormorant

27 nests were counted on the eastern side, extending from the summit to within 15 metres of the water, on the bare rock ledges. They were built from seaweed and small amounts of *Tetragonia*.

East side of the Coffee Pot.

VEGETATION

Only one species of plant, *Tetragonia implexicoma,* was found scattered sparsely among crevices and small areas of jumbled boulders.

OTHER SEABIRD SPECIES

Fairy Prion – an adult was found in a crevice near the summit and 3 dead adults were also found.

No mammals, reptiles or other birds were recorded.

COMMENTS

Severe storms, which would disrupt breeding birds, are reported to wash over most or all of the rock. Such storms also prevent Fairy Prions from using suitable rock crevices and boulders for nesting.

West Pyramid

(Trumpeter Islets Group, page 584)

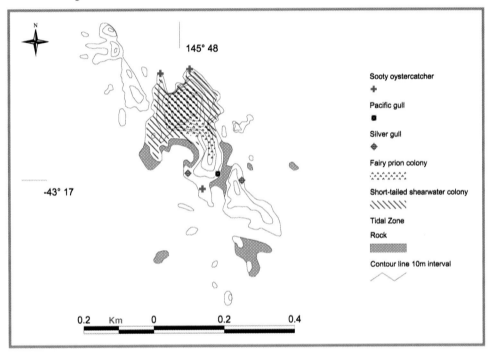

Location: 43°18'S, 145°49'E
Survey date: 4/12/81
Area: 2.5 hectares
Status: National Park

An elongate island, West Pyramid is steep sided with its long axis running north-west to south-east. The southern end of the island, which is about a quarter of the total island area, is bare and rocky. This is connected to the pyramid-shaped northern end via a steep, boulder-strewn ridge. The northern end is surrounded by vertical cliffs that descend to the water and the vegetated summit of the island is located at the apex. There is a small pebble beach on the north-west side at the base of a cliff.

BREEDING SEABIRD SPECIES

Short-tailed Shearwater

Up to 3000 pairs were located, restricted in their distribution by suitable burrowing habitat, to the northern section of the island. The major concentration was found on the western slopes where *Senecio lautus* was dominant. Lower concentrations of birds were found along the ridge and down the slope on the eastern side.

Fairy Prion

An estimated 1000 to 2000 nests were found in crevices and burrows on the northern section of the island. The highest densities of burrows occur

on the lower slopes, comprising about 50% of the total number. There is intense competition between Fairy Prions and Short-tailed Shearwaters for burrows, resulting in low occupancy rates for the prions. The burrows are restricted in depth by the lack of soil and follow the strike and dip of the base rock, resulting in many burrows interconnecting. During this survey, most birds were incubating eggs.

Pacific Gull

One pair, with 2 addled eggs and one downy chick, was recorded at the southern end of the island.

Silver Gull

44 pairs, most at new nests or with eggs were located at two colonies on the southern section of the island. One colony was on bare rock, with nests of dry *Poa poiformis*, while the other was in a patch of *P. poiformis*.

Sooty Oystercatcher

One pair, with 2 eggs, was nesting in a gull's nest at the southern end of the island and another, also with 2 eggs, was at the northern end of the pebble beach. The other pair was displaying territorial behaviour at the northern tip of the island.

VEGETATION

The northern section is vegetated with *Leptospermum scorparium, Rhagodia candolleana* and *Tetragonia implexocoma* which dominate the summit and the ridge. Scattered patches of *Carpobrotus rossii* and *S. lautus*, surrounded by bare patches of rock, dominate the rest of the island.

No mammals or other birds were recorded.

REPTILES

Tasmanian Tree Skink

COMMENTS

The gulls and Sooty Oystercatchers were breeding in areas vulnerable to high seas. Soils are extremely fragile.

Trumpeter Islets

(Trumpeter Islets Group, page 584)

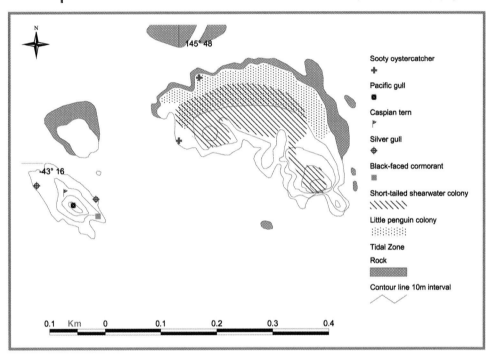

Location: 43°17'S, 145°48'E
Survey date: 4/12/81
Area: Main islet: 0.75 hectares, smaller islet: 0.29 hectares
Status: National Park

There are two islets. The main islet, with two peaks and flat ground to the north, has soft sandy peat. There are also a number of bare earth patches, especially on the flat ground. The smaller islet is steep-sided, with bare sheer rock on the western face and steep rock ledges on the eastern side.

BREEDING SEABIRD SPECIES

Little Penguin

Up to 1000 pairs inhabit the main islet, the majority nesting on the flat ground to the north, interspersed with shearwaters.

Short-tailed Shearwater

An estimated 1000 pairs primarily inhabit the relatively flat ground to the north end of the main islet but also extending into the scrub. There were a number of unoccupied areas especially in the scrub around the peaks.

Pacific Gull

One pair was located on the summit of the small islet.

Silver Gull

136 pairs at nests were located on the rock ledges beneath the overhanging *Carpobrotus rossii* on both the east and west sides of the small islet.

East side.

Sooty Oystercatcher
2 pairs, with 2 eggs each, were located, one on the north side of the main islet and one on the west side.

Black-faced Cormorant
35 nests, constructed of dry grass and seaweed, were concentrated in a small area of bare rock ledges at the south-east end of the small islet.

Caspian Tern
1 pair, with 1 egg and 2 downy chicks, was located 4 metres to the north of the Pacific Gull nest on the summit of the small islet.

VEGETATION
On the main islet, *Senecio* sp. and *C. rossii* were dominant with scrub confined largely to the two peaks which have little soil on the upper parts. There are small *C. rossii* patches between the rocks on the smaller islet.

OTHER SEABIRD SPECIES
Fairy Prion – 1 dead, with no evidence of breeding

No mammals were recorded.

BIRDS
Native:
Masked Lapwing - 2
Superb Fairy-wren
Striated Fieldwren
Little Grassbird

REPTILES
Tasmanian Tree Skink

COMMENTS
This is a pristine island with high diversity and abundance of seabirds.

Hobbs Island

(Trumpeter Islets Group, page 584)

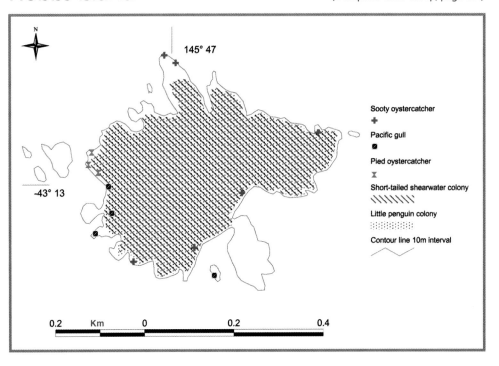

Location: 43°13'S, 145°47'E
Survey date: 3/12/81
Area: 9.73 hectares
Status: National Park

Sometimes referred to as Green Island, it is composed of sandy soil with peat occurring in scattered patches. It has flat topography, rising to a summit of only 30 – 40 metres in the east, a rocky shoreline with a south-facing pebble beach and good vegetation cover throughout.

BREEDING SEABIRD SPECIES
Little Penguin

An estimated 11 000 pairs breed all over the island, the higher densities of $1.52/m^2$ occurring in the sandy soil, where *Poa* predominates, with lower densities of $0.84/m^2$ in peat soils.

Short-tailed Shearwater

Interspersed with Little Penguins, an estimated 11 000 occur in similar densities.

Pacific Gull

4 pairs, 2 with eggs and 2 with downy chicks were present. 3 pairs were on the western shore between the rock and the vegetation and one pair with small downy chicks was on the small islet off the southern coast.

Sooty Oystercatcher

6 pairs, with eggs, were recorded. 2 pairs were on the north north-east tip on nest scrapes, 2 pairs were on the eastern shore, one pair with one egg and 2 pairs, each with 2 eggs, were on the southern shore.

Pied Oystercatcher
3 pairs with 2 eggs each, were nesting on the western shore on the border of the rock and vegetation.

OTHER SEABIRD SPECIES
Fairy Prion – a dead bird was found on the beach

Silver Gull – 25 were present, but not breeding

MAMMALS
Water Rat – one dead on the south-east beach

BIRDS
Native:

White-faced Heron

Orange-bellied Parrot – 5 separate sightings – a pair was feeding on *Poa* seeds

Striated fieldwren

Forest Raven – 60+

Little Grassbird – numerous

REPTILES
Tasmanian Tree Skink

VEGETATION
Poa poiformis on sandy soil is dominant over the island. *Senecio* sp. and fern species dominate the peat soils. *Senecio pinnatifolius* and *Carpobrotus rossii* are also widespread, with the gully areas predominantly vegetated by *Carex appressa*.

COMMENTS
This island is one of the more important breeding sites for the Little Penguin in the south-west of the State.

Hibbs Pyramid Group

(Region 7, page 501)

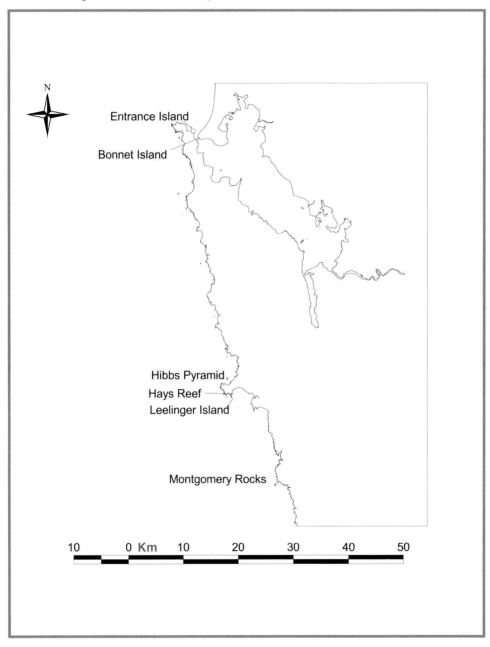

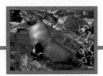

Montgomery Rocks

(Hibbs PyramidGroup, page 593)

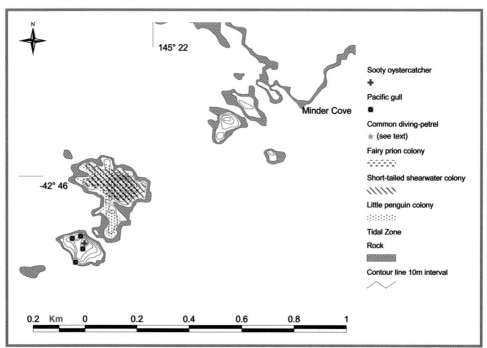

Location: 42°47'S, 145°23'E
Survey date: 22/12/88
Area: 3.69 hectares
Status: Conservation Area

This group consists of a main rock of 2.74 hectares and a smaller rock of 0.95 hectares. The main pyramid-shaped rock of 50 metres is composed of dolerite and limestone with high steep sides and a central ridge.

BREEDING SEABIRD SPECIES
Little Penguin

60 pairs were located over the main island mainly in matted *Poa* and rock crevices right to the summit. There were two main access points, one in the south-east and one on the west coast.

Short-tailed Shearwater

560 pairs were nesting sparsely in 50% *Poa poiformis* and 50% *Senecio lautus*. Burrows were either very shallow or just under matted *Poa* clumps.

Fairy Prion

760 burrows were found all over the upper slopes of the main rock, chiefly in association with the shearwater colonies on the south-east side. Many were also scattered around the island in rock crevices. There may also be small numbers on the western rock.

Common Diving-Petrel

Approximately 100 burrows were found on the ledges of the west to north-west steep slopes of the main rock in low matted *Poa*.

Montgomery Rocks

Pacific Gull
4 pairs were defending off the western rock.

Sooty Oystercatcher
One pair was defending the north shore of the western rock.

VEGETATION
The west and north-west sides have steep cliffs and lightly *Poa*-covered ledges. The south-east side is steep and well-vegetated. The smaller rock to the west is almost barren with only a small area of vegetation, mainly *Carpobrotus rossii* and *Rhagodia candolleana*, which are not present on the main rock.

No mammals or other birds were recorded.

REPTILES
Metallic Skink

COMMENTS
The isolated nature, seabird diversity and pristine nature of the rocks make them significant for the region.

Leelinger Island

(Hibbs Pyramid Group, page 593)

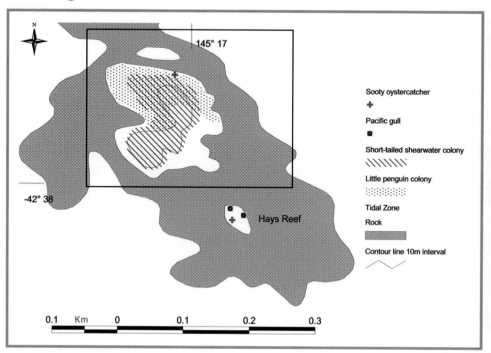

Location: 42°38'S, 145°17'E
Survey date: 19/12/88
Area: 1.54 hectares
Status: Conservation Area

It is a flat, gently undulating, dolerite island with vegetation dominated by *Poa poiformis*. The south-west is bare rock protruding from the sea covered with stipa. There is a small shingle and boulder cove at the south-east end. Elsewhere, the shoreline is rocky with a gentle slope to the sea. It is separated from the mainland by a 20 metre wide permanent channel.

BREEDING SEABIRD SPECIES
Little Penguin
Up to 400 pairs were located. 90% of the nests are under very dense *P. poiformis* with scattered patches of *Correa* sp. particularly at the north end. Burrows are scattered amongst the shearwater colony.

Short-tailed Shearwater
6000 pairs were located throughout the central part of island in dense *P. poiformis* and scattered patches of *Correa* sp., particularly at the north end.

Pacific Gull
2 pairs, one with half grown chicks and one newly-hatched dead chick were found on Hays Reef.

Sooty Oystercatcher
One pair was sighted on Hays Reef and one pair was on the north-east tip on shingle with one newly-hatched chick.

VEGETATION
90% very dense *P. poiformis* with scattered patches of *Correa* sp. particularly at the north end. *Carpobrotus rossii* is dominant in some coastal areas away from shearwater colonies.

Coralline algae colours the rocks.

OTHER SEABIRD SPECIES
Black-faced Cormorant

Great Cormorant

Pied Oystercatcher – a pair not nesting

Crested Tern – 4 fly by + 24 in high altitude sky chasing, possibly courtship.

Fairy Tern – 2 fly by and 2 pairs possibly nesting. – mildly aggressive.

No mammals or reptiles were recorded

BIRDS
Native:
White-bellied Sea-Eagle

Grey Teal

Superb Fairy-wren

Tasmanian Scrubwren

Pink Robin

Black Currawong

Forest Raven

Kelp forests on the seaward side.

Hays Reef

(Hibbs PyramidGroup, page 593)

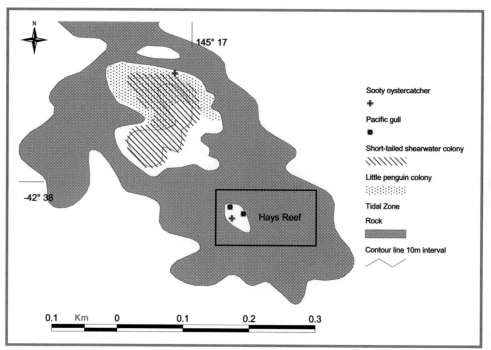

Location: 42°38'S, 145°17'E Check
Survey date: 19/12/88
Area: 1.54 hectares
Status: Conservation Area

Connected at low tide to mainland Tasmania, it is a small rocky reef with small pockets of vegetation dominated by *Carpobrotus rossii*.

BREEDING SEABIRD SPECIES
Pacific Gull
2 pairs, one with half grown chicks and one newly hatched dead chick were nesting in association with the *C. rossii*.

Sooty Oystercatcher
One pair defending a nest scrape was recorded.

OTHER SEABIRD SPECIES
Black-faced Cormorant

Great Cormorant

Pied Oystercatcher – a pair not nesting

Crested Tern – Up to 24 in high altitude sky chasing, possibly courtship.

Fairy Tern – 4 seen acting mildly aggressive.

No other fauna was recorded.

COMMENTS
It would be useful to ascertain whether the two recorded tern species actually breed here.

Hibbs Pyramid

(Hibbs PyramidGroup, page 593)

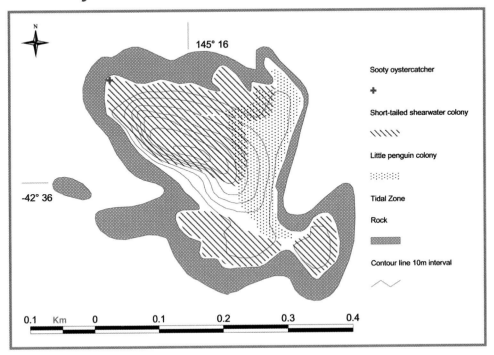

Location: 42°36'S, 145°16'E.
Survey date: 20/12/88
Area: 4.65 hectares
Status: Conservation Area

This is a vegetated, steep-sided pyramid-shaped dolerite island, the upper third of which is mostly bare rock with light *Poa* and scattered scrub species. The west side of the island is dominated by steep cliffs to the sea. The island's shoreline is bare rock but for a small pebble and boulder beach near the east end.

BREEDING SEABIRD SPECIES
Little Penguin

820 pairs were located in colonies, concentrated in from the beach area amongst the shearwaters. Most of the Little Penguins were in burrows with some under dense, matted *Poa* and in rock crevices.

Classic island pyramid shape.

Pebble cove on east side.

Short-tailed Shearwater
An estimated 9055 pairs occur in four distinct areas. The main colony in the north of the island has an estimated 3960 burrows in an area dominated by *Carpobrotus rossii* and *Senecio* sp. with 20% bare soil. The central colony, with an estimated 2600 burrows, is dominated by *Poa* (90%) and *Senecio* sp. and the two colonies which follow the ridge line are in very dense, matted *Poa*.

Sooty Oystercatcher
1 pair was defending on the north-west coast.

VEGETATION
The north of the island is dominated by *C. rossii* and *S. lautus* and the eastern section is dominated by *Poa poiformis* (90%) and *Senecio lautus*. Dense matted *Poa* occurs around the ridge line with an extensive shrub area on the upper eastern summit slopes, especially at the north end.

OTHER SEABIRD SPECIES
Silver Gull – 10 offshore

Common Diving-Petrel – a wing was found

No mammals were recorded although Fur Seals haul out on the adjacent rock.

BIRDS
White-bellied Sea-Eagle

Forest Raven

REPTILES
Metallic Skink

Tasmanian Tree Skink

Entrance Island

(Hibbs PyramidGroup, page 593)

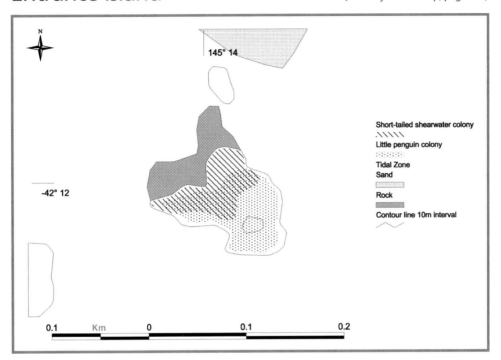

Location: 42°12'S, 145°14'E
Survey date: 7/12/89
Area: 6.1 hectares
Status: Non-allocated Crown Land

Situated at the entrance to Macquarie Harbour, this small island rising to only 5 metres in height has rocky shores, black, sandy soil and numerous rocky outcrops. Signs of past inhabitation include an old house site, a light beacon and jetty.

BREEDING SEABIRD SPECIES
Little Penguin

An estimated 100 pairs occur all over the islet in crevices, under rocks and under old building materials, the majority in short, shallow burrows in dark black sandy soil.

Short-tailed Shearwater

An estimated 1900 pairs occur, concentrated at the northern end and elsewhere lightly scattered amongst Little Penguins. Their habitat is dominated by bracken, particularly in the south, introduced grasses, *Poa poiformis,* kangaroo apple and scattered shrubs of *Myoporum* sp.

VEGETATION

The vegetation is dominated by bracken in the south, introduced grasses, *P. poiformis*, kangaroo apple and *Myoporum* species. Other introduced plants include radiata pine, thistles and fuscias.

Light beacon and jetty.

No mammals or other seabird species were recorded.

BIRDS
Native:
Superb Fairy-wren

Introduced:
Common Starling

REPTILES
Metallic Skink

COMMENTS
Considering its size, this island supports a diverse array of fauna and flora. Despite the tide flow at times being particularly strong past this and Bonnet Island, it does not pose too great an obstacle for the Little Penguins that come here to breed.

Bonnet Island

(Hibbs PyramidGroup, page 593)

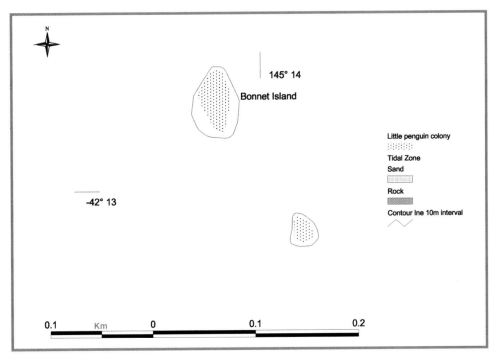

Location: 42°13'S, 145°14'E
Survey date: 7/12/89
Area: 2.21 hectares
Status: Non-allocated crown land

Situated in Kelly Channel at the entrance to Macquarie Harbour, this small islet rises to a height of 6 metres with a central summit ridge running north to south with steep slopes descending to rocky shores. Fifty metres further south, there is another very small, scrub-covered, rocky islet, which has little soil depth. This islet appears to be unnamed.

BREEDING SEABIRD SPECIES
Little Penguin

An estimated 250 pairs are concentrated along the central summit ridge on steep slopes. Their burrows are short and shallow with some under dense vegetation, either grasses, sedges or *Tetragonia implexicoma*. 7 pairs were in rock crevices on the small southerly islet nesting under matted *Tetragonia*.

VEGETATION

The vegetation is dominated by bracken fern, introduced grasses, *Poa poiformis*, *Tetragonia*, kangaroo apple and *Myoporum* species. Other introduced plants include radiata pine, fruit trees, thistles and fuscias.

COMMENTS

An island of relatively low flora and fauna significance but of important historical value.

Historical light beacon.

No mammals or other bird species were recorded.

REPTILES
Metallic Skink

Appendix 1
Status of Fauna under *Tasmanian Threatened Species Protection Act 1995*

Under the *National Parks and Wildlife Service Act 1970*, all species listed in Table 1 are wholly protected in that it is an offence to trade, collect, possess or disturb them unless a permit is obtained from the Director of the Parks and Wildlife Service.

Under the Tasmanian *Threatened Species Protection Act 1995*, the Little Tern is listed as endangered, the Shy Albatross as vulnerable and the Fairy Tern and White-fronted Tern as rare.

Under the Commonwealth *Endangered Species Protection Act 1992*, the Little Tern is listed as endangered and the shy albatross as vulnerable.

The following definitions apply under the Tasmanian *Threatened Species Protection Act 1995*:

Endangered species
- are in danger of extinction because long-term survival is unlikely while the factors causing the species to be endangered continue operating or
- are presumed extinct on the grounds that no occurrence of the species in the wild can be confirmed during the past 50 years.

Vulnerable species
- are likely to become endangered while the factors causing them to be vulnerable continue operating.

Rare species
- are those which have small populations in Tasmania that are not endangered or vulnerable but are at risk.

Appendix 2
ISLANDS WITH NO BREEDING SEABIRDS

ISLAND NAME	ISLAND GROUP	MAIN FEATURES
Reid Rocks	New Year Island group	A nature reserve, Reid Rocks is the only breeding site for Australian Fur Seals in western Bass Strait, with the annual mean seal pup production over the past ten years being 1500. It is regularly wave-washed and has no vegetation.
Mackerel Islets	Tasman Island group	The main eastern islet has dense blackwood stands in the central area and introduced grasses and *Carpobrotus rossii* over the remainder of the island. It has insufficient soil for burrowing seabirds. The west islet is a flat rocky reef with no vegetation. It is used as a casual Black-faced Cormorant roost site.
Dart Island	Tasman Island group	Extensive stands of blackwoods, *Allocasuarina* sp and some *Eucalyptus* sp. dominate the vegetation. Rabbit browsing damage is severe. There are shallow soils, the vegetation at the west end is adequate for concealing little penguins, but none were present.
Clydes Island	Tasman Island group	Being accessible from mainland Tasmania, this island is a popular destination for visitors. Seabirds are discouraged from using the site because of its cliff perimeter, thick scrub and insufficient soil.
King George Island	Sloping Island group	Having been destroyed by burning and clearing, bracken is dominant with remnant patches of *Allocasuarina* and *Eucalyptus* spp, Metallic and White's Skinks were recorded. Rabbits were present. A Brown Falcon and Superb Fairy-wren were seen. The remnants of 2 buildings are on the east side. Gulls and Oystercatchers may breed at this locality in some years as is common on the nearby causeway. Of particular importance is the dependence of waders using this islet as a high tide refuge.
Woody Island	Sloping Island group	There was obvious rabbit browsing damage over the whole island, very sandy soil and in many parts, sparse to no ground cover. 80+ *Eucalyptus* sp. 2 *Allocasuarina* sp., 2 large radiata pine and several seedlings were recorded. White's Skinks were present.

ISLAND NAME	ISLAND GROUP	MAIN FEATURES
Snake Island	Betsey Island group	With a low rocky shoreline and flat terrain it is unsuitable for seabirds. The vegetation was dominated by *Allocasuarina* and *Eucalyptus* spp. Pied Oystercatchers were recorded at the northern end. No snakes were seen. A Pacific Gull pair was unsuccessfully attempting to breed January 2001. (Tim Reid pers. comm.)
Satellite Island	Partridge Island group	Its rocky shore, hard ground and steep cliffs make it unsuitable for seabirds. Rabbits and tiger snakes were seen.
Garden Island	Partridge Island group	Dry eucalyptus forest covers much of the island. It is very stony with little soil, making it unsuitable for burrowing seabirds.
Hope Island	Partridge Island group	Used for agriculture, it has a rocky shoreline and is mostly cleared with small patches of eucalypts. Rabbits were present.
Faith Island	Partridge Island group	Two gravestones are concealed by scrub in the north-east. It is a low flat islet with non-breeding Pacific Gulls, Pied Oystercatchers, Black-faced Cormorants and Kelp Gulls.
Seagull rock	Partridge Island group	A small islet dominated by *Carpobrotus rossii* and introduced grasses. There is no burrowing habitat. A patch of over 60 orchids was recorded.
Ketchem Island	Mutton Bird Island group	Ketchem Island has no seabirds. On the islet to the east a pair of Sooty Oystercatchers was recorded. The vegetation is confined to the razorback ridge and comprises *Melaleuca* sp, *Carpobrotus rossii* and stipa.
Munday Island	Breaksea Island group	Thick *Melaleuca* scrub dominates the island. Black Cormorants and Sooty Oystercatchers were present. A raven's nest with the remains of a Fairy Prion were found in the scrub. There were no burrows.
Black Island	Hibbs Pyramid group	Composed of 80% bare jagged rock, it is wave-washed apart from several small patches of *Senecio* sp. at the north-west end. Crested Terns and Silver Gulls were present. There is insufficient soil for burrowing birds.

Appendix 3
Definition of the Land Tenure Status of Offshore Islands

Schedule 3 of the Tasmanian *National Parks and Wildlife Act, 1970* gives the following definitions of national parks, nature reserves, game reserves and conservation areas - the major categories of land tenure of the offshore islands.

National Park

A national park is a large natural area of land containing a representative or outstanding sample of major natural regions, features or scenery. Its purpose is to protect and maintain the natural and cultural values of the area of land while providing for ecologically sustainable recreation consistent with conserving those values.

The management objectives are:

a) to conserve natural biological diversity;

b) to conserve geological diversity;

c) to preserve the quality of water and protect catchments;

d) to conserve sites or areas of cultural significance;

e) to encourage education based on the purpose of reservation and the natural or cultural values of the national park, or both;

f) to encourage research, particularly that which furthers the purpose of reservation;

g) to protect the national park against, and rehabilitate the national park following, adverse impacts such as those of fire, introduced species, diseases and soil erosion on the national park's natural and cultural values and on assets within and adjacent to the national park;

h) to encourage and provide for tourism, recreational use and enjoyment consistent with the conservation of the national park's natural and cultural values;

i) to encourage cooperative management programs with Aboriginal people in areas of significance to them in a manner consistent with the purpose of reservation and the other management objectives;

j) to preserve the natural, primitive and remote character of wilderness areas.

Nature Reserve

A nature reserve is an area of land that contains natural values that:

a) contribute to the natural biological diversity or geological diversity of the area of land, or both; and

b) are unique, important or have representative value.

Its purpose is to conserve the natural biological diversity or geological diversity of the area of land, or both, and the conservation of the natural values of that area of land that are unique, important or have representative value.

The management objectives are:

a) to conserve natural biological diversity;

b) to conserve geological diversity;

c) to preserve the quality of water and protect catchments;

d) to conserve sites or areas of cultural significance;

e) to encourage education based on the purposes of reservation and the natural or cultural values of the nature reserve, or both;

f) to encourage research, particularly that which furthers the purposes of reservation;

g) to protect the nature reserve against, and rehabilitate the nature reserve following, adverse impacts such as those of fire, introduced species, diseases and soil erosion on the nature reserve's natural and cultural values and on assets within and adjacent to the nature reserve;

h) to encourage cooperative management programs with Aboriginal people in areas of significance to them in a manner consistent with the purpose of reservation and the other management objectives.

Conservation Area

A conservation area is an area of land predominantly in a natural state. Its purpose is to protect and maintain the natural and cultural values of the area of land and the sustainable use of natural resources of that area of land.

The management objectives are:

a) to conserve natural biological diversity;

b) to conserve geological diversity;

c) to preserve the quality of water and protect catchments;

d) to conserve sites or areas of cultural significance;

e) to provide for the controlled use of natural resources;

f) to provide for exploration activities and utilisation of mineral resources subject to appropriate controls;

g) to provide for the taking, on an ecologically sustainable basis, of designated game species for commercial or private purposes or both;

h) to provide, in special circumstances, for other small-scale commercial or industrial uses;

i) to encourage education based on the purposes of reservation and the natural or cultural values of the conservation area, or both;

j) to encourage research, particularly that which furthers the purposes of reservation;

k) to protect the conservation area against, and rehabilitate the conservation area following, adverse impacts such as those of fire, introduced species, diseases and soil erosion on the conservation area's natural and cultural values and on assets within and adjacent to the conservation area;

l) to encourage appropriate tourism, recreational use and enjoyment consistent with the conservation of the conservation area's natural and cultural values;

m) to encourage cooperative management programs with Aboriginal people in areas of significance to them in a manner consistent with the purposes of reservation and the other management objectives.

Game Reserve

A game reserve is an area of land containing natural values that are unique, important or have representative value particularly with respect to game species.

Its purpose is the conservation of the natural values of the area of land that are unique, important or have representative value, the conservation of the natural biological diversity or geological diversity of that area of land, or both,

and the ecologically sustainable hunting of game species in that area of land.

The management objectives are:

a) to conserve natural biological diversity;
b) to conserve geological diversity;
c) to preserve the quality of water and protect catchments;
d) to conserve sites or areas of cultural significance;
e) to conserve sites or areas of cultural significance;
f) to encourage appropriate tourism, recreational use and enjoyment, particularly sustainable recreational hunting;
g) to encourage education based on the purposes of reservation and the natural or cultural values of the game reserve, or both;
h) to encourage research, particularly that which furthers the purposes of reservation;
i) to protect the game reserve against, and rehabilitate the game reserve following, adverse impacts such as those of fire, introduced species, diseases and soil erosion on the game reserve's natural and cultural values and on assets within and adjacent to the game reserve;
j) to encourage cooperative management programs with Aboriginal people in areas of significance to them in a manner consistent with the purposes of reservation and the other management objectives.

Appendix 4
Table of common and scientific names for fauna species found in the book

BREEDING SEABIRDS

COMMON NAMES	SCIENTIFIC NAMES
Little Penguin	*Eudyptula minor*
Common Diving-Petrel	*Pelecanoides urinatrix*
Fairy Prion	*Pachyptila turtur*
Sooty Shearwater	*Puffinus griseus*
Short-tailed Shearwater	*Puffinus tenuirostris*
Shy Albatross	*Thalassarche cauta*
White-faced Storm-Petrel	*Pelagodroma marina*
Soft-plumaged Petrel	*Pterodroma mollis*
Australasian Gannet	*Sula serrator*
Black-faced Cormorant	*Phalacrocorax fuscescens*
Australian Pelican	*Pelecanus conspicillatus*
Pied Oystercatcher	*Haemotopus longirostris*
Sooty Oystercatcher	*Haemotopus fuliginosus*
Pacific Gull	*Larus pacificus*
Kelp Gull	*Larus dominicanus*
Silver Gull	*Larus novaehollandiae*
Caspian Tern	*Sterna caspia*
Crested Tern	*Sterna bergii*
White-fronted Tern	*Sterna striata*
Little Tern	*Sterna albifrons*
Fairy Tern	*Sterna nereis*

NON-BREEDING SEABIRDS

COMMON NAMES	SCIENTIFIC NAMES
Southern Fulmar (Antarctic Fulmar)	*Fulmarus glacialoides*
Southern Giant-Petrel	*Macronectes giganteus*
Northern Giant-Petrel	*Macronectes halli*
Cape Petrel	*Daption capense*
Grey Petrel	*Procellaria cinerea*
Fluttering Shearwater	*Puffinus gravia*
Black-browed Albatross	*Diomedea melanophyrs*
Buller's Albatross	*Diomedea bulleri*
Little Pied Cormorant	*Phalacrocorax melanoleucos*
Pied Cormorant	*Phalacrocorax varius*
Little Black Cormorant	*Phalacrocorax sulcirostris*
Great Cormorant	*Phalacrocorax carbo novaehollandiae*
Great Skua	*Catharacta skua*
Arctic Tern	*Sterna paradisaea*

TERRESTRIAL BIRDS

COMMON NAMES	SCIENTIFIC NAMES
Brown Quail	*Coturnix ypsilophora*
Red Junglefowl (Chicken)	*Gallus gallus*
Common Pheasant	*Phasianus colchicus*
Indian Peafowl (Peacock)	*Pavo cristatus*
Wild Turkey	*Maleagris gallopavo*
Black Swan	*Cygnus atratus*
Cape Barren Goose	*Cereopsis novaehollandiae*
Australian Shelduck (shelduck)	*Tadorna tadornoides*
Pacific Black Duck	*Anas superciliosa*
Grey Teal	*Anas gracilis*
Chestnut Teal	*Anas castanea*
White-faced Heron	*Egretta novaehollandiae*
Cattle Egret	*Ardeola ibis*
Great Egret	*Ardea alba*

COMMON NAMES	SCIENTIFIC NAMES
Straw-necked Ibis	*Threskiornis spinicollis*
White-bellied Sea-Eagle	*Haliaeetus leucogaster*
Swamp Harrier	*Circus approximans*
Brown Goshawk	*Accipiter fasciatus*
Grey Goshawk	*Accipiter novaehollandiae*
Wedge-tailed Eagle	*Aquila audax*
Little Eagle	*Hieraaetus morphnoides*
Brown Falcon	*Falco berigora*
Peregrine Falcon	*Falco peregrinus*
Nankeen Kestrel	*Falco cenchroides*
Lewin's Rail	*Rallus pectoralis*
Tasmanian Native-Hen	*Gallinula mortierii*
Latham's Snipe	*Gallinago hardwickii*
Whimbrel	*Numenius phaeopus*
Eastern Curlew	*Numenius madagascariensis*
Common Redneck	*Tringa totanus*
Common Sandpiper	*Actitis hypoleucos*
Ruddy Turnstone	*Arenaria interpres*
Great Knot	*Calidris tenuiroistris*
Red Knot	*Calidris canutus*
Sanderling	*Calidris alba*
Red-necked Stint	*Calidrus ruficollis*
Red-capped Plover	*Charadrius ruficapillus*
Black-fronted Dotterel	*Elseyornis melanops*
Hooded Plover	*Charadrius rubricollis*
Banded Lapwing	*Vanellus tricolor*
Masked Lapwing	*Vanellus miles*
Rock Dove (Pigeon)	*Columba livia*
Common Bronzewing	*Phaps chalcoptera*
Brush Bronzewing	*Phaps elegans*
Yellow-tailed Black-Cockatoo	*Calyptorhynchus funereus*

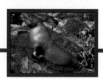

COMMON NAMES	SCIENTIFIC NAMES
Sulphur-crested Cockatoo	*Cacatua galerita*
Green Rosella	*Platycercus caledonicus*
Eastern Rosella	*Platycercus eximus*
Swift Parrot	*Lathamus discolor*
Blue-winged Parrot	*Neophema chrysostoma*
Orange-bellied Parrot	*Neophema chrysogaster*
Pallid Cuckoo	*Cuculus pallidus*
Fan-tailed Cuckoo	*Cuculus pyrrhophanus*
Southern Boobook	*Ninox novaeseelandiae*
Masked Owl	*Tyto novaehollandiae*
Barn Owl	*Tyto alba*
White-throated Needletail	*Hirundapus caudacutus*
Azure Kingfisher	*Alcedo azurea*
Laughing Kookaburra (Kookaburra)	*Dacelo novaeguineae*
Superb Fairy-Wren	*Malurus cyaneus*
Spotted Pardalote	*Pardalotus punctatus*
Forty-spotted Pardalote	*Pardalotus quadragintus*
Striated Pardalote	*Pardalotus striatus*
Tasmanian Scrubwren	*Sericornis humilis*
Scrubtit	*Acanthornis magnus*
Striated Fieldwren (Calamanthus)	*Calamanthus fuliginosus*
Brown Thornbill	*Acanthiza pusilla*
Tasmanian Thornbill	*Acanthiza ewingii*
Yellow Wattlebird	*Anthochaera paradoxa*
Noisy Miner	*Manorina melanocephala*
Yellow-faced Honeyeater	*Lichenostomus chrysops*
Yellow-throated Honeyeater	*Lichenostomus flavicollus*
Strong-billed Honeyeater	*Melithreptus validirostris*
Black-headed Honeyeater	*Melithreptus affinis*
Crescent Honeyeater	*Phylidonyris pyrrhoptera*
New Holland Honeyeater	*Phylidonyris novaehollandiae*

COMMON NAMES	SCIENTIFIC NAMES
Tawny-crowned Honeyeater	*Phylidonyris melanops*
Eastern Spinebill	*Acanthorhynchus tenuirostris*
White-fronted Chat	*Epthianura albifrons*
Scarlet Robin	*Petroica multicolor*
Flame Robin	*Petroica phoenicea*
Pink Robin	*Petroica rodinogaster*
Dusky Robin	*Melanodryas vittata*
Olive Whistler	*Pachycephala olivacea*
Golden Whistler	*Pachycephala pectoralis*
Grey Shrike-thrush	*Colluricincla harmonica*
Satin Flycatcher	*Myiagra cyanoleuca*
Grey Fantail	*Rhipidura fuliginosa*
Willie Wagtail	*Rhipidura leucophrys*
Black-faced Cuckoo-shrike	*Coracina novaehollandiae*
Grey Butcherbird	*Cracticus torquatus*
Black Currawong	*Strepera fuliginosa*
Grey (clinking) Currawong	*Strepera versicolor*
Forest Raven	*Corvus tasmanicus*
Little Raven	*Corvus mellori*
Skylark	*Alauda arvensis*
Richard's Pipit	*Anthus novaeseelandiae*
House Sparrow	*Passer domesticus*
Beautiful Firetail (Firetail Finch)	*Stagonopleura bella*
European Greenfinch	*Carduelis chloris*
European Goldfinch	*Carduelis carduelis*
Welcome Swallow	*Hirundo neoxena*
Tree Martin	*Hirundo nigricans*
Little Grassbird	*Megalurus gramineus*
Silvereye	*Zosterops lateralis*
Common Blackbird	*Turdus merula*
Common (European) Starling	*Sturnus vulgaris*

REPTILES

COMMON NAMES	SCIENTIFIC NAMES
DRAGONS	
Mountain Dragon	*Tympanocryptis diemensis*
SKINKS	
Three-lined Skink	*Bassiana duperreyi*
She-oak Skink	*Cyclodomorphus casuarinae*
White's Skink	*Egernia whitii*
Southern Water Skink	*Eulamprus tympanum*
Bougainvilles Skink	*Lerista bougainvillii*
Mountain Skink	*Niveoscincus orocrypta*
Ocellated Skink	*Niveoscincus ocellata*
Pedra Branca Skink	*Niveoscincus palfreymani*
Tasmanian Tree Skink	*Niveoscincus pretiosa*
Metallic Skink	*Niveoscincus metallica*
Southern Grass Skink	*Pseudemoia entrecasteauxii*
Blue-tongue Lizard	*Tiliqua nigrolutea*
SNAKES	
Copperhead Snake	*Austrelaps superbus*
White-lipped Whip Snake	*Drysdalia coronoides*
Tiger Snake	*Notechis ater (Notechis scutatus)*

AMPHIBIANS

COMMON NAMES	SCIENTIFIC NAMES
Green and Gold Bell Frog	*Litoria raniformis*
Eastern Banjo Frog	*Limnodynastes dumerilii*

MAMMALS

COMMON NAMES	SCIENTIFIC NAMES
Echidna	*Tachyglossus aculeatus*
Swamp Antechinus	*Antechinus minimus minimus*
Eastern Quoll	*Dasyurus viverrinus*
Tasmanian Devil	*Sarcophilus harrisii*
White-footed Dunnart	*Sminthopsus leucopus leucopus*
Southern Brown Bandicoot	*Isoodon obesulus affinis*
Eastern Barred Bandicoot	*Perameles gunnii*
Common Wombat	*Vombatus ursinus tasmaniensis*
Common Ringtail Possum	*Pseudocheirus peregrinus convoluter*
Brushtail Possum	*Trichosurus vulpecula fuliginosus*
Tasmanian Bettong	*Bettongia gaimardi*
Long-nosed Potoroo	*Potorous tridactylus apicalis*
Eastern Grey (Forester) Kangaroo	*Macropus giganteus tasmaniensis*
Red-necked (Bennett's) Wallaby	*Macropus rufogriseus rufogriseus*
Tasmanian Pademelon	*Thylogale billardierii*
Water Rat	*Hydromys chrysogaster*
House Mouse	*Mus musculus*
New Holland Mouse	*Pseudomys novaehollandiae*
Swamp Rat	*Rattus lutreolus*
Black Rat	*Rattus rattus*
New Zealand Fur Seal	*Arctocephalus forsteri*
Australian Fur Seal	*Arctocephalus pusillus doriferus*
Leopard Seal	*Hydrurga leptonyx*
Southern Elephant Seal	*Mirounga leonina macquariensis*
Southern Right Whale	*Eubaleaena australis*
Common Dolphin	*Delphinus delphis*
Rabbit	*Oryctolagus cuniculus cuniculus*
Fallow Deer	*Dama dama dama*
Cat	*Felis catus*
Horse	*Equus asinus*
Pig	*Sus scrofa*
Cattle	*Bos taurus*
Goat	*Capra hircus*
Sheep	*Ovis aries*
Hare	*Lepus capensis*

Appendix 5
PLANT SPECIES RECORDED ON THE ISLANDS

SCIENTIFIC NAMES	COMMON NAMES
AIZOCEAE	
Carpobrotus rossii	pigface
Disphyma crassifolium	rounded noon-flower
Tetragonia implexicoma	iceplant
APIACEAE	
Apium prostratum	sea celery
ASTERACEAE	
Bracteantha bracteata	
Helichrysum leucopsideum	satin everlasting
Olearia axillaris	coast daisy bush
Olearia persoonioides	
Olearia phlogopappa	daisy bush
Ozothamnus turbinatus	coast everlasting
Senecio lautus	coast fireweed
BRASSICACEAE	
Lepidium foliosum	
CAMPANULACEAE	
Wahlenbergia gracilis	bluebell
CASUARINACIAE	
Allocasuarina verticillata	drooping she-oak
CHENOPODIACEAE	
Atriplex cinerea	grey saltbush
Chenopodium glaucum	
Rhagodia candolleana	coast saltbush
Sclerostegia arbuscula	shrubby glasswort
Sarcocornia quinqueflora	beaded glasswort

SCIENTIFIC NAMES	COMMON NAMES
EPACRIDACEAE	
Cyathodes abietina	
Cyathodes juniperina	mountain berry
Epacris myrtifolia	
Epacris stuartii	
Leucopogon parviflorus	coastal beard heath
EUCRYPHIACEAE	
Eucryphia lucida	leatherwood
EUPHORBIACEAE	
Euphorbia paralias	sea spurge
FAGACEAE	
Nothofagus cunninghamii	myrtle, myrtle beech
GERANIACEAE	
Geranium potentilloides	mountain geranium
Pelargonium australe	wild geranium
LAMIACEAE	
Westringia brevifolia	
MALVACEAE	
Lavatera plebeia	Austral hollyhock
MIMOSACEAE	
Acacia mucronata var. longifolia	
Acacia melanoxylon	blackwood
Acacia sophorae	coast wattle
Acacia stricta	
Acacia verticillata	prickly moses
Paraserianthes lophantha	Cape Leuwin wattle
MYOPORACEAE	
Myoporum insulare	boobyalla

SCIENTIFIC NAMES	COMMON NAMES
MYRTACEAE	
Eucalyptus globulus	blue gum
Eucalyptus nitida	
Eucalyptus ovata	black gum
Eucalyptus viminalis	white gum
Kunzea ambigua	
Leptospermum scoparium	manuka, tea tree
Leptospermum laevigatum	coast tea tree
Melaleuca armillaris	
Melaleuca ericifolia	paperbark
Melaleuca squarrosa	scented paperbark
PINACEAE	
Pinus radiata	radiata pine
PITTOSPORACEAE	
Pittosporum bicolor	
PROTEACEAE	
Banksia marginata	
RHAMNACEAE	
Pomaderris apetala	
ROSACEAE	
Aceana novae-zelandiae	buzzy
RUBIACEAE	
Coprosma repens	shining mirror bush
RUTACEAE	
Correa alba	white correa
Correa backhouseana	common correa
SOLANACEAE	
Lycium ferocissimum	African boxthorn
Solanum vescum	kangaroo apple

SCIENTIFIC NAMES	COMMON NAMES
STYLIDIACEAE	
Stylidium graminifolium	trigger plant
THYMELAEACEAE	
Pimelea drupacea	
Pimelea linifolia	
WINTERACEAE	
Tasmannia lanceolata	native pepper
LILIACEAE	
Bulbine bulbosa	
Bulbine semibarbata	leek lily
Dianella tasmanica	
Drymophila cyanocarpa	
POACEAE	
Poa labillardierei	
Poa poiformis	blue tussock grass
Austrostipa stipoides	stipa/coastal spear grass
CUPRESSACEAE	
Callitris rhomboidea	Oyster Bay pine
ASPLENIACEAE	
Asplenium obtusatum	shore spleen wort
DICKSONIACEAE	
Dicksonia antarctica	treefern, manfern
DENNSTAEDTIACEAE	
Pteridium esculentum	
Histiopteris incisa	bat's wing fern

References and Bibliography

AUSTRALIAN BROADCASTING COMMISSION (Ed.) 1969; **Bass Strait Australia's Last Frontier;** A.B.C., Sydney.

ALEXANDER, K., ROBERTSON, G. & GALES, R. 1997; **The Incidental Mortality of Albatrosses in Longline Fisheries;** Australian Antarctic Division, Tasmania.

BAXTER, A.S. 1992; **The Management of Whale and Dolphin Watching Kaikoura New Zealand;** Department of Conservation, New Zealand.

BIODIVERSITY GROUP - ENVIRONMENT AUSTRALIA 1998; **Threat Abatement Plan for the Incidental Catch of Seabirds During Oceanic Longline Fishing Tasmanian Islands;** Earth Science Section, Parks and Wildlife Service, Tasmania.

BROTHERS, N., GALES, R. & PEMBERTON, D. 1993; **Prey Harvest of the Australasian Gannet (Sula serrator) in Tasmania;** in Wildlife Reserarch Vol. 20, CSIRO, Melbourne.

BROTHERS, N. & HARRIS, S. 1999; **The Effects of Fire on Burrow-nesting Seabirds Particularly Short-tailed Shearwaters (Puffinus tenuirostris) and their Habitat in Tasmania;** Papers and Proceedings of the Royal Society of Tasmania. Vol. 133(1), Hobart.

BRYANT, S. & JACKSON, J. 1999; **Tasmania's Threatened Fauna Handbook, What, Where and How to Protect Tasmania's Threatened Animals;** Threatened Species Unit, Parks and Wildlife Service, Tasmania.

BUCHANAN, A.M. (ed.) 1999; **A Census of the Vascular Plants of Tasmania & Index to the Students' Flora of Tasmania;** Tasmanian Museum and Art Gallery, Hobart.

CAMERON, M. (ed.) 1996; **A Guide to Flowers and Plants of Tasmania:** Launceston Field Naturalists Club, Tasmania.

CHRISTIDIS, LESLIE & BOLES, WALTER E. 1994; **The Taxonomy and Species of Birds of Australia and the Territories;** Royal Australian Ornithologists Union, Melbourne.

Corella Journal of the Australian Bird Study Association: seabird island series, New South Wales.

CORNELL, C. (Translator) 1974; **The Journal of Post Captain Nicolas Baudin 1800 – 1803;** Library Board of South Australia, Adelaide.

CUMPSTON, J.S. 1978; **The Furneaux Group Bass Strait, First Visitors 1797 – 1810;** Roebuck Society, Canberra.

DIXON, G. 1996; **A Reconnaissance Inventory of Sites of Geoconservation Significance on Tasmanian Islands;** unpublished report, Parks and Wildlife Service, Tasmania.

EDGECOMBE, J. 1986; **Flinders Island and Eastern Bass Strait;** J. M. Edgecombe, Sydney.

FELTON, H. (ed.) 1984 **Return to the Islands, Aborigines in Tasmania;** Education Department, Tasmania.

FLANNERY, T. (ed.) 2000; **Matthew Flinders Great Adventures in the Circumnavigation of Australia Terra Australis**, Text Publishing, Melbourne.

FOWLER, R.M. 1980; **The Furneaux Group Bass Strait, A History;** Roebuck Society, Canberra.

GALES, R. 1993; **Co-operative Mechanisms for the Conservation of Albatross;** ANCA and Australian Antarctic Foundation, Hobart.

GALES, R., PEMBERTON, D., LU, C.C & CLARKE, M.R. 1993; **Cephalopod diet of the Australian Fur Seal: Variation Due to Location, Season and Sample Type** in Australian Journal of Marine and Freshwater Research, Victoria.

GREEN, R.H. & RAINBIRD, J.L 1993; **Reptiles from the Islands of Tasmania**; technical report, Queen Victoria Museum and Art Gallery, Launceston.

HARRIS, S. & MCKENNY, H. 1999; **Preservation Island, Furneaux Group: Two Hundred Years of Vegetation Change** in Papers and Proceedings of the Royal Society of Tasmania, 133 (1), Hobart.

HIGGINS, P.J. & DAVIES, S. J. 1996; **Handbook of Australian, New Zealand and Antarctic Birds Vol 3;** Oxford University Press, Melbourne.

HUME, F. 2000; **Princess Melikoff Trust Marine Mammal Program Report 1999-2000**; Nature Conservation Branch, Department of Primary Industries Water and Environment, Hobart.

HUME, F. & GALES R. 1999; **Princess Melikoff Trust Marine Mammal Program Report 1998-1999**; Marine Conservation Branch, Tasmanian Parks and Wildlife Service, Hobart.

JOHNSTONE, G.W., MILLEDGE, D. & DORWARD, D.F. 1975; **The White-capped Albatross of Albatross Island: Numbers and Breeding Behaviour** from The Emu Vol. 75 Part 1, Melbourne.

KIRKPATRICK, J.B., MASSEY J.S. & PARSONS R.F. 1971; **Natural History of Curtis Island, Bass Strait 2: Soils and Vegetation with notes on Rodondo Island** in Papers and Proceedings of the Royal Society of Tasmania, Volume 107, Hobart.

MARCHANT, S. & HIGGINS, P.J. 1990; **Handbook of Australian, New Zealand and Antarctic Birds Vol 1 & 2;** Oxford University Press, Melbourne.

Maria Island National Park and Ile Des Phoques Nature Reserve Management Plan 1998; Parks and Wildlife Service, Department of Environment and Land Managment, Hobart.

MARMION P., 1997; **Seal Ecotourism Potential in Tasmania,** Parks and Wildlife Service, Department of Environment and Land Managment, Hobart.

NELSON, B. 1980; **Seabirds: Their Biology and Ecology;** Hamlyn Publishing Group Ltd., London.

PEMBERTON D., GALES, R. & HUME F. 1998; **Princess Melikoff Trust Marine Mammal Program Report 1997-1998;** Tasmanian Parks and Wildlife Service, Hobart.

PEMBERTON, D. & KIRKWOOD, R.J. 1994; **Pup Production and Distribution of the Australian Fur Seal** *Arctocephalus pusillus doriferus* **in Tasmania**, in *Wildlife Research.*, Vol. 21, Hobart.

PLOMLEY, N.J.B. 1983; **The Baudin Expedition and the Tasmanian Aborigines;** Blubberhead Press, Australia.

RYAN, L. 1981; **The Aboriginal Tasmanians;** University of Queensland Press, Brisbane.

SERVENTY, D.L., SERVENTY, V. & WARHAM, J. 1971; **The Handbook of Australian Sea-birds;** A.H. and A.W. Reed Ltd., Sydney.

SIM, R. AND STUART I., 1991; **The Outer Furneaux Island Archaeological Survey**, Tasmanian Environment Centre Inc. Hobart.

SHAUGHNESSY, P.D. 1999; **The Action Plan for Australian Seals;** Environment Australia, Canberra.

SMITH, S.J. 1990; **Checklist of the Vertebrate Animals of Tasmania;** St. David's Park Publishing, Hobart.

STANBURY, P & PHIPPS, G. 1980; **Australia's Animals Discovered;** Peragamom Press, Australia.

STRATON, M., CLAYTON, M. SCHODDE, R., WOMBEY, J. AND MASON, .I. 1998; **CSIRO Lists of Australian Vertebrates: A reference with conservation status;** CSIRO Publishing, Australia.

WATTS, D. 1999; **Field Guide to Tasmanian Birds;** New Holland Publishers Pty. Ltd., Australia.

WBM OCEANICS AUSTRALIA AND CLARIDGE, G. 1997; **Guidelines for Managing Visitation to Seabird Breeding Islands;** Great Barrier Reef Marine Park Authority, Townsville.

WHITE, G. 1980; **Islands of South-west Tasmania;** G. White, Sydney.

General Subject Index
Photograph page references in *italic*
See also Index to Islands and Index to Seabirds and Terrestrial Birds

A

A Reconnaissance Inventory of Sites of Geoconservation Significance on Tasmanian Islands x

Abel Tasman viii

Aborigines viii, 59, 531, 533

aerial photography 73

African Boxthorn see boxthorn

agricultural activites 435, 461, 478

airforce bombing practice 194

airports 135

airstrips 77, 96, 151, *153*, 199, 218, 237, 241, 361, 380, 397

Ansons Bay 404

archaeological material 147

artificial nest burrows 146

Atlantic Ocean 6, 12

Australian Bird Study Association x

Australian Bush Heritage Fund 148

Australian Fur Seal *141, 142, 143, 166*
see also island groups: 46, 56, 140, 171, 180, 188, 196, 311, 374, 410, 439, 467, 486, 502, 512, 563

Australian Maritime Safety Authority 319

Australian Sea Lion 82

B

banding 291

barley grass 348

Bering Sea 5

blackberries 492, 500

blowholes 425, 534

Blue-tongue Lizard 52, 54, 160, 204, 220, 238, 277, 300, 314, 436, 483

botanical specimens x
see also vegetation

boating activities 142

Bougainville's Skink 146, 152, 160, 187, 242, 258, 308, 364, 381, 398

Boullanger/Robbins Island wetland site 128, 130

boxthorn 92, 219, 250, 251, 363, 381, 393, 430, 457, 459, 462, 464, 470

Brushtail Possum 148, 152

buildings 312
see also houses; huts; monument; museum; sheds

burning practice 220, 331, 399, 416, 451

burnt land 82, 257, 284, 285, 298, 299, 406, 483, 489, 496

C

calcarenite sand 158
see also limestone outcrops

campers 416

campsites 147

Cape butterflies 52

Cape Lewin wattle 470

Cape Portland 384

Captain Cook 510

cat eradication 277, 303, 449

caterpillars 175

cattle 79, 128, 148, 152, 160, 201, 219, 238, 256, 308, 348, 352
see also grazing

cattle watering troughs 158

cattleyards 147

cats 79, 152, 238, 242, 277, 285, 303, 314, 326, 348, 381, 398, 418, 436, 449, 465, 488
behaviour & predation studies 451

caves 57, 80, 90, 93, 95, 147, 185, 425, 427, 479, 521, 524, 537

colonies of seabirds, guidelines for visiting 42-3
see also visits

conifers 381

Copperhead Snake 228, 277, 291, 326, 436

coralline algae 597

crayfish 160, 555

D

dams 147, 151, 531

decoy systems 319

deer 382, 436

density of birds, measurement of ix

Department of Primary Industry, Water & Environment 201, 291, 376, 418
see also management plans

description & distribution of seabirds 1-39

Devonian granite 218

dogs 418

dolerite 268, 358, 361, 367, 447, *448*

dolphins 520

E

Eastern Barred Bandicoot 324, 436

Eastern Grey (Forester) Kangaroo 79, 436

echidnas 308, 436

ecotourism 41, 417, 427, 451, 488

endangered species 31

Endangered Species Protection Act 1992 (Cwlth) 82, 509

endemic seabirds 14, 20, 32

environmental pollutants 3

Eucalyptus nitida 78

Eucalyptus ovata 348
see also individual islands—vegetation recorded

F

farm buildings 307

farmed land 298

farming practices 269, 285, 299, 326, 382

fences 77, 147, 255, 257, 261, 262, 380

ferries 185

fire management 221

fires 64, 78, 83, 148, 204, 258, 264, 267, 269, 276, 303, 307, 309, 319, 379, 454, 483, 522, 555
see also burnt land

fireweed 377

fish processing 281, 306

fisheries & Short-tailed Shearwaters 5

fishers 164

fishing 142
overfishing 86
set nets 471

flowstone (phosphatic) 427

forests 276

Fortescue Bay 442

fossil cliffs 435

fossils 479

Franklin Sound 309

Freycinet Peninsula 411, 414

Frederick Henry Bay 33

freshwater soaks 144, 255, 227, 257, 261, 377

frogs 272, 308

G

garden plants 381

geoconservation 173, 248, 238

Geographic Information Systems (GIS) ix, xi

geographical features 147, 544

geological unique features 92, 95, 427, 539

geomorphological features 148, 159, 480, 495, 498, 533

geomorphology 59, 382

George Rocks Nature Reserve 401

goats 219, 228, 398, 488

granite 238, 358, 505

grasshoppers 426

graves & gravestones 147, *148*, 218, 249, 251, *307*, 481

grazing 465, 483 by cattle 128, 159, 160, 201, 220-1, 237, 255, 307, 309, 347, 352, 398
by fallow deer 382 by horses 352
by pademelons 237, 347
by rabbits 406, 437
by Red-Necked Wallabies 237
by sheep 227, 228, 237, 255, 285, 324, 326, 331, 347, 352, 358, 382, 398, 416, 451

by stock 79, 258, 264, 268, 272, 298, 303, 343, 363, 379, 449
practices 465

Green & Gold Bell Frog 272

guano 26, 33, 88, 281 collectors 553

H

habitat destruction of seabirds 2, 4, 6, 8, 10, 12, 28

hares 324

harvesting, of seabirds viii illegal 75, 319, 497
Short-tailed Shearwaters 57, 64, 98, 120, 123, 125, 225, 240, 241, 274, 276, 285, 303, 315, 321, 324, 326, 331

helipad 144, *146*

historical sites 221, 251, 347, 364, 533, *604*

holiday makers 393
see also tourists

Holocene dune system 52, 380, 397

horehound 242

horses 352

House Mouse 79, 201, 238, 242, 285, 314, 326, 348, 359, 381, 398

houses 147, 151, 218, 227, 237, 261, 298, 323, 324, 361, 380, 397, 460, 462, 472, 477, 531, 601

huts 57, 77, 96, 199, 218, 241, 255, 271, 284, 290, 330, 358, 361, *363*, 377

I

Indian Ocean 12

introduced species 83, 152, 219, 238, 324, 381, 382, 399, 435, 464

Iron Baron oil spill 3, 379

island industry 41

islands, by region **44**

J

jetties 151, 237, 380, 435, 477, 601, *601*

Jurassic dolerite 450

K

karst system 349, 355

Kelly Channel 603

kelp 98, 597

Killiecrankie 332

L

Lady Barron 281, 284, 292

lagoons 77, 144, 307, 434

landslip 416

light beacon 432, 601, *602*, *604*

lighthouses 77, 79, 151, 249, *250*, 397, 449, *448*, 510, 531 keepers 249, 251, 449, 531, 533
see also navigation lights

limestone 238, 268, 361, 367, 520, 518, 524

lizards 281, 446
see also Blue-tongue Lizard

logging 526

Long-nosed Potoroo 148, 436, 523, 526

M

Macquarie Harbour 601, 603

mallow 135, 366

mammals xii, 617
see also individual islands—mammals recorded

Mannalargenna Cave 347

management plans 142, 418, 471

maps & mapping ix

Maria Island Marine Reserve 433

Maria Island National Park 432

market garden (Chinese) 51, 52

Metallic Skink see all island groups except North Coast, Pedra Branca and Petrel Island

mice 52, 204, 242, 277, 436
see also House Mouse; New Holland Mouse

middens *75*

migration, migratory seabird species 4, 5, 6, 8, 12, 36
partly migratory seabird species 16, 26, 28, 30
see also movement

military target practice 179
artillery shells & missiles 192, *193*

mining activities 414, 435

monument 312

Mountain Dragon 238, 424

movement, circumpolar seabirds 10
dispersive seabirds 2, 8, 14, 16, 18, 24, 26, 28, 32, 34, 36
see also migration; resident seabirds; sedentary seabirds

Mt William National Park 404

mudflats 129, 255, 261, 266, 284, 286, 298, 299, 300, 307

Muscovite granite 505

museum 151

muttonbirders 92
see also under harvesting

Myoporum insulare 348

Myxoma virus 471

N

Narawntapu National Park 136, 137

National Estate Grants Program Report on Seabirds x

National Oil Spill Response Atlas x

Nature Conservation Branch, Department of Primary Industry, Water & Environment 291, 379

naval bombing practice 194

navigation lights 144, *146*, 158, 319, 472, *473*

navigation obelisk 499, *499*

netting practice 428

New Holland Mouse 49

New Zealand Fur Seal 82, 490, 508, 529, 533, 536

New Zealand 6, 7, 28, 29

Ninepin Point Marine Reserve 479

Nomenclature Board of Tasmania 233, 296, 301, 304, 549

O

oceanic cormorant 32

Ocellated Skink 228, 238, 242, 251, 258, 277, 381, 421, 424, 428, 436, 449

overfishing 86

Oyster Bay Pine 307, 309, 423

P

Pacific Ocean 4, 6, 12

permits 75, 79, 83, 92, 95, 142, 183, 376, 393

Pedra Branca Skink 508

Phytophthora cinnamomi 43, 492

pigface *81*

pigs 52, 256

plastic piton 511

Pleistocene dunes 261

poaching 79, 331, 355, 399, 404, 442, 451, 461, 465

poisoning programs 17, 135

pollutants 3

population monitoring study 509

populations of seabirds 40

Port Davey 564

Precambrian conglomerate boulders 80

predators of seabirds 6, 9

Q

quail hunting & shooters 221, 262

Queen Victoria Museum and Art Gallery x

quolls see Spotted-tailed Quoll

R

rabbit, silver-haired 471
  see also island groups 56, 116, 140, 196, 374, 410, 431, 439, 454, 468, 476, 486, 574

rabbit control 67, 201

radiata pine 457

rare species 29, 31

rat eradication & control 287, 364, 416, 438

rats 138, 152, 160, 201, 204, 242, 272, 277, 287, 291, 364, 403, 497
  see also Swamp Rat; Water Rat

Red-necked (Bennett's) Wallaby 52, 238, 303, 308, 314, 436

refuse disposal sites see tip sites

reptiles 67, 125, 276, 283, 616
  see also skinks

resident seabirds 14, 16, 18, 20, 24, 26

Ringtail Possum 461

roads 77, 151

rock samples x

Ross Sea 6

Royal Australasian Ornithologists Union xii

S

sea spurge 153, 221

seal harvesting viii, 218 iron pot *284*

seal ladder 141

seal watching 43, 376, 490

seals *see* Australian Fur Seal; Australian Sea Lion New Zealand Fur Seal; Southern Elephant seal

sedentary seabirds 22, 26, 32, 38

serrated tussock 461

shacks 275, 404
   see also houses; sheds

sheds 151, 199, 227, 237, 255, 261, 284, 324, 380, 531

sheep viii, 79, 98, 152, 228, 238, 256, 324, 326, 331, 343, 348, 352, 382, 397, 451
   see also grazing

She-oak Skink 436, 449, 471

shipwrecks 218, 327, *352*, 377, *378*

skinks 303
   see also Bougainville's Skink; Metallic Skink; Mountain Dragon; Ocellated Skink; She-oak Skink; Southern Grass Skink; Southern Water Skink; Tasmanian Tree Skink; Three-lined Skink; White's Skink

snails 152, 160, 539

snakes 381
   see also Copperhead Snake; Tiger Snake; White-lipped Whip Snake

soils, acid 78 peat 159, 547

South Bruny Island National Park 487, 489

South West World Heritage Area xii

Southern Brown Bandicoot 148, 152, 436

Southern Elephant Seal 82, *532*, 533

Southern Grass Skink 291, 308, 379

Southern Water Skink 183

Spotted-tailed Quoll 52

Stack Island Nature Reserve 69, 71

status of islands xii

stockyards 255, 284, 307

Storm Bay 33

surveying techniques ix-x

surveys 178, 509, 533, 549

Swamp Antechinus 520, 529, 533

Swamp Rat 152, 204, 436, 526

Sydney Cove 219

synchrony 4

T

Tamar River 375

Tasman Sea 309

Tasmanian Bettong 436

Tasmanian Devil 238

Tasmanian Museum & Art Gallery 379

Tasmanian Pademelon 52, 128, 228, 238, 308, 314, 324, 326, 348, 436, 523, 526

Tasmanian Tree Skink, *see* island groups on pp. 56, 431, 502, 512, 545, 563, 574, 584, 593

thistles 106, 221, 348, 363, 462

Threatened Species Protection Act 1995 (State) xi, 25, 29, 31, 82, 368, 509

Three-lined Skink 148, 160, 228, 242, 246, 277, 314, 324, 348, 381, 436, 458, 520, 533

thistles 359

tidal delta 309

Tiger Snake, *see* island groups on pp. 46, 56, 89, 116, 196, 215, 236, 270, 293, 311, 322, 346, 357, 374, 431

tip sites 14, 16, 19

tourism 375, 376, 404, 418, 442, 577
see also ecotourism

tramways 414

transport on islands 241

U

uncooperative species x

V

vandalism 331, 399

vegetation xii, 618-21
diversity 64
endangered heath 492
native 221, 449
rehabilitation program 134
see also individual islands—vegetation

visits to islands 42-3, 396, 428, 454, 456, 467, 471, 472, 475, 509, 511
disturbance by visitors 404
guidelines for minimal impact 42
management & research 92

volcanic features 86

vulnerable species 13, 23, 25, 27
to disturbance 35

W

walls 249, 298, *299*, 380, 425

Water Rat 66, 277, 291, 319, 433, 436, 465, 467, 592

weed control 283

weed eradication 134

wetlands 115, 128, 130, 218, 221

whaling, iron pot *284*

wharf 77
see also jetties

white settlement viii

White-footed Dunnart 156

Wild Wind, trading ketch iv, 549

White's Skink, *see* island groups on pp. 46, 56, 140, 157, 180, 196, 236, 270, 311, 322, 330, 335, 357, 357, 374, 410, 431, 439, 468

White-lipped Whip Snake 147, 152, 220, 238, 283, 297, 324, 326, 341, 352, 368, 372, 436

windmills 237

Wilsons Promontory 181

Wombat 52, 436

World Heritage Area 505

Index to Islands
Map page references in **bold**; photograph page references in *italic*

A

Actaeon Island **486**, 495-7, **495**, *496*
Actaeon Island Group **409**, 486-500, **486**
Albatross Island 11, 38, **56**, 80-3, **80**, *81*, *83*
Anderson Island **254**, 255-6, **255**
Apple Orchard Point Islet **293**, 296-7, **296**
Arch Rock **476**, 479-80, **479**, *479*

B

Babel Island **311**, 312-5, **312**, *313*, *315*
Babel Island Group **195**, 311-21, **311**
Badger Island **236**, 237-8, **237**
Badger Island Group **195**, 236-52, **236**
Barren Island **452**, 457-8, **457**, *458*
Barrenjoey *see* Tenth Island
Bass Pyramid **188**, 192-4, *192*
Bass Pyramid Group **139**, 188-94, **188**
Battery Island **196**, 213-4, **213**, *214*
Baynes Island **374**, 387-8, **387**
Beagle Island **236**, 252-3, **252**
Bears Island **56**, 76, **76**
Betsey Island & Little Betsey Island 3, **468**, 469-71, **469**, *471*
Betsey Island Group **409**, 468-75, **468**, 607
Big Black Reef **226**, 229-30, **229**
Big Caroline Rock **563**, 564-5, **564**, *565*
Big Dog Island *see* Great Dog Island
Big Green Island **357**, 361-3, **361**, *362*, *363*
Big Green Island Group **195**, 357-72, **357**

Big Sandy Petrel Island **116**, 124-5, **124**, 126
Big Stony Petrel Island **116**, 121-3, **121**, *122*
Billy Goat Reefs **270**, 282-3, **282**
Bird Island (Hunter Island Group) 21, **56**, 62-4, **62**
Bird Island (Prime Seal Island Group) **346**, 355-6, **355**, *356*
Bird Rock **374**, 400-01, **400**, 404
Black Island 607
Black Pyramid Rock 37, 84-6, *84*, *85*
Black Rocks *see* Western Rocks
Blanche Rock **486**, 493-4, **493**, *494*
Bonnet Island **593**, 603-4, **603**, *604*
Boxen Island **226**, 234-5, **234**
Breaksea Island Group **501**, 574-92, **574**, 607
Breaksea Islands **574**, 575-7, **575**, *576*
Briggs Islet **270**, 280-1, **280**
Bruny Island xi, 482

C

Cape Barren Island xi, 35
Cat Island 27, 37, **311**, 316-9, **316**, *317*
Chalky Island 31, **357**, 367-8, **367**
Charity Island **476**, 481, **481**
Chicken Island **512**, 513-5, **513**, *514*, 517
Christmas Island **46**, 47-9, **47**, *48*
Clarke Island xi, 352
Clydes Island 606
Cone Island **171**, 174-5, **174**
Cooties Reef **293**, 310, **310**

Councillor Island 3, **46**, 53-5, **53**
Courts Island **486**, 487-8, **487**, 488
Craggy Island **188**, 189-90, *189*, **190**
Curlew Island **476**, 484-5, **484**
Curtis Island **171**, 172-3, **172**
Curtis Island Group **139**, **171**
Cygnet Island **374**, 389-90, **389**

D

Dart Island 606
De Witt Island **512**, 524-6, **524**, *525*
Deal Island **140**, 151-3, **151**, *151*, *153*
Devils Tower **171**, 178-9, **178**, *178*, *179*
Diamond Island **410**, 417-8, **417**
Doughboy Island **254**, 268-9, **268**
Doughboy Island East **89**, 90-2, **90**, *91*
Doughboy Island West **89**, 93-5, **93**, *94*
Doughboy Islands 11
Dover Island **140**, 149-50, **149**
Dugay Islet **56**, 67, 68-9, **68**

E

East Coast Islands – Region 6 **44**, 409-500
East Island **157**, 169-70, **169**
East Kangaroo Island 27, **357**, 358-60, **358**, *360*
East Moncoeur Island **180**,186-7, **186**
East Pyramids xi, **545**, 557-8, **557**, *558*
Eddystone Rock **502**, 510-12, **510**
Edwards Islet **56**, 67, 69, 70-1, **70**
Egg Island, Horseshoe Reef **132**, 135, **135**
Entrance Island **593**, 601-2, **601**, *602*
Erith Island **140**, 147-8, **147**

F

Faith Island **476**, 481, 607
Fisher Island 17, 290-1, **270**, **290**
Fisher Island Reef **270**, 292, **292**
Fitzroy Islands **574**, 580-1, **580**
Flat Island *see* Mutton Bird Island
Flat Top Island **512**, 527-9, **527**, *528*
Flat Witch Island **512**, 527-9, **527**
Flinders Island xi, 285, 331, 355
Forsyth Island **196**, 202-4, **202**
Foster Islands **374**, 391-3, **391**
Fulham Island **452**, 464-5, **464**
Furneaux Islands 13, 29

G

Garden Island 607
George Rocks **374**, 402-4, **402**, *403*, *404*
Goose Island **236**, 249-51, **249**, *250*
Gossys Reef **329**, 334, **334**
Governor Island **410**, 419-20, **419**
Great Dog Island **270**, 275-6, **275**, *276*
Great Dog Island Group **195**, 270-92, **270**
Green Island (D'Entrecasteaux Channel) 19
Green Island *see* Hobbs Island
Gull Island **196**, 197-8, **197**, *198*
Gun Carriage Island *see* Vansittart Island 307

H

Half Tide Rock **116**, 126, **126**
Harbour Islets **89**, 105-7, **105**, *106*
Hay Island **563**, 568-9, **568**
Hays Reef **593**, 598, **598**

Hen Island **512**, 516-7, **516**, *517*
Henderson Islets **89**, 103-4, **103**
Hibbs Pyramid **593**, 599-600, **599**, *599*, *600*
Hibbs Pyramid Group, **501**, 593-604, **593**, 607
Hippolyte Rocks **439**, 440-2, **440**, *441*, *442*
Hobbs Island **585**, 591-2, **591**
Hog Island **452**, 459, **459**
Hogan Group **139**, **157**, 168
Hogan Island **157**, 158-9, **158**, 168
Hope Island **476**, 481, 607
Horseshoe Reef *see* Egg Island
Howie Island **116**, 129-30, **129**
Hunter Island xi, 67
Hunter Island Group 21, **45**, **56**
Huon Island **476**, 477-8, **477**

I

Ile des Phoques **410**, 425-7, **425**, *425*
Ile du Golfe 9, **512**, 518-20, **518**, *519*, *520*
Ile du Nord **431**, 432-3, **433**, *433*
Inner Little Goose Island **236**, 245-6, **245**
Inner Rocks, New Harbour **545**, 546-7, **546**, *547*
Inner Sister Island **322**, 323-4, **323**
Iron Pot **468**, 472-3, **472**, *473*
Isabella Island **357**, 370, 371-2, **371**, *372*
Judgement Rocks **140**, 141-3, **141**, *143*

K

Kangaroo Island **116**, 127-8, **127**
Kathleen Island **574**, 578-9, **578**, *579*
Kent Group xi, **139**, **140**
Ketchem Island 607

Key Island **226**, 231-2, **231**, *232*
Key Reef **226**, 233, **233**, *234*
King George Island 606
King Island xi, 2, **46**

L

Lachlan Island **431**, 437-8, **437**, *438*
Leelinger Island **593**, 596-7, **596**
Little Anderson Island **254**, 257-8, **257**
Little Badger Island **236**, 239-40, **239**, *242*, 354
Little Betsey Island *see* Betsey Island
Little Chalky Island **357**, 365-6, **365**
Little Christmas Island **410**, 429-30, **429**, *429*
Little Dog Island **270**, 271-2, **271**
Little Dog Island (north rock) **270**, 273-4, **273**
Little Goose Island **236**, 247-8, **247**, 354
Little Green Island **270**, 272, 284-5, **284**, *285*
Little Island **329**, 332-3, **332**
Little Spectacle Island 452, 455-6, 455
Little Stony Petrel Island **116**, 119-20, **119**, *119*, *125*
Little Swan Island 35, **374**, 393, 394-6, **394**, *395*, *396*
Little Trefoil Island **89**, 99-100, **99**, *100*
Little Witch *see* Flat Witch Island
Little Waterhouse Island **374**, 383-4, **383**
Long Island (Region 3) **157**, 165-6, **165**, **166**
Long Island (Region 4) **226**, 227-8, **227**
Long Island Group **195**, 226-35, **226**
Louisa Island **512**, 521-3, **521**, *522*, *523*
Lourah Island **563**, 572-3, **572**
Low Islet Group 352

Low Islets (east island) **196**, 205-6, **205**
Low Islets **346**, 350-2, **350**, *351*
Low Islets (west Island) **207**

M

Maatsuyker Island **512**, 531-3, **531**
Maatsuyker Island Group viii, 501, 512-44, **512**
Mackerel Islets 606
Maclean Island **374**, 385-6, **385**
Maria Island **431**, 434-6, **434**
Maria Island Group **409**, 431-8, **431**
Marriot Reef **335**, 344-5, **344**, *345*
Mavourneen Rocks **574**, 582-3, **582**
Mewstone 38, **502**, 503-5, **503**, *504*
Mid Woody Islet **254**, 259-60, **259**
Middle Pasco Island (north & south) **335**, 338-9, **338**
Mile Island **357**, 369-70, **369**, *370*
Montgomery Rocks **593**, 594-5, **594**, *595*
Moriarty Rocks **196**, 209-10, **209**
Mount Chappell Island **236**, 241-2, **241**, *242*
Munday Island 607
Murkay Islet (east) **89**, 108-9, **108**
Murkay Islets (middle) **89**, 110-11, **110**
Murkay Islets (west) **89**, 112-3, **112**
Mutton Bird Island xi, **545**, 553-5, **553**, *554*
Mutton Bird Island Group xi, **501**, 545-62, **545**, 607

N

Nares Rocks **56**, 60-1, **60**, *61*
Neds Reef **254**, 266-7, **266**

Needle Rocks **512**, 542-4, **542**, *543*, *544*
New Year Island **46, **47, **50**, *51*
New Year Island Group **45, 46**, 606
Night Island **215**, 216-7, **216**, *217*
Ninth Island 3, 27, **374**, 377-9, **377**, *378*, *379*
North Bass Strait Islands – Region 3 **44**, **139**
North Coast Group 132-8, **132**
North Coast Islands – Region 2 **44**, 131-8, **131**
North East Isle **140**, 154-6, **154**, *155*, *156*
North East Islands – Region 5 **44**, 373-408, **373**
North Pasco Island **335**, 340-1, **340**
North West Mount Chappell Islet **236**, 243-4, **243**, 354
North West Islands – Region 1 **44, 45**, 45-130

O

Outer Sister Island **322**, 325-6, **325**, *326*
Oyster Rocks **254**, 263-4, **263**, *264*
Oyster Rocks West **254**, 265, **265**

P

Paddys Island **374**, 407-8, **407**
Partridge Island **476**, 482-3, **482**
Partridge Island Group **409**, 476-85, **476**, 607
Pasco Island Group **195**, 335-45, **335**
Passage Island **196**, 199-201, **199**, *200*
Passage Island Group **195**, 196-214, **196**
Pedra Branca 11, 37, 38, **502**, 506-9, **506**, 506, *509*
Pedra Branca Group **501**, 502-09, **502**
Pelican Island **293**, 302-3, **302**, 304
Pelican Island Reef **293**, 304, **304**

Penguin Island **132**, 136, **136**
Penguin Islet 35, **56**, 72-5, **72**, *73*, *74*
Petrel Island Group **45**, 116-30, **116**, 125
Picnic Island **410**, 421-2, **421**
Preservation Island **215**, 218-21, **218**, *219*
Preservation Island Group **195**, 215-25, **215**
Preservation Islets **215**, 222-3, **222**, *223*
Prime Seal Island **346**, 346-9, **347**, *349*
Prime Seal Island Group **195**, 346-56, **346**
Puncheon Island **293**, 298-9, **298**, *299*
Puncheon Islets **293**, 300-1, **300**

R
Ram Island **293**, 294-5, **294**, *295*
Refuge Island **410**, 428, **428**
Reid Rocks 606
Robbins Island xi
Rodondo Island **180**, 181-3, **181**, *182*
Rodondo Island Group **139**, **180**
Round Island **157**, 167-8, **167**
Round Top Island **512**, 540-1, **540**, *541*
Roydon Island **336**, 336-7, **336**, *337*
Rum Island **215**, 224-5, **224**, *225*

S
Satellite Island 607
Schouten Island **410**, 414-6, **414**
Schouten Island Group **409**, 410-30, **410**
Seacrow Islet **89**, 101-2, **101**
Seagull Rock 607
Sentinel Island **329**, 330-1, **330**
Sentinel Island Group **195**, 329-34, **329**

Shag Reef **322**, 327-8, **327**
Shanks Islands **563**, 570-1, **570**, *571*
Shell Islets **89**, 114-5, **114**
Sidmouth Rock **502**, 510-12, *511*
Sisters Island **132**, 133-4, **133**
Sisters Island Group **195**, 322-8, **322**
Sloping Island & Sloping Reef **452**, 460-1, **460**
Sloping Island Group **409**, 452-67, **452**, 606
Smooth Island **452**, 462-3, **462**
Snake Island 607
South & West Islands – Region 7 **44**, 501-604, **501**
South Black Rock **56**, 87-8, *87*
South East Great Dog Islet **270**, 278-9, **278**
South East Mutton Bird Islet **545**, 550-1, **550**
South Pasco Island **335**, 342-3, **342**
South West Isle **140**, 144-6, **144**, 145
South West Mutton Bird Islet **545**, 550-1, **550**
South West Petrel Island **116**, 117-8, **117**, *118*
Southport Island **486**, 491-2, **491**
Southport Island Reef **486**, 491-2, **491**
Spectacle Island **452**, 453-4, **453**, *454*
Spences Reefs **270**, 286-7, **286**, *287*
Spike Island **196**, 211-2, **211**
St Helens Island **374**, 405-6, **405**, *406*
Stack Island **56**, 65-7, **65**, *67*
Steep Head *see* Steep Island
Steep Island **56**, 57-9, **57**, *58*, *59*
Sterile Island **486**, 498-9, **498**, *499*
Storehouse Island **311**, 320-1, **320**, *321*

Sugarloaf Rock (Curtis Island Group) xi, **171**, 176-7, **176**, *176*, *177*

Sugarloaf Rock (Mutton Bird Island Group) **545**, 559-60, **559**

Sugarmouse Island **545**, 556, **556**

Swainson Island **563**, 566-7, **566**, *567*

Swainson Island Group **501**, 563-73, **563**

Swan Island **374**, 397-9, **397**

T

Taillefer Rocks **410**, 423-4, **423**

Tasman Island 9, **439**, 447-9, **447**, *448*

Tasman Island Group **409**, 439-51, **439**, 606

Tenth Island **374**, 375-6, **375**, *376*

The Carbuncles **132**, 137-8, **137**

The Coffee Pot **58**4, 585-6, **585**, *585*

The Friars **486**, 489-90, **489**, *490*

The Images **486**, 500, **500**

The Lanterns **439**, 443-4, **443**

The Nuggets **410**, 411-3, **411**, *411*, *412*

The Thumbs **439**, 445-6, **445**, *446*

Three Hummock Island **56**, 77-9, **77**, *78*

Tin Kettle Island **254**, 261-2, **261**, *262*

Tin Kettle Island Group **195**, 254-69, **254**

Trefoil Island **89**, 96-8, **96**

Trefoil Island Group **45**, 89

Trumpeter Islets **584**, 589-90, **589**, *590*

Trumpeter Islets Group **501**, 584-92, **584**

Tucks Reef **293**, 305-6, **305**

Twin Islets (north) **157**, 161-2, **161**, *162*

Twin Islets (south) **157**, 163-4, **163**

V

Vansittart Island **293**, 307-9, **307**

Vansittart Island Group **195**, 293-310, **293**

Visscher Island 13, 19, **452**, 466-7, **466**, *467*

W

Walker Island xi

Walker Island (Maatsuyker Island Group) **512**, 534-6, **534**, *535*, *536*

Waterhouse Island **374,** 380-2, **380**, *381*

Waterhouse Island Group **373**, 374-408, **374**

Wedge Island **439**, 450-1, **450**

Wendar Island xi, **545**, 561-2, **561**, *562*

West Moncoeur Island **180**, 184-5, **184**, *185*

West Pyramid **584**, 587-8, **587**

Western Rocks **512**, 530, **530**

Wild Wind Islets **545**, 548-9, **548**, *549*

Woody Island 606

Wright Rock **188**, 191, *191*

Wybalenna Island **346**, 353-4, **353**, *354*

Index to seabirds and terrestrial birds
Map page references in **bold**; photograph page references in *italic*

A

Arctic Tern 442

Australasian Gannet
  breeding sites 37, **37**, 318, 507, 511
  description & habitat 36, *36*, *85*, *318*
  see also island groups on pp. 56, 140, 180, 195, 373, 410, 439, 486, 502

Australian Pelican
  breeding sites 34, **34**, 74, 206, 208, 392, 395
  description & habitat 34-5, *34*
  see also island groups on pp. 56, 196, 215, 226, 270, 293, 322, 335, 346, 357, 374

Australian Shelduck 228, 308

Azure Kingfisher 194

B

Banded Lapwing 348

Barn Owl 82, 160

Beautiful Firetail 148, 150, 152, 308, 348, 415, 424, 444, 529

Black-browed Albatross 198, 348

Black Currawong
  see island groups on pp. 56, 311, 346, 454, 512, 545, 563, 574, 593

Black-faced Cuckoo Shrike 82

Black-faced Cormorant xii
  breeding sites 33, **37**
  see also island groups on pp. 48, 56, 89, 116, 132, 215, 226, 236, 254, 357, 294, 311, 322, 346, 374, 410, 439, 468, 486, 502, 512, 545, 584
  description & habitat 32-3, *33*, *328*
  see also island groups on pp. 56, 89,116, 140, 157, 171, 180, 196, 215, 226, 236, 270, 293, 322, 330, 335, 357, 375 410, 431, 439, 454, 476, 492, 486, 512, 549, 574, 593

black-faced shag *see* Black-faced Cormorant

Black-fronted Dotterel 74

Black-headed Honeyeater 471, 483

Black Swan 66, 228, 251, 303, 308, 573

Blackbird *see* Common Blackbird

blue penguin *see* Little Penguin

Blue-winged Parrot 152, 204

Brown Goshawk 66, 406, 449, 471, 483

Brown Falcon
  see island groups on pp. 56, 89, 140, 157, 171, 196, 215, 236, 254, 270, 311, 322, 335, 346, 357, 379, 410, 439, 468, 512, 545

Brown Quail
  see island groups on pp. 56, 116, 140, 160, 157, 180, 196, 215, 226, 236, 254, 311, 322, 335, 346, 357, 374, 410, 454, 468, 476

Brown Thornbill 107, 201, 204, 238, 308, 324, 348, 526, 567, 569

Brush Bronzewing 277, 348, 415, 523, 533

Bullers Albatross 348, 505

C

Calamanthus *see* Striated Fieldwren

Cape Barren Goose *360*
  management program 242
  protection of 360

see also island groups on pp. 56, 89, 116, 140, 157, 171, 180, 196, 215, 226, 236, 254, 270, 293, 319, 322, 330, 335, 346, 357, 374

Cape Petrel 442, 444

Caspian Tern
breeding sites 25, **25**
see also island groups pp. 46, 58, 89, 116, 196, 215, 226, 236, 254, 270, 293, 322, 330, 335, 346, 357, 374, 410, 431, 454, 468, 476; 512, 563, 574, 584
description & habitat 24-5, *24*
see also pp. 66, 118, 135, 146, 152, 220, 228, 262, 354, 372, 406, 465

Cattle Egret 508

Chatham Island Albatross 82

Chestnut Teal 66, 220, 235, 308, 393

chicken see Red Junglefowl

Common Blackbird
see island groups on pp. 56, 140, 157, 171, 180, 196, 215, 226, 236, 254, 270, 293, 311, 335, 346, 357, 374, 410, 439, 476, 512

Common Diving-Petrel
breeding sites 11, **11**, 46, 56, 89, 116, 140, 157, 171, 180, 180, 330, 374, 410, 439, 486, 512, 545
description & habitat 10-11, *10*
co-existing birds 5
see also island groups on pp. 64, 72, 75, 113, 146, 214, 253, 415, 436, 444, 449, 508, 558, 600

Common Starling see island groups on pp. 56, 89, 116, 140, 157, 171, 196, 215, 226, 236, 254, 270, 293, 311, 322, 335, 346, 357, 374, 439, 454, 545, 593

Cormorants see Black-faced Cormorant; Great Cormorant; Little Pied Cormorant

Crescent Honeyeater
see island groups on pp. 56, 140, 196, 322, 410, 439, 454, 486, 512, 574

Crested Tern xii
breeding sites 27, **27**
see also island groups on pp. 56, 196, 215, 311, 346, 374, 410, 454, 486
description & habitat 26-7, *26*
see also island groups on pp. 56, 116, 132, 140, 157, 171, 196, 226, 236, 254, 311, 330, 335, 346, 374, 410, 439, 476, 593

cuckoos see Fan-tailed Cuckoo; Pallid Cuckoo

curlew see Eastern Curlew

currawongs see Black Currawong; Grey (Clinking) Currawong

D

Domestic Pigeon 97, 194

Dominican Gull see Kelp Gull

Dusky Robin 308

E

Eastern Curlew 128, 235, 289, 364, 415, 467

Eastern Spinebill 82, 348, 451, 533, 577

European Goldfinch
see island groups on pp. 89, 140, 157, 171, 196, 215, 226, 236, 254, 270, 293, 346, 439, 512, 563

European Greenfinch 238, 567

F

fairy penguin see Little Penguin

Fairy Prion 10, 11
breeding sites 8, **8**
see also island groups on pp. 46, 58, 89, 486, 584, 140, 157, 171, 180, 188, 410, 410, 439, 502, 512, 545, 563

co-existing seabirds 5
description & habitat 8-9, *8*, *83*

Fairy Tern
breeding sites 30, **30**, 54, 66, 368
description & habitat 30-1, *30*
see also island groups on pp. 46, 56, 89, 116, 215, 226, 270, 357, 374, 593

Fan-tailed Cuckoo 148, 150, 183, 324, 348, 415

Flame Robin 152, 194, 201, 228, 238, 324, 449

Fluttering Shearwater 426

Forest Raven 9, 11
see island groups on pp. 56, 92, 116, 140, 171, 180, 188, 196, 215, 226, 254, 270, 293, 311, 322, 335, 346, 359, 410, 439, 454, 471, 476, 486, 502, 512, 555, 563, 574, 584, 593

Forty-spotted Pardalote 483

G

Giant Petrel 235, 426, 508

Giant Skua 508

gannet *see* Australasian Gannet

Golden Whistler 415

Goldfinch *see* European Goldfinch

Great Egret 308

Great Knot 289

Great Cormorant
see island groups on pp. 56, 89, 374, 410, 431, 454, 476, 486, 545, 574

Green Rosella 152, 415, 444, 483, 526, 529, 533, 577, 579

Greenfinch *see* European Goldfinch

Grey Butcherbird 471

Grey (Clinking) Currawong 415

Grey Fantail
see island groups on pp. 56, 171, 196, 215, 226, 236, 256, 270, 293, 311, 322, 346, 410, 512

Grey Petrel 539

Grey Shrike-thrush 415, 483

Grey Teal 220, 228, 235, 238, 244, 256, 308, 388, 597

gulls 11, 29
see also Kelp Gull; Pacific Gull; Silver Gull

H

honeyeaters *see* Black-headed Honeyeater; Crescent Honeyeater; New Holland Honeyeater; Yellow-faced Honeyeater

Hooded Plover
see island groups on pp. 46, 56, 89, 116, 180, 196, 215, 226, 236, 254, 311, 335, 346, 357, 357, 374

House Sparrow 97, 251, 256, 308, 314, 319, 324, 326, 337, 359, 449, 465, 533

I

Indian Peafowl (Peacock) 262, 348

K

Kelp Gull ix, 15, 210, 413, 415, 433, 456
breeding sites 19, *19*
see also island groups on pp. 431, 454, 468, 476, 374, 502
description & habitat 18-9, *18*

king muttonbird *see* Sooty Shearwater

L

Latham's Snipe 289

Laughing Kookaburra 415

Lewin's Rail 82, 97, 326, 341, 406, 533

Little Black Cormorant 415, 444, 446, 485

Little Eagle 173

Little Grassbird
see island groups on pp. 46, 56, 89, 116, 140, 157, 188, 196, 215, 226, 236, 254, 293, 311, 322, 335, 357, 374, 486, 563, 584

Little Penguin xi, *2*, 5, 82, 242, 348, 378
breeding sites 3, **3**
see also all island groups except Pedra Branca
co-existing seabirds 5
description & habitat 2

Little Pied Cormorant 66, 260, 289, 436, 462, 523

Little Raven 175, 179

Little Tern 31, *31*
breeding site 31

M

Masked Lapwing
see island groups on pp. 46, 56, 89, 140, 157, 196, 215, 226, 236, 254, 270, 293, 311, 322, 330, 335, 346, 359, 357, 374, 410, 431, 584

mollymawk see Pacific Gull; Shy Albatross

moonbird see Short-tailed Shearwater

Mother Carey's chicken see White-faced Storm-Petrel

muttonbird see Short-tailed Shearwater

N

Nankeen Kestrel 59, 92, 97, 152, 160, 314, 319, 471, 533

New Holland Honeyeater 64, 201, 228, 308, 449, 471, 526, 533, 579

New Zealand muttonbird see Sooty Shearwater

Noisy Miner 471

Northern Giant-Petrel 497

O

oil bird see Fairy Prion

Olive Whistler 148, 150, 152, 173, 183, 201, 204, 220, 228, 262, 308, 324, 526

Orange-bellied Parrot 175, 533, 592

P

Pacific Black Duck 66, 97, 228, 523

Pacific Gull ix, 9, 19, 21
breeding sites 14, **14**
see also all island groups except Tasman Island, Breaksea Island & Hibbs Pyramid
description & habitat 14-5, *14*
see also island groups on pp. 116, 132, 196, 242, 254, 374, 410, 431, 439, 486, 512, 545, 563, 574

Pallid Cuckoo 97, 152

Peacock see Indian Peafowl

pelican see Australian Pelican

petrel species, breeding site 525

Peregrine Falcon 9
see island groups on pp. 56, 140, 157, 171, 180, 188, 196, 236, 311, 410, 439, 454, 476, 486, 502, 512, 563

Pied Oystercatcher ix, 20, 22-3, *22*
breeding sites **23**
see also island groups on pp. 46, 89, 116, 226, 236, 254, 270, 293, 311, 322, 357, 374, 431, 454, 468, 476, 512
description & habitat 22
see also 129, 220, 242, 269, 285, 348, 359, 366, 456, 458, 481, 572, 597

Pink Robin 82, 526, 533, 597

R

Red-capped Plover 49, 64, 256, 277, 285, 308, 352, 366, 398

Red-necked Stint 114

Red Junglefowl (chicken) 97, 314, 381

robins see Dusky Robin, Pink Robin; Scarlet Robin; Flame Robin

Richard's Pipit 82, 97, 173, 194, 319, 426, 449

Ruddy Turnstone 49, 64, 66, 82, 208, 214, 235, 238, 289, 310, 319, 348, 352, 359, 366, 372, 401

S

Satin Flycatcher 508

seagull see Silver Gull

Sanderling 49, 102, 114

Sandpiper sp. 289

Satin Flycatcher 88

Scarlet Robin 152, 256, 262, 308, 415

Scrubtit 599

Scrubwren see Tasmanian Scrubwren

Short-tailed Shearwater xi, 7, 9, 77, 82, 86, 217, 483, 508
  breeding sites 5, **5**
  see also all island groups except Long Island and Pedra Branca
  co-existing birds 5
  description and habitat 4-5, *4*

Shy Albatross 82, *82*, 83, *83*, 198, 415, 426, 449, 451, 497, 539
  breeding sites 39, **39**, 81, 504, 508
  description & habitat 38-9, *38*

Silver Gull ix, 27
  breeding sites 17, **17**, 419, 441, 457
  see also island groups except: North Coast, Curtis Island, Preservation Island, Great Dog, Pasco Island, Maria Island, Partridge Island and Hibbs Pyramid
  description and habitat 16-7, *16*
  see also island groups on pp. 56, 89, 132, 140, 157, 171, 180, 196, 215, 226, 236, 254, 346, 374, 410, 431, 439, 454, 476, 486, 512

Silvereye see island groups on pp. 46, 56, 116, 140, 157, 171, 180, 188, 196, 215, 226, 236, 254, 293, 311, 322, 330, 335, 346, 357, 374, 410, 439, 468, 486, 502, 512, 545, 563

Skylark
  see island groups on pp. 46, 56, 89, 140, 157, 171, 196, 215, 226, 236, 254, 270, 293, 330, 335, 346, 374, 410, 439

snipe see Latham's Snipe

Soft-plumaged Petrel, breeding site 532

Sooty Oystercatcher ix, 22, 88, 128, 129, 135, 210, 242, 401, 406, 421, 456, 467, 488, 492, 558, 562
  breeding sites 21, **21**
  see also all island groups except Babel Island, Pedra Branca and Hibbs Pyramid
  description & habitat 20-1, *20*

Sooty Shearwater breeding sites 7, **7**, 448, 487, 532
  description & habitat 6-7, *6*

Southern Boobook 82, 415, 533

Southern Fulmar 299, 436

Southern Giant-Petrel 497

spinebill see Eastern Spinebill

Spotted Pardalote 471, 526

Starling see Common Starling

Straw-necked Ibis 152

Striated Fieldwren 555, 565, 590, 592

Strong-billed Honeyeater 483, 526

Sulphur-crested Cockatoo 555

Superb Fairy-wren 414, 577, 579, 590, 597, 602

Swamp Harrier
see island groups on pp. 46, 56, 89, 116, 157, 171, 196, 215, 236, 270, 311, 335, 374, 410, 431, 439, 454, 468, 512, 545

Swift Parrot 526

T

Tasmanian Native Hen 415, 451, 462

Tasmanian Scrubwren 183, 324, 348, 415, 444, 483, 520, 526, 533, 569, 579, 597

Tasmanian Thornbill 148, 152, 190, 415, 520, 533, 579

Tawny-crowned Honeyeater 488

terns ix
see also Crested Tern; Fairy Tern; Little Tern; White-fronted Tern

Tree Martin 204, 415, 449, 467, 492, 526, 577

Turkey 97

W

waders 49, 107, 114, 115, 128, 256, 289

Wedge-tailed Eagle 183, 187, 324, 415, 449, 471

Welcome Swallow 54, 61, 64, 66, 82, 86, 92, 97, 123, 146, 152, 156, 160, 173, 201, 220, 228, 242, 256, 285, 415, 533

whale bird see Fairy Prion

Whimbrel 341

White-bellied Sea-Eagle
see island groups on pp. 56, 140, 157, 171, 180, 188, 196, 236, 270, 311, 322, 346, 374, 410, 439, 468, 476, 512, 545, 595

white-capped albatross see Shy Albatross

White-faced Heron
see island groups on pp. 46, 56, 89, 116, 226, 293, 311, 322, 346, 357, 374, 410, 431, 439, 454, 468, 512, 545, 584

White-faced Storm-Petrel 54, 55, 72, 75, 107, 108, 123, 253, 279, 341, 433
breeding sites 13, **13**
see also island groups on pp. 56, 89, 116, 132, 196, 236, 270, 293, 311, 331, 330, 335, 346, 357, 374, 410, 454
description & habitat 12-3, *12*

White-fronted Chat
see island groups on pp. 46, 56, 89, 116, 196, 215, 226, 236, 254, 270, 393, 322, 330, 335, 346, 357, 374, 454

White-fronted Tern 210, 220, 223, 235, 262, 264, 274, 283, 273, 281, 354, 508
breeding sites 29, **29**, 208, 231, 263, 259, 267, 273
description & habitat 28-9, *28*

White-throated Needletail 66

Willie Wagtail 97

Y

Yellow-faced Honeyeater 183

Yellow-tailed Black-Cockatoo 201, 308, 415, 449

Yellow-throated Honeyeater 415, 471, 523

Yellow Wattlebird 415

yolla see Short-tailed Shearwater

Notes